Also in the Variorum Collected Studies Series:

BRUCE M.S. CAMPBELL
The Medieval Antecedents of English Agricultural Progress

PAMELA NIGHTINGALE
Trade, Money, and Power in Medieval England

DAVID M. PALLISER
Towns and Local Communities in Medieval and Early Modern England

STEPHEN D. WHITE
Re-Thinking Kinship and Feudalism in Early Medieval Europe

RAYMOND VAN UYTVEN
Production and Consumption in the Low Countries, 13th–16th Centuries

E.M.C. VAN HOUTS
History and Family Traditions in England and the Continent, 1000–1200

ANNE J. DUGGAN
Thomas Becket: Friends, Networks, Texts and Cult

RICHARD W. PFAFF
Liturgical Calendars, Saints and Services in Medieval England

HELMUT GNEUSS
Language and History in Early England

H.S. OFFLER (Ed. Dr A.I. Doyle and A.J. Piper)
North of the Tees
Studies in Medieval British History

DAVID NICHOLAS
Trade, Urbanisation and the Family: Studies in the History of Medieval Flanders

PATRICK SIMS-WILLIAMS
Britain and Early Christian Europe: Studies in Early Medieval History and Culture

ADRIAAN VERHULST
Rural and Urban Aspects of Early Medieval Northwest Europe

JOHN H. MUNRO
Bullion Flows and Monetary Policies in England and the Low Countries, 1350–1500

DAVID N. DUMVILLE
Histories and Pseudo-Histories of the Insular Middle Ages

VARIORUM COLLECTED STUDIES SERIES

Field Systems and Farming Systems in Late Medieval England

In memory of my farming forebears,
the Hall family of Manor Farm, Marsworth, Buckinghamshire,
and especially Minnie Kate Hall, 1883–1959

Bruce M.S. Campbell

Field Systems and Farming Systems in Late Medieval England

ASHGATE
VARIORUM

Published in the Variorum Collected Studies Series by

Ashgate Publishing Limited
Wey Court East
Union Road
Farnham, Surrey
GU9 7PT
England

Ashgate Publishing Company
Suite 420
101 Cherry Street
Burlington, VT 05401–4405
USA

Ashgate website: http://www.ashgate.com

ISBN 978–0–7546–5946–4

British Library Cataloguing-in-Publication Data
Campbell, B. M. S.
Field systems and farming systems in late medieval England.
– (Variorum collected studies series ; no. 903)
1. Agricultural systems – Great Britain – History 2. Field crops – Great Britain – History 3. Middle Ages
I. Title
630.9'42'0902

ISBN 978–0–7546–5946–4

US Library of Congress Cataloging-in-Publication Data
Campbell, B. M. S.
Field systems and farming systems in late medieval England / Bruce M.S. Campbell.
p. cm. – (Variorum collected studies series)
Includes bibliographical references and index.
ISBN 978–0–7546–5946–4 (alk. paper)
1. Agricultural systems – Great Britain – History. 2. Field crops – Great Britain – History. 3. Middle Ages.
I. Title. II. Series: Collected studies.

S453.C36 2008
630.942'0902–dc22 2007050603

The paper used in this publication meets the minimum requirements of the American National Standard for Information Sciences – Permanence of Paper for Printed Library Materials, ANSI Z39.48–1984. ∞ ™

Printed and bound in Great Britain by TJ International Ltd, Padstow, Cornwall

VARIORUM COLLECTED STUDIES SERIES CS903

CONTENTS

This volume contains xiv + 316 pages

PUBLISHER'S NOTE

The articles in this volume, as in all others in the Variorum Collected Studies Series, have not been given a new, continuous pagination. In order to avoid confusion, and to facilitate their use where these same studies have been referred to elsewhere, the original pagination has been maintained wherever possible.

Each article has been given a Roman number in order of appearance, as listed in the Contents. This number is repeated on each page and is quoted in the index entries.

INTRODUCTION

The thirteen essays reproduced in this volume and researched and written between the mid 1970s and late 1990s share several common concerns. First and most fundamentally, they aim to reconstruct late-medieval agrarian arrangements from the wealth of extant contemporary documentary sources preserved in a wide array of public and private archives. In the process they pioneer fresh ways of using and analysing that information, for one of the challenges presented by these sources is how to reconstruct an aggregate picture from evidence which is essentially local and particular in nature. Manorial extents, surveys, accounts, and court rolls naturally lend themselves to micro-scale analysis of individual demesnes, manors, vills, and estates. To employ these same sources to make valid generalisations about all farms, manors, vills, and estates is quite another matter, not least because substantial as is the written record, it is neither complete nor unbiased in its coverage. The almost infinitely varied character of agriculture presents a further obstacle to effective historical generalisation above the level of the individual production unit. One way forward is to develop objective methods of classification which identify and differentiate between the key variant types of field systems, farming systems, and land-use. This is easier said than done and is an explicit concern of five of the essays in this volume. Moreover, classification, as anyone who has attempted it will confirm, can easily become an end in itself, each result inspiring an iterative quest for more and better data and a further refinement of method. Nevertheless, in the final analysis, all classifications are no more than artefacts whose real value lies in the assistance they lend to effective description and the tool provided to analysis of the respective roles of environmental, institutional, and economic factors in shaping how agricultural resources were combined and used.

Classification is certainly indispensable to reconstructing the agricultural geography of the period, an exercise whose results imply that by the close of the thirteenth century factor costs, transaction costs, and commodity prices were already an important determinant of land-use and farm enterprise. At this stage markets did not determine everything, far from it, but they accounted for far more than was appreciated 30 years ago when the first of these essays were being researched. Collectively, the latter endorse the view that the English agrarian economy was already highly commercialized by the end of the thirteenth century as also the revisionist verdict that by-and-large landlords were economically rational in their utilisation and management of demesne resources.[1] Equally clearly, late-medieval English agriculture emerges as neither unvarying nor unchanging. All of the essays in this volume stress the diverse and evolutionary nature of agriculture – its fields, farms, enterprise, and land use – and the challenges and opportunities created by, first, the expansion and, then, the contraction of population and market demand on either side of the demographic watershed represented by the Black Death of 1348–9.

In the mid 1960s Joan Thirsk rekindled debate on the origin of English commonfields – a form of cooperative agrarian management that prevailed in many English townships for the better part of a millennium of more – by questioning whether the commonfield system 'was in full working order before our documents begin', as was the then prevailing orthodoxy, and proposing instead that 'the common-field system was a gradual growth, co-ordinated and systematized by practical necessity as populations grew and land became more and more subdivided'.[2] This is the hypothesis addressed by the first five essays in this volume, which are all concerned with commonfields, their physical evolution, forms of regulation, and the demographic and economic circumstances most likely to promote and perpetuate their systematisation and regularisation. Close examination of the evolution of fields and holdings in eastern Norfolk, probably the single most densely populated and intensively cultivated locality in rural England by the close of the thirteenth century,[3] provides explicit evidence of the progressive morcellation of once consolidated plots and holdings as a result of rising tenant numbers under conditions of *de facto* peasant proprietorship. Significantly, given current debate about the key contribution of factor markets to long-term economic growth and the role of land markets in creating agrarian capitalism, active buying and selling of customary and freehold land reinforced rather than counteracted these tendencies towards morcellation.[4]

---

[1] On commercialisation see R.H. Britnell, *The commercialisation of English society 1000–1500*, Cambridge, 1993; R.H. Britnell and B.M.S. Campbell, eds., *A commercialising economy: England 1086 to c.1300*, Manchester, 1995. On decision making, see B.M.S. Campbell, *English seigniorial agriculture 1250–1450*, Cambridge, 2000; D. Stone, *Decision-making in medieval agriculture*, Oxford, 2005.

[2] J. Thirsk, 'Preface to the third edition', v–xv in C.S. Orwin and C.S. Orwin, *The open fields*, 3rd edn., Oxford, 1967, xv. See also, J. Thirsk, 'The origin of the common fields', *Past and Present* 29 (1964), 3–25.

[3] B.M.S. Campbell, 'Agricultural progress in medieval England: some evidence from eastern Norfolk', *Economic History Review*, 2nd series, 36 (1983), 26–46, reprinted in B.M.S., Campbell, *The medieval antecedents of English agricultural progress*, Variorum Collected Studies Series, Aldershot, 2007.

[4] B. van Bavel, T. de Moor, J.L. van Zanden, eds., 'The rise, organization, and institutional framework of factor markets', *Continuity and Change*, special issue. Also, C.T. Bekar and C.G. Reed, *Distributional dynamics in a stochastic environment with tradable assets: medieval English land markets*, unpublished working paper, 2007; B.M.S. Campbell, 'Land markets and the morcellation of holdings in Pre-Plague England and Pre-Famine Ireland', in G. Beaur and P. Schofield, eds., *Property rights, the market in land and economic growth in Europe (13th–19th centuries)*, forthcoming, 2008/9. For a fuller discussion of the problems of this period see B.M.S. Campbell, *Land and people in late medieval England*, Variorum Collected Studies Series, Aldershot, forthcoming 2009.

The upshot was a mounting complexity and irregularity of field and holding layout which, far from promoting cooperation, positively inhibited adoption of anything more complex to implement than common pasturage on the harvest stubble (sometimes extended to include land destined to remain fallow until the spring).

At the climax of medieval population growth and agricultural expansion, irregular commonfields were probably the agrarian norm. Analysis of information contained in the *inquisitiones post mortem* reveals that by the first half of the fourteenth century common rights and common management of pastoral resources were territorially far more extensive than the distribution of regular two- and three-field systems with their symmetrical layout of holdings and regular rotation of crops designed to consolidate for grazing purposes all land remaining fallow throughout the year.[5] Very likely, therefore, at this point in time, irregular commonfields of one sort or another – for functionally, as essay IV shows, there were many variant types – accounted for the greater share of agricultural land use and production and, as in eastern Norfolk, were typically associated with the highest rural population densities and most intensive and developed mixed-farming systems. Only in a broad swathe of country stretching across the midlands of England were regular commonfield systems, and the nucleated villages which typically accompanied them, the norm.[6] Both are now regarded, on mainly archaeological and morphological evidence, as the product of a large-scale rationalisation and reorganisation of rural settlement that took place over the course of the ninth to twelfth centuries.[7] Why this should have occurred and by what precise agency it was brought about – collective action, seigniorial imposition, or collaboration between lords and tenants – nevertheless remain matters of speculation. What is clear, however, is that with a national population in 1086 of no more than 2½ million, it was labour rather than land that was in relatively short supply. As the two essays, 'Commonfield origins – the regional dimension' and 'Commonfield agriculture: the Andes and medieval England compared', both argue, such a circumstance provided an incentive to coordinate cropping so that pastoral resources could be pooled and labour economized upon. Moreover, the fewer the cultivators and the slacker the demand for land, the easier it would have been to achieve the kind of comprehensive rationalisation of field and holding layout required by adoption of the two- and three-field system.

Once created, of course, regular field systems proved to be self-perpetuating and remarkably enduring. In contrast, their irregular counterparts succumbed more readily to processes of consolidation, engrossment, and piecemeal and private en-

---

5 B.M.S. Campbell and K. Bartley, *England on the eve of the Black Death: an atlas of lay lordship, land and wealth, 1300–49*, Manchester, 2006, 55–68.

6 For the distribution of villages see, B.K. Roberts and S. Wrathmell, *An atlas of rural settlement in England*, London, 2000.

7 Current thinking on this subject is reviewed by C.C. Dyer and P.R. Schofield, 'Recent work on the agrarian history of medieval Britain', 21–55 in I. Alfonso, ed., *The rural history of medieval European societies: trends and perspectives*, Turnhout, 2007, 25–9.

closure, set in train by the prolonged demographic decline that followed the Black Death and promoted by the very *post mortem* and *inter vivos* transfers of land between individuals which under earlier and opposite circumstances had been a source of so much morcellation. Again, detailed analysis of eastern Norfolk traces the point at which consolidation and engrossment became the predominant net outcomes of land transfers and emphasizes the exceptionally long time frame required to reverse the extreme morcellation that had come into being by the early fourteenth century. When agrarian capitalism eventually emerged, in the form of substantial owner-occupied farms hiring labour in order to produce surpluses for profit, this was the product of the cumulative effect over more than two centuries of the selective exchange and inheritance of land between tenants.[8] In other words, in eastern Norfolk capitalist farms were created not by the intervention of landlords but by processes internal to peasant society. By the sixteenth century these processes had gained sufficient momentum to be reinforced rather than reversed by the renewal of population growth, a development which had momentous consequences for the long-term development of English agriculture but whose exact causes continue to challenge detailed diagnosis.[9]

Within the institutional framework provided by field systems and their associated rights and regulations decisions about what crops and livestock to produce were taken at the level of the individual production unit. Many factors shaped those decisions: the precise factor endowments of land, labour, and capital; the levels and forms of rent that had to be paid; the specific consumption requirements of the household; the costs of delivering commodities to market and the prices they commanded once delivered there; and, in the case of the seigniorial sector, the place of the individual demesne within the overall estate structure. Medievalists continue to debate the relative importance of these factors and have approached the subject via detailed case studies of individual demesnes, analysis of specific urban hinterlands, and – as in the case of essays VI to XIII in this volume – classification and mapping of farm enterprise, farming systems, and land use at regional and national scales.[10]

---

[8] For a detailed analysis of the latter stages of this process in eastern Norfolk see J. Whittle, *The development of agrarian capitalism: land and labour in Norfolk 1440–1580*, Oxford, 2000.

[9] Many of the issues raised in R.H. Tawney, *The agrarian problem in the sixteenth century*, London, 1912, remain unresolved.

[10] D. Stone, 'Medieval farm management and technological mentalities: Hinderclay before the Black Death', *Economic History Review*, 2nd series, 54 (2001), 612–38; Stone, *Decision making*; B.M.S. Campbell, J.A. Galloway, D.J. Keene, and M. Murphy, *A medieval capital and its grain supply: agrarian production and its distribution in the London region c.1300*, Historical Geography Research Series 30, 1993; J.S. Lee, *Cambridge and its economic region, 1450–1560*, Studies in Regional and Local History 3, Hatfield, 2005; Campbell, *English seigniorial agriculture*. See also, H. Kitsikoupolos, 'Urban demand and agrarian productivity in pre-plague England: reassessing the relevancy of von Thünen's model', *Agricultural History* 77 (2003), 482–522.

The potential of surviving written records for reconstructing the agricultural geography of England before and after the Black Death was convincingly demonstrated by John Langdon in his methodologically innovative study *Horses, oxen and technological innovation* published in 1986.[11] The database of crop and livestock information assembled by Langdon from a national sample of manorial accounts has provided the empirical basis of all subsequent work, amplified by data from a fuller sample of accounts for ten counties around London (created in conjunction with the two 'Feeding the City' projects) and all accounts extant for the exceptionally well-recorded county of Norfolk. Impressive as is this body of information it is beyond doubt that a further systematic search of all available public and private archives would add significantly to it. Essays VI and VII – 'Towards an agricultural geography of medieval England' and 'The diffusion of vetches in medieval England' – demonstrate the utility of such geographically comprehensive datasets for reconstructing the diffusion of specific farming technologies, notably the adoption of horses for draught power and cultivation of vetches as a nitrifying fodder crop and partial substitute for hay where the latter was scarce. Both also highlight the receptiveness of demesne managers to new technological opportunities when their adoption offered practical and economic advantages and could be integrated without undue difficulty into the prevailing technological complex.

Analysing crops and animals in combination presents a significantly greater methodological challenge due to the larger number of variables involved. This problem is most effectively addressed by the development of classifications. It was Mark Overton who in the mid 1980s identified cluster analysis as a particularly effective statistical technique for deriving farming classifications from historical data.[12] Accordingly, it is this technique that is employed to derive the series of classifications of arable husbandry, pastoral husbandry, mixed husbandry, and land use presented in essays VIII to XIII.[13] The methods and results set out in 'Mapping the agricultural geography of medieval England' constitute the first systematic attempt to develop a national classification of medieval husbandry types using cluster analysis. A strength of this paper is its comparison between the results obtained from applying several different clustering techniques (each based upon different statistical principles) in order to establish the optimum number of husbandry types and identify differences that are intrinsic to the data rather than a product of the choice of technique. Its main limitation is the use of closed numbers to specify the component crop and livestock variables (i.e. individual crops are expressed as

[11] J.L. Langdon, *Horses, oxen and technological innovation: the use of draught animals in English farming from 1066–1500*, Cambridge, 1986.

[12] M. Overton, 'Agricultural regions in early modern England: an example from East Anglia', University of Newcastle upon Tyne, Department of Geography, Seminar Paper 42, 1985.

[13] Statistical and computational credit for developing these classifications belongs to my co-authors: John Power and Ken Bartley.

a percentage of the total cropped area and individual livestock as a percentage of total livestock units) since this results in variables which are highly correlated (e.g. a higher percentage of one crop inevitably results in a lower percentage of others). Nor at this stage had a solution been found to the problem of how to compare data on crops with data on livestock. Although sub-optimal, the results obtained are nonetheless revealing, as is demonstrated by their application to a discussion of the livestock on the Norfolk demesne of Bawdeswell as enumerated by Geoffrey Chaucer in his portrayal of Oswald the reeve.

This approach is refined and taken a stage further in essay X – 'Cluster analysis and the classification of medieval demesne-farming systems' – with reference to data drawn from the period 1250–1349. Here the component variables are specified in such a way as to avoid the correlations inherent to closed numbers, primarily by relying upon logged ratios with statistically determined upper-bound limits combined with percentages calculated using independent bases (e.g. wheat as a percentage of winter grains and oats as a percentage of spring grains). Ratios enable crops and livestock to be specified in much the same way and, by including the ratio of livestock to crops, it becomes possible to generate a typology which accommodates both the arable and pastoral components of farm enterprise. Again, three different clustering techniques are applied to the data but this time the final typology is determined by those 'core' demesnes identically classified by each method. The remaining 'non-core' demesnes are then allocated to whichever of these core types they most closely resemble using function coefficients obtained from discriminant analysis. The result, which should be as statistically robust as the available data and choice of variables permit, is an eight-category classification of mixed-farming types with potentially 56 sub-types, a solution which satisfactorily encapsulates the many subtle gradations of farm enterprise which existed in practice. Since location is not included as one of the component variables it is all the more significant that, when mapped, each of the core mixed-farming types displays its own distinctive geographical distribution. The role of commercial factors in shaping the overall configuration of mixed-farming types and determining where the most integrated and intensive mixed-farming systems were developed is considered in the essay XI – 'Economic rent and the intensification of English agriculture'.

This same exercise is then repeated with respect to the period 1350–1449 in essay XII – 'The demesne-farming systems of post Black Death England'. This yields a farming classification for the period of population decline and economic contraction that followed the Black Death which is methodologically exactly comparable with that obtained for the preceding period of growth and expansion. Unsurprisingly, given the radically altered supply and demand conditions which prevailed following the plague, neither the attributes nor the number – seven rather than eight – of the core farming types exactly replicate those which had prevailed before. The extent of the match between the eight pre-Black Death mixed-farming types and the seven post-Black Death mixed-farming types is established by applying discriminant functions to the mean characteristics of each farming type. This provides a statistically objective method of linking cross-sectional with temporal analysis and thereby reconstructing the spatial dynamic of agrarian change. The same essay also demonstrates the benefits of deriving classifications at a national rather than regional or county level. Applying the same methodology to the Norfolk dataset and comparing the results obtained with the corresponding national classification shows that the latter captures a significantly wider range of farming types. Paradoxically, a national perspective may be required to highlight what is most distinctive at a local level.

The final essay in this volume – '*Inquisitiones post mortem*, GIS, and the creation of a land-use map of pre Black Death England' – develops this method of classification to its logical extreme and, with reference to the wealth of land-use information contained in the extents attached to *inquisitiones post mortem* (*IPM*s), offers the most ambitious and comprehensive national classification of all. The *IPM*s provide a density of geographical coverage which manorial accounts, abundant as they are, can never match. Consequently, classification yields not a patchy scatter of discrete points but continuous territorial coverage of the greater part of the country: a veritable land-use map of early fourteenth-century England. The picture that emerges is kaleidoscopic. Broad geographical contrasts in land use turn out, upon closer inspection, to have been characterized by much subtle and localized variation. The result demonstrates what can be achieved by applying modern statistical methods to data derived from medieval sources and the potential for further work of this nature. More importantly (and in conjunction with the classifications of field systems and farming systems similarly reproduced in this volume), it provides a systematic spatial framework within which future research into and discussion of the late medieval agrarian economy can be conducted.[14] It also serves as a salutary reminder of the benefits to be derived from combining micro and macro scales of analysis.

BRUCE M.S. CAMPBELL

Belfast
Michaelmas 2007

---

[14] As exemplified by M. Bailey, *Medieval Suffolk: an economic and social history 1200–1500*, Woodbridge, 2007.

ACKNOWLEDGEMENTS

For their kind permission to reproduce the essays included in this volume grateful acknowledgement is made to my co-authors Ricardo Godoy (for essay V); John Power (VIII, X, XII) and Ken Bartley (XII, XIII). Grateful acknowledgement is made to the following persons, institutions and publishers for their kind permission to reproduce the essays included in this volume: Wiley-Blackwell Publishing Ltd, Oxford (I, VII, X, XIII); Norfolk Archaeology, Wymondham (II); Agricultural History Review, Exeter (III, VI, XII); Thomson Publishing Services, Andover (IV); Daniel W. Bromley and Institute for Contemporary Studies, San Francisco, CA (V); Elsevier, London (VIII); Medieval Institute Publications, Kalamazoo, MI (IX); Brill, Leiden (XI); and Manchester University Press for the maps reproduced in XIII (taken from Bruce M.S. Campbell and Ken Bartley, *England on the eve of the Black Death: an atlas of lay lordship, land and wealth, 1300–49*, Manchester, Manchester University Press, 2006). Thanks are also due to Gill Alexander for redrawing and in the process making corrections to many of the figures that are reproduced here.

I

Population Change and the Genesis of Commonfields on a Norfolk Manor[1]

IN the Middle Ages eastern Norfolk possessed a social structure which was unusually flexible and free. The institution of lordship appears to have been retarded by the area's comparatively late settlement (a substantial settled population not being attracted until the late ninth century) and by a strong free-Danish element in the population;[2] so much so that as late as 1086 over half of the population was of free or socage status.[3] As was usual, some extension of seigneurial authority followed the Norman Conquest but this locality remained an area of comparatively weak manorialization throughout the Middle Ages, with an almost total lack of coincidence between manor and vill.[4] In the thirteenth century partible inheritance was still the predominant custom on most manors and by this date most villeins had already established the right to alienate land.[5] This relative weakness of the feudal nexus seems to have fostered above-average rates of population growth, a trend which was reinforced by the area's physical advantages of subdued relief, soils for the most part fertile and easily tilled, and abundant local supplies of peat for fuel. Domesday Book shows that eastern Norfolk was already the most populous district in England in the late eleventh century and it probably remained so for at least the next two and a half centuries.[6]

This general laxity of institutional constraints in matters of land tenure and social structure was also to be seen in the organization of agriculture: field systems in this area allowed individuals maximum initiative in matters of cultivation. The only clearly documented common right which applied to subdivided fields in eastern Norfolk was the right of common grazing on the aftermath of the harvest known as harvest shack.[7] Even the foldcourse system of western Norfolk, which

[1] I am grateful to Dr A. R. H. Baker and Dr H. S. A. Fox for comments on an early draft of this article.

[2] J. M. Lambert, J. N. Jennings, C. T. Smith, C. Green, and J. N. Hutchinson, *The Making of the Broads: A Reconsideration of their Origin in the Light of New Evidence* (Royal Geographical Society Research Series, III, 1960); R. R. Rainbird-Clarke, *East Anglia* (1960).

[3] B. Dodwell, 'The Free Peasantry of East Anglia in Domesday', *Norfolk Archaeology*, XXVII (1939), 145–57; H. C. Darby, *The Domesday Geography of Eastern England* (Cambridge, 1952).

[4] D. C. Douglas, *The Social Structure of Medieval East Anglia* (Oxford Studies in Social and Legal History, IX, 1927). For an example of the complexity of manorial arrangements which could be found within townships of this area see B. M. S. Campbell, 'Field Systems in Eastern Norfolk during the Middle Ages: A Study with Particular Reference to the Demographic and Agrarian Changes of the Fourteenth Century' (unpublished Ph.D. thesis, University of Cambridge, 1975), pp. 253–74.

[5] Campbell, thesis, pp. 328–9 and 189–219.

[6] H. C. Darby, ed. *A New Historical Geography of England* (Cambridge, 1973), pp. 45–7. For post-Domesday demographic trends in eastern Norfolk see Campbell, thesis, pp. 325–8. See also J. Sheail, 'The Distribution of Taxable Population and Wealth in England during the Sixteenth Century', *Transactions of the Institute of British Geographers*, LV (1972), 145–57.

[7] The documentary evidence is fully discussed in Campbell, thesis.

allowed the folding of communal flocks upon full-year fallows,[1] is not recorded here. The latter arrangement depended upon the close juxtaposition of extensive sheep-walks to the arable fields, a relatively high frequency of fallowing, and the utilization of irregular cropping shifts to concentrate fallow strips together. Yet in eastern Norfolk, except in the immediate vicinity of the coastal dunes, Broadland marshes, and occasional extensive heaths, sheep were almost everywhere of secondary importance to cattle, and many townships suffered from a serious shortage of pasture. Moreover, because of the pressure upon land and the generally high fertility of the soil, methods of cultivation were very intensive and fallowing was correspondingly infrequent; indeed, on some demesnes methods of cultivation had become so intensive by the late thirteenth century that annual fallows had been virtually eliminated.[2] All this served to reduce the opportunities for, whilst increasing the difficulties of, operating a foldcourse system. But practical objections apart, the foldcourse system could never have been important in eastern Norfolk for it revolved around the superior flock right of the manorial lord whose prerogative did not include the numerous freemen and sokemen of this locality. Only tenants of villein status and below were obliged to place their sheep in the lord's custody between Pentecost and Martinmas and they numbered less than half the population. Thus cultivators were left to make their own arrangements for fallow grazing with the result that these subdivided arable fields only qualify for the epithet "commonfields" on the single ground that they were subject to the right of harvest shack,[3] the most elementary of common rights.

The general irregularity and lack of systematization exhibited by eastern Norfolk's commonfields, coupled with the area's late settlement and subsequent rapid growth of population, raise a number of important questions concerning current notions about the origin of the commonfield system and suggest that this locality may provide an important testing ground for the principal theories. After two generations of research and debate these divide into two groups: those which maintain that the commonfield system was a deliberate creation—either introduced from abroad as argued by H. L. Gray[4] or established by pioneer colonists of Anglo-Saxon and Scandinavian origin as deduced by C. S. and C. S. Orwin[5]—and those which postulate that commonfields evolved gradually as holdings became increasingly fragmented and co-operation between cultivators

[1] H. L. Gray, *English Field Systems* (Cambridge, Mass., 1915), pp. 305–54; K. J. Allison, 'The Sheep-Corn Husbandry of Norfolk in the Sixteenth and Seventeenth Centuries', *Agricultural History Review*, x (1962), 80–101; M. R. Postgate, 'Field Systems of East Anglia', in A. R. H. Baker and R. A. Butlin, eds. *Studies of Field Systems in the British Isles* (Cambridge, 1973), pp. 281–324.

[2] Fallows consistently occupied less than 10 per cent of the arable area on the demesnes at Halvergate, South Walsham, and Acle, and less consistently so at Martham, Hemsby, Flegg, and Plumstead. A full discussion of demesne farming in this area is given in Campbell, thesis, pp. 337–57.

[3] The term "subdivided fields" is used morphologically, to denote arable fields in which the lands of different cultivators lay unenclosed and intermixed. The term "commonfields" is used functionally, to denote subdivided fields subject to common rules of management.—A. R. H. Baker, 'Some Terminological Problems in Studies of British Field Systems', *Agric. Hist. Rev.* XVII (1969), 136–40; I. H. Adams, *Agrarian Landscape Terms: A Glossary for Historical Geography* (Institute of British Geographers Special Publication, 9, 1976), p. 79. Subdivided fields which were not commonfields have been identified in both Kent and the Lincolnshire fenland during the Middle Ages.—A. R. H. Baker, 'Field Systems in Southeast England', in Baker and Butlin, op. cit. pp. 386–9; H. E. Hallam, *Settlement and Society: A Study of the Early Agrarian History of South Lincolnshire* (Cambridge, 1965), pp. 137–61.

[4] Gray, op. cit. [5] C. S. and C. S. Orwin, *The Open Fields* (Oxford, 1938).

increasingly necessary, the principal advocate here being J. Thirsk.[1] At present, although each of these theories has gained a measure of support, none commands widespread acceptance.[2] Recent detailed research into the development of field systems in Kent, Yorkshire, and Devon during the Middle Ages has partly endorsed Thirsk's arguments and given valuable insight into this much neglected but most formative period in the history of the commonfield system,[3] but has failed to resolve the basic issue. The exact relationship between population growth, the parcellation of holdings and fields, and the co-ordination and systematization of cropping and grazing practices remains in question and requires much closer observation. It is on this issue that this article endeavours to cast light.

I

Within eastern Norfolk clearest evidence of the early history of the commonfield system is provided by the township of Martham in the coastal district of Flegg. Like so many townships in this area, Martham was divided among several different manors, the largest of which belonged to the Prior of Norwich and comprised slightly more than half the arable area. It is to this manor that the bulk of surviving documentation relates, the absence of comparable data for the other manors in the township preventing any full appraisal of the topographical extent of the township's commonfields and the size and layout of peasant holdings. Notwithstanding this limitation, though, available evidence (of which the most important is a detailed extent of 1292)[4] is sufficient to show that Martham's commonfields covered an extensive area and were characterized by extreme subdivision. The Prior's manor alone contained 846 acres (equivalent to 1,057 statute acres)[5] of commonfield land and it is clear from abuttals given in the extent that on the south and east the arable fields extended right up to the parish boundary where they merged with those of the neighbouring townships of Rollesby, Hemsby, and Somerton. Only in the north of the township, where the soils were too ill-drained for cultivation, do other land uses appear to have predominated, although a small pocket of marsh and fen also existed in the extreme south of the township. Internally the commonfields were made up of a number of unequally sized fields and of a myriad of small strips. It is impossible to be certain as to the precise number and size of Martham's commonfields, but it is clear that Estfeld, Westfeld, and Suthfeld were the largest and that these were probably equalled in extent by a plethora of smaller fields such as Morgrave, Tomeres, Hiltoftes, and Welletoft. The internal subdivision of these fields was,

[1] J. Thirsk, 'The Common Fields', *Past & Present*, xxix (1964), 3–25. Also 'The Origin of the Common Fields', *Past & Present*, xxxiii (1966), 142–7, and 'Preface to the Third Edition', in C. S. and C. S. Orwin, *The Open Fields* (Oxford, 3rd edn. 1967).

[2] For the most up-to-date and comprehensive survey of the literature see Baker and Butlin, op. cit.

[3] A. R. H. Baker, 'Open Fields and Partible Inheritance on a Kent Manor', *Economic History Review*, 2nd ser. xvii (1964), 1–23; J. A. Sheppard, 'Pre-Enclosure Field and Settlement Patterns in an English Township: Wheldrake near York', *Geografiska Annaler*, xlviii B (1966), 59–77; H. S. A. Fox, 'The Chronology of Enclosure and Economic Development in Medieval Devon', *Econ. Hist. Rev.* 2nd ser. xxviii (1975), 181–202.

[4] B.M. Stowe MS 936.

[5] A perch of 18½ft was used throughout (ibid. fo. 37), hence each Martham acre is equivalent to approximately 1¼ statute acres. Unless otherwise stated, all acreages are as given in the extent.

in every case, extreme. The 846 acres surveyed in the extent were divided into no less than 2,122 separate strips of which only 129 were larger than 1 acre and 14 larger than 2 acres.[1] In 1292 mean parcel size stood at 1·6 rods (0·5 statute acre) and modal parcel size at 1 rod (0·31 statute acre).

Just as arable land at Martham was subdivided into a multitude of separate strips, so too it was divided amongst a host of different owners. A total of 376 tenants held land of the Prior of Norwich in 1292 and in each case the amounts of land held were small—never more than 18 acres (22·5 statute acres) and in 90 per cent of cases less than 5 acres. The mean and modal amounts held were just over 2 acres and just under 1 acre respectively. It is improbable that such small acreages of land represented the sum total held by these tenants, but how much additional land they held of other manors and in neighbouring townships cannot be ascertained. The general laxity of manorial controls and the emergence by the close of the thirteenth century of an active peasant land market would certainly have provided both the opportunity and the means for peasants to have built up holdings of considerable tenurial complexity. Nevertheless, if peasants held as little land from other manors as they did of this manor their aggregate holdings would still not have been large, so that a mean holding size no greater than 5 acres is entirely plausible.

More important to the operation of the commonfields than holding size was holding layout. Here again available evidence is imperfect, but there are grounds for supposing that Martham was no different from townships elsewhere in East Anglia and that holding layout was irregular.[2] Such land as tenants held from the Prior of Norwich, for example, was distributed with the utmost irregularity. In some cases strips were widely scattered among the commonfields whereas in others they were relatively concentrated in their distribution. Moreover, landownership was maintained in a constant state of flux by the twin action of partible inheritance and an active peasant land market[3] which would have militated against the persistence of a regular scheme of holdings. Nor was there any special reason why holdings should have been regular, for common rotations were absent and fallow grazing rights were confined to harvest shack. The only function which a scattered distribution of strips may have served may have been to minimize the risk of harvest failure,[4] although it is unlikely that this is a sufficient reason for the intermixture of holdings. Indeed, their haphazard form suggests an origin which was spontaneous rather than planned, an inference which is verified by an analysis of the 1292 extent and related documents.

At Martham, as on other manors in East Anglia, rents and services were

[1] The size distribution of these strips was as follows:

Under ½ rod:	76 strips	3–under 4 rods:	108 strips
½–under 1 rod:	271 strips	4–under 6 rods:	94 strips
1–under 2 rods:	1,057 strips	6–under 8 rods:	21 strips
2–under 3 rods:	475 strips	8 rods and over:	14 strips

[2] This conclusion is endorsed by later topographical surveys of several townships in this locality, e.g. those of Hemsby (Norfolk Record Office (hereafter N.R.O.), Middleton, Killin & Bruce 19.11.68), Westwick (N.R.O. Pet. MS 584), Coltishall (King's College Cambridge E28), Hempstead with Lessingham (King's College Cambridge P34), Blofield (N.R.O. NRS 16646 37G).

[3] See below, pp. 183–90.

[4] D. N. McCloskey, 'English Open Fields as Behaviour towards Risk', in Paul Uselding, ed. *Research in Economic History*, 1 (Greenwich, Connecticut, 1976), 124–70.

affixed to land rather than to tenure, with the result that there was little or no need for the standardized virgates and bovates which were such a characteristic feature of much of central and southern England. In this East Anglia was closely analogous to Kent, where rents and services were attached to blocks of land known as *iuga*.[1] At Martham, however, rents and services had not always been allocated in this way and its system of territorially based rents appears to have superseded and been adapted from an earlier tenurial scheme. The fiscal units involved were known as *tenementa*, did not generally form contiguous blocks of land, and once comprised individual landholdings. In the case of villein tenants these holdings—known as *eriungs*—were of a standardized 12 acres (15 statute acres) in size and all owed the same rents and services:

> *Thomas Knight tenuit quondam xij acras terre de villenagio que vocatur j eriung et reddit inde* . . . (rents and services follow). *Sciendum est quod xij acre de villenagio vocantur unum Eriung. Et quilibet tenens unum Eriung faciet in omnibus sicud predictum est de tenemento Thome Knight. Et habentur in Martham xxij Eriung et iij acre de villenagio et omnes isti herciabunt totam ville exceptis terris quesitis ad siliginem, avenam, et falihes.*[2]

The creation of these standardized holdings was almost certainly the work of the manorial authorities and was probably associated with the general downgrading in status of the manorial population which took place some time after 1086 (when the population comprised 36 freemen, 27 sokemen, seven villeins, three bordars, and one serf).[3] As such it marks an important step in the development of Martham's fields, for it would have involved a substantial reallocation of land. Once created, however, the integrity of these holdings evidently proved impossible to preserve and it was then, as their ownership fragmented, that the changeover to a territorial system of rents and services occurred, with the *tenementa* becoming fossilized as fiscal units. The same was true of the holdings of freemen and sokemen, although in their case there was no standardization of size before their somewhat lighter rents and services were assessed and fixed. These fiscal units were still in use when the Prior of Norwich's manor was surveyed in 1292, the extent of that year recording the *tenementa* along with the names of the tenants who then held portions of them. Thus for each villein and socage *tenementum* (each of which had once formed a single holding) this document records the name of the original owner, the rents and services which he owed, the names of the tenants currently responsible for these dues, and details of the amount and location of land which each held. This information is sufficient to allow a partial reconstruction of the original layout of the *tenementa* which, when compared with their layout in 1292, can provide answers to several important questions concerning the antecedents of Martham's commonfields.

[1] Baker, 'Open Fields'.

[2] B.M. Stowe MS 936, fo. 39*v*. For a full discussion and possible interpretation of this arrangement see W. Hudson, 'Traces of Primitive Agricultural Organization as Suggested by a Survey of the Manor of Martham, Norfolk (1101–1292)', *Transactions of the Royal Historical Society*, 4th ser. I (1918), 28–58 and 'The Anglo-Danish Village Community of Martham, Norfolk', *Norfolk Arch.* xx (1919–21), 273–316. For vestiges of similar systems in other East Anglian townships see Douglas, op. cit. Within the vicinity of Martham evidence of a structured system of landholdings is also to be found at Lessingham, where a 7½-acre holding was in use in the mid-thirteenth century (B.M. Add. MS 24, 316, fos. 51–6), on the manor of Crictot Hall in Hevingham, where villein holdings were once based on a 6 acre unit (N.R.O. NRS 19280 33 F9), and—most strikingly—at Cawston, where a three-tiered system of 18 acre, 4 acre, and 2 acre holdings had once prevailed (P.R.O. SC 11 Roll 471).

[3] H. A. Doubleday and W. Rye, eds. *The Victoria History of the County of Norfolk*, II (1906), 118.

Precisely when the *eriungs* were created and the *tenementa* became fossilized is difficult to gauge. W. Hudson, in his pioneering study of this manor,[1] surmised that the *eriungs* were created shortly after 1101 when Norwich Cathedral Priory acquired its property at Martham, but if so the establishment of the *tenementa* as fiscal units probably took place some time later by which time several *eriungs* had already undergone subdivision. A date for the fossilization of the *tenementa* towards the middle of the twelfth century and possibly even as late as the beginning of the thirteenth century would thus seem plausible. Such a date would accord with the great increase in tenant numbers which subsequently occurred (numbers rising by 350 per cent from 107 to 376 in 1292, an increase which suggests the passage of several generations) and the fact that many tenants already possessed hereditary surnames when the *tenementa* became fossilized[2] (60 per cent of these surnames still being current in 1292). More important than the precise date at which the *tenementa* were created, though, are the changes which their layout subsequently underwent.

During the thirteenth century pressure upon land steadily mounted and, in the absence of any real expansion in the cultivated area, Martham's rising population was largely accommodated by the subdivision of existing holdings (see Table 1). That this had a profound effect upon field layout is revealed by a reconstruction of the original layout of the *tenementa* using the abuttals given in the 1292 extent (of which there are two for each strip), this reconstruction being

Table 1. *Martham, Norfolk: Number and Size of the* Tenementa *in the Twelfth Century and of Landholdings in 1292*

	Twelfth century			*1292*		
Size of holding (acres)	*No. of holdings*	*% of total no.*	*% of total area*	*No. of holdings*	*% of total no.*	*% of total area*
18 and over	5	4·6	15·4	0	0	0
17–under 18	0	0	0	1	0·3	2·0
16–under 17	5	4·6	9·6	0	0	0
15–under 16	1	0·9	1·8	0	0	0
14–under 15	4	3·7	6·6	1	0·3	1·7
13–under 14	2	1·9	3·1	0	0	0
12–under 13	5	4·6	7·1	3	0·8	4·4
11–under 12	2	1·9	2·6	2	0·5	2·7
10–under 11	6	5·6	7·1	3	0·8	3·7
9–under 10	4	3·7	4·3	1	0·3	1·1
8–under 9	6	5·6	5·7	6	1·6	6·0
7–under 8	13	12·0	11·1	4	1·1	3·5
6–under 7	14	13·0	10·7	8	2·1	6·2
5–under 6	8	7·4	5·1	19	5·1	12·3
4–under 5	10	9·3	4·9	18	4·8	9·6
3–under 4	5	4·6	1·8	25	6·7	10·3
2–under 3	6	5·6	1·7	47	12·4	14·1
1–under 2	8	7·4	1·2	85	22·6	13·8
Under 1	3	2·8	0·2	153	40·7	8·6
Total	107	100%	100%	376	100%	100%

[1] Hudson, 'Traces'.

[2] In Norfolk, according to R. A. McKinley, "such evidence as there is for the surnames of the peasant population would suggest that before 1200 hereditary surnames were rare amongst them".—R. A. McKinley, *Norfolk Surnames in the Middle Ages* (English Surnames Series, II, 1975), p. 7. At Martham, however, the early development of private property rights in land combined with a very high density of population may have led many peasants to adopt hereditary surnames well before the thirteenth century.

based on the assumption that where adjacent strips belonged to the same *tenementum* subdivisions had taken place. By its nature this exercise is involved and at times somewhat arbitrary and consequently has been confined to a random sample of 26 *tenementa* comprising 170 acres and 419 strips. This sample is sufficiently large to provide results—summarized in Table 2—which are statistically significant, although such are the vagaries of the data that they are more reliable as a guide to relative than absolute change. Nevertheless, even allowing for a wide margin of error it is plain that during the thirteenth century the subdivision of holdings frequently entailed the parcellation of land. In the case of the sampled *tenementa* the number of separate parcels increased from approximately 283 at the end of the twelfth century to 419 in 1292, which, if representative, indicates an increase for the whole commonfield area of 50 per cent (this compares with a 350 per cent increase in the number of landholders). In this development tenurial considerations seem to have carried but little weight; on the other hand, geographical factors do seem to have had some effect, some portions of the arable fields experiencing much greater subdivision than others. Whereas in the late thirteenth century there was great uniformity of parcel size in all parts of the commonfields, in the late twelfth century this had been much less the case. During the thirteenth century the rate of subdivision tended to be

Table 2. *Martham, Norfolk: The Subdivision of* Tenementa

Original owner of Tenementum	*Area (acres)*	*Approx. original no. of parcels*	*No. of parcels in 1292*	*Increase in no. of parcels by 1292*	*No. of tenants in 1292*
Amable de Hemesby	0·75*	1	3	2	3
Cruchestoft	3·5	7	12	5	7
Agnes Keneman	1·5*	1	4	3	4
Matilda Edmund	1·25	4	4	0	4
John Wymere	7·0	6	12	6	5
Roger de Hil	10·17†	18	41	23	11
John Sunof	9·0†	12	24	12	19
Roger Starlyng	5·75	16	22	6	18
Ralph le Greyve	12·06†	23	27	4	15
Syward Stiward	6·11†	22	28	6	12
Roger Mome	6·0	11	12	1	4
William Goshey	6·0*	7	18	11	11
John le Dekene	15·75†	29	30	1	14
Robert le Clerk	25·13	39	41	2	10
Nicholas Haral	6·5†	8	11	3	3
Gunnilda Frone	3·0†	6	10	4	6
William Haylston	4·25	7	9	2	4
Humfrey Hisbald	2·0†	2	2	0	1
William Leve	7·09	16	19	3	16
Syware Blakeman	6·0†	1	10	9	10
Robert Curtman	5·56	14	21	7	7
Barth. son of Matild.	7·13	11	14	3	12
Wluiua Bo	0·5	1	3	2	3
John Godknape	5·0†	4	11	7	5
Robert le Longe	10·38†	13	26	13	8
William Chaplain	2·5	4	5	1	4
Total	169·875	283	419	136	217

* Wholly copyhold. † Wholly socage.

higher where the scope for subdivision was greatest. Thus Estfeld and Westfeld, which were already characterized by a high degree of subdivision, together experienced only a 37 per cent increase in parcellation, while Suthfeld and several of the smaller arable fields, particularly those with names containing the suffix "toft", manifest much higher rates of increase.[1] The explanation for this may be that Estfeld and Westfeld were the oldest and most central of Martham's arable fields and consequently were already highly fragmented when the *tenementa* were were established, whereas Suthfeld and other of the smaller fields were the products of more recent colonization and hence had been liable to subdivision for a much shorter period of time. In their case complete parcellation and with it final absorption into the commonfields did not come until the twelfth and thirteenth centuries. This progression from arable intake to subdivided commonfield can be illustrated by the example of the block of land known as Blakemannestoft.

The 6 acres of socage land comprising Blakemannestoft were situated on the edge of the cultivated area adjacent to the common pasture of Sco, a position which suggests a relatively late origin as an assart. Indeed, when the *tenementa* became fossilized it was still in the ownership of one man—Syware Blakeman—from whose surname the field name Blakemannestoft (the element *toft* being Old Danish for curtilage or close)[2] is plainly derived. By 1292, however, the physical unity of this block of land had broken down: where there had originally been but one tenant there were now ten and the land itself had been subdivided into as many separate parcels. In short, Blakemannestoft had been incorporated into, and become indistinguishable from, the rest of the commonfields. Only its name survived as a reminder that it had once formed a single landholding. Other portions of the commonfields bore very similar names—a personal name as prefix combined with the element *toft* as suffix (thus Hardyngestoft, Hilltoftes, Morgrimestoft, and Godknapestoft)—and closer inspection indicates that these, too, shared much the same history. Each seems to have begun as a separate close, probably created by the men whose surnames provide the prefixes Hardynge, Hil, Morgrime, and Godknape (surnames which were still current at Martham in 1292), but subsequently to have become subdivided and assimilated into the commonfields.

These few cases prove that Martham's commonfields expanded by a process of secondary subdivision[3] but analysis of the *tenementa* also shows that commonfields were by no means a creation of the thirteenth century. As early as the late twelfth century the majority of holdings lay fragmented and intermixed. Of the 26 *tenementa* examined only five were made up of land in just one field, all the others comprised a much wider scatter of strips, nine containing land in at least four different fields. In the distribution of their land within the township they anticipated the haphazard holding layout of a century later. The changes in field layout which can be identified at Martham during the thirteenth century were therefore changes of degree rather than of kind, with parcellation increasing in

[1] The relative increase in the subdivision of these fields was as follows: Estfeld 30 per cent, Westfeld 50 per cent, Suthfeld 90 per cent, Tomeres and Morgrave 60 per cent, various tofts 85 per cent.

[2] A. H. Smith, *English Place-Name Elements* (Cambridge, 1956).

[3] The term "secondary subdivision" is used to denote the parcellation of existing landholdings, as opposed to the creation of intermixed holdings from the first.

intensity and spreading in extent. Like the great contemporary colonizing movement of which the growth of these commonfields was in a sense part, the developments of the twelfth and thirteenth centuries represent but the tail-end of a much older process. Nevertheless, they do serve to show—as Thirsk has claimed—that secondary subdivision could perform a prominent and perhaps even decisive role in the formation of commonfields.

II

As already noted, the subdivision of holdings and parcellation of fields which took place at Martham during the course of the thirteenth century were closely associated with the great increase in population which was general at that time[1] and which was represented on the Prior of Norwich's manor by a 350 per cent increase in tenant numbers. Yet population growth, whilst a necessary precondition for subdivision, is not of itself a sufficient explanation of this phenomenon. Socio-economic factors were also important. It was these which determined whether increases in population were accommodated by the modification of existing households or the creation of new ones, only the latter being associated (where opportunities for further colonization were either slight or non-existent) with the subdivision of holdings and hence the parcellation of land. Of particular importance in determining which of these paths was followed were the nature of the prevailing tenurial regime and the extent to which the manorial authorities were prepared to condone the subdivision of holdings and alienation of land. Where tenure was inflexible, impartible inheritance prevailed, and the alienation of land was actively discouraged, population growth was unlikely to have much effect on holding size and layout, but where, as at Martham, the converse applied, then the potential for change was considerable.[2] Thus, in this township the fragmentation of land was as much a result of the operation of a custom of partible inheritance and the manorial authorities' preparedness to control rather than curb the alienation of land as of population growth *per se*. Any full explanation of the evolution of Martham's commonfields must therefore rest upon an understanding of these two processes—inheritance custom and the land market.

The operation of inheritance custom, as historians are becoming increasingly aware, could have far-reaching implications for many aspects of rural society.[3] However, it is only comparatively recently that it has been recognized as of importance to the growth of field systems. Thirsk first drew attention to this fact,

[1] For evidence of which see H. E. Hallam, 'Some Thirteenth Century Censuses', *Econ. Hist Rev.* 2nd ser. x (1957), 340–61; J. B. Harley, 'Population Trends and Agricultural Developments from the Warwickshire Hundred Rolls of 1279', *Econ. Hist. Rev.* 2nd ser. xi (1958), 8–18; J. Z. Titow, 'Some Evidence of the Thirteenth Century Population Increase', *Econ. Hist. Rev.* 2nd ser. xiv (1961), 218–23; M. M. Postan, ed. *The Cambridge Economic History of Europe*, i, *The Agrarian Life of the Middle Ages* (Cambridge, 2nd edn. 1966); M. M. Postan, *The Medieval Economy and Society* (1972).

[2] For a classic analysis of this contrast see G. C. Homans, *English Villagers of the Thirteenth Century* (Cambridge, Mass. 1941), also Harley, loc. cit.; J. Z. Titow, 'Some Differences between Manors and their Effects on the Condition of the Peasant in the Thirteenth Century', *Agric. Hist. Rev.* x (1962), 1–13.

[3] Homans, op. cit.; J. Thirsk, 'Industries in the Countryside', in F. J. Fisher, ed. *Essays in the Economic and Social History of Tudor and Stuart England* (1961), pp. 70–88; R. J. Faith, 'Peasant Families and Inheritance Customs in Medieval England', *Agric. Hist. Rev.* xiv (1966), 77–95; B. Dodwell, 'Holdings and Inheritance in Medieval East Anglia', *Econ. Hist. Rev.* 2nd ser. xx (1967), 53–66; J. Goody, J. Thirsk, and E. P. Thompson, eds. *Family and Inheritance: Rural Society in Western Europe, 1200–1800* (Cambridge, 1976).

and evidence of parcellation resulting from partible inheritance has since been found in counties as widely separate as Kent, Cambridgeshire, Yorkshire, and Devon.[1] It has even been claimed that partible inheritance was once ubiquitous and that the predominance of impartible inheritance in so much of England in the Middle Ages (particularly the heavily manorialized commonfield communities of the midland and southern counties) was a relatively recent development.[2] Yet despite the interest which different inheritance customs have aroused and the claims which have been made for them, empirical studies of their operation are few. Most historians have been content merely to identify which custom prevailed and only rarely have gone on and followed through how that custom worked in practice. This is unfortunate, for there could be a wide divergence between what happened in theory and what took place in practice. Subdivision was not an invariable concomitant of partible inheritance nor was it entirely precluded by impartible inheritance. In the former case there might be only one surviving heir, or co-heirs might agree on a settlement which did not require subdivision,[3] or even if subdivision did take place it might be annulled by subsequent land sales and exchanges. In short, to establish that partible inheritance was promoting the formation of subdivided fields it is necessary to do more than just establish its existence as a custom.

At Martham a substantial body of evidence testifies to the fact that inheritance custom was partible. In the late thirteenth and early fourteenth centuries this evidence is mostly indirect—the abuttals of strips recorded in the 1292 extent[4] and cases of co-heirship recorded in obituaries in the court rolls—but from the end of the fourteenth century, when co-heirs were fewer and customary inheritance was in danger of falling into abeyance, direct statements of partibility begin to appear. Thus in 1390 it was recorded in the court rolls that *tenementa tenta de domine in bondagio secundum consuetudinem manerii . . . inter heredes masculos sunt partibilis*,[5] whilst in the following century details are sometimes given of the manner in which holdings were to be divided.[6] The latter provide explicit testimony of the parcellation which could result from partible inheritance, but, unfortunately, are too few and appear too late to provide any real measure of the relative

[1] Thirsk, 'The Common Fields'; Baker, 'Open Fields'; M. Spufford, *A Cambridgeshire Community, Chippenham from Settlement to Enclosure* (Occasional Paper, Department of English Local History, University of Leicester, XX, 1965); Sheppard, loc. cit. Fox, loc. cit.

[2] Faith, loc. cit.

[3] Two types of division may be envisaged, those in which parcellation did occur (here termed "subdivision") and those in which it was avoided (termed "partition").—A. R. H. Baker, 'Some Fields and Farms in Medieval Kent', *Archaeologia Cantiana*, LXXX (1965), 160; R. A. Dodgshon, 'The Land-Holding Foundations of the Open-Field System', *Past & Present*, LXVII (1975), 3–29.

[4] Partible inheritance tended to produce a distinctive pattern of landownership whereby related tenants held adjacent and identical strips. The 1292 extent yields many examples of this including one in which a threefold subdivision between the brothers Robert de Hil senior, Robert de Hil junior, and John de Hil had led to the creation of 18 additional strips.

[5] N.R.O. NNAS 5928 20 D3.

[6] For example: *Thomas Mundes et Willelmus Mundes filii et heredes Margaret Mundes admissi fuerunt ad j acram terre native in Martham tanquam ad hereditatem suam iuxta consuetudinem huius manerii eo quod dicta Margareta mater eorum inde obiit seisita. Et admittus est modo ad hanc curiam veniant dicti heredes in plena curia coram senescalle et ostendant curiam quod ipsi de particione terre predicte in formam sequentie convenerunt et agreaverunt. Scilicet quod dictus Thomas habeat et teneat pro parte sua inde dimidiam acram que est orientalis pars predicte acre. Et dictus Willelmus habeat et teneat residuam dimidiam acram inde scilicet occidentalem partem predicte acre pro parte sua. Et dant domino de fine pro aggreamento predicto intrando etc.*—N.R.O. NNAS 5952 20 D5.

importance of subdivision and partition.[1] Nevertheless, it is plain that both occurred and, moreover, that the latter was relatively frequent. This is evident from the analysis of the *tenementa* described in the 1292 extent (see Tables 1 and 2), which shows that the proliferation of strips lagged well behind the multiplication of holdings, for whereas the number of holdings on the Prior of Norwich's manor increased by 350 per cent during the thirteenth century the number of strips increased by only 50 per cent during the same period. Either the increase in the number of holdings owed very little to partible inheritance (which seems improbable although some tenants certainly acquired holdings by purchase) or, as seems more likely, large numbers of holdings were being partitioned rather than subdivided. So intense was the morcellation of Martham's commonfields by the end of the thirteenth century that parcellation was rapidly approaching a threshold beyond which it was neither desirable nor necessary to proceed further. In other words, the rate of subdivision decreased as the amount of parcellation increased. Yet although partition gradually gained favour over subdivision the further alternative of co-parcenage never became popular. Of 2,122 strips recorded in the 1292 extent only 56 were held by co-parceners. Clearly, in matters of landownership, as in those of cultivation, emphasis was consistently upon the individual rather than the group.

Before a holding could qualify for subdivision, however, there had to be more than one heir. Co-heirship was never general and even under the most favourable circumstances of high fertility and low mortality there were always some tenants who left only one heir. The size of this proportion set an absolute ceiling to the amount of subdivision which could thus take place and this ceiling was sometimes very low: how low is revealed by an analysis of obituaries recorded in Martham's court rolls. Thus between 1351 and 1400, a period for which 142 obituaries are extant, only 37 tenants left more than one heir[2]—barely a quarter of the total. Moreover, there were only three cases in which the number of co-heirs exceeded three, which limited the amount of subdivision taking place still further. Admittedly, this was an exceptional period, the shortage of co-heirs bearing out the dismal picture which is generally painted of demographic developments at this time,[3] but even well before the Black Death when the population was still growing the proportion of cases in which tenants left co-heirs does not appear to have been large. The data which survive from this earlier period are much more fragmentary, but again they suggest that tenants who left only one heir were in the majority.[4]

This conclusion is endorsed and amplified by an analysis of the pattern of heirship on the manor of Hakeford Hall in the nearby township of Coltishall. Here, as at Martham, partible inheritance was practised, and in fact socio-economic conditions were closely analogous in the two townships, both being

[1] See above, p. 183, n. 3. [2] N.R.O. NRS 11312 26 B3, NNAS 5924 29 D3, NNAS 5930 20 D4.

[3] J. Saltmarsh, 'Plague and Economic Decline in England in the Later Middle Ages', *Cambridge Historical Journal*, VII (1941–3), 23–41; M. M. Postan, 'Some Economic Evidence of Declining Population in the Later Middle Ages', *Econ. Hist. Rev.* 2nd ser. II (1950), 221–46; J. M. W. Bean, 'Plague, Population and Economic Decline in England in the Later Middle Ages', *Econ. Hist. Rev.* 2nd ser. XV (1962–3), 423–37; J. A. Raftis, 'Changes in an English Village after the Black Death', *Mediaeval Studies*, XXIX (1967), 158–77; J. Hatcher, *Plague, Population and the English Economy, 1348–1530* (1977).

[4] N.R.O. Dean and Chapter Muniments, Martham Court Rolls, 1288–99.

characterized by high densities of population, small holdings, and complex and flexible systems of land tenure. At Coltishall, however, the sequence of surviving court rolls is much fuller and provides particularly good coverage of the last quarter of the thirteenth and first half of the fourteenth centuries.[1] Moreover, nominative listings and other evidence of a quasi-demographic nature contained in the court rolls allow the chronology of population change to be reconstructed, thereby providing the demographic context within which it is essential to view the incidence of co-heirship. This evidence confirms that on this manor the increase in population continued without serious interruption until well into the second quarter of the fourteenth century.[2] In 1349, however, bubonic plague precipitated a mortality of at least 55 per cent and subsequent outbreaks almost certainly reduced the population further so that in contrast to the first half of the fourteenth century the second half of the century was a period of almost unrelieved population decline. This contrast is reflected in the pattern of heirship (see Table 3), the incidence of co-heirs being twice as frequent during the period of population growth as during the period of population decline. Not only this, but cases involving three, four, or even five co-heirs, although never at any time

Table 3. *Martham and Coltishall: The Incidence of Co-heirship*

Years	*Total no. of deceased tenants*	*No. of tenants with at least:* 2 co-heirs *no.*	%	3 co-heirs *no.*	%	4 co-heirs *no.*	%	5 co-heirs *no.*	%
Martham:									
1289–99	10	3	30·0	0	0	0	0	0	0
1350–1400	142	37	26·1	6	4·2	3	2·1	0	0
1401–50	70	19	27·1	6	8·6	2	2·9	0	0
Coltishall:									
1275–1348	163	64	39·3	35	21·5	12	7·4	5	3·1
1349	92	18	19·6	2	2·2	0	0	0	0
1350–1400	95	19	20·0	8	8·4	3	3·2	1	1·1

very common, were more numerous during the former than the latter period. Nevertheless, although conditions of population growth were plainly more conducive to the division of holdings than those of decline, at no time did the number of tenants succeeded by co-heirs comprise more than a substantial minority of the total. Hence, even under the relatively favourable conditions which prevailed before the Black Death it could have taken up to six generations before each holding had undergone subdivision at least once, the equivalent period after the Black Death being 11 generations or more. If the parcellation of land was the outcome of partible inheritance alone it must therefore have been a much more protracted process (and one which was much more dependent upon population growth) than has often been appreciated. For it to have produced the profound fragmentation of land which existed in Martham in 1292 would have taken several centuries.

Apart from the changes induced by partible inheritance, however, changes in

[1] King's College, Cambridge, E29–38.

[2] For a full discussion of this evidence and its implications see B. M. S. Campbell, 'Population Pressure, Inheritance and the Land Market in a Fourteenth Century Peasant Community', in R. M. Smith and E. A. Wrigley, eds. *Land, Kinship and Life-cycle* (forthcoming).

field and holding layout could also arise from the alienation of land during a tenant's lifetime. How long tenants had enjoyed this privilege at Martham is not clear, but it had certainly received the sanction of the manorial authorities by the second half of the thirteenth century when the 1292 extent and surviving court rolls show that it was well established and frequently exercised. Indeed, an active peasant land market seems to have been an established feature of most manors in this area by this time.[1] The existence of this market has important implications for the evolution of Martham's commonfields, for it was potentially the most dynamic of all agents of change. Its potential lay in its ability to sustain a much higher volume of turnover and regularly involve a much larger proportion of tenants than could ever result from inheritance custom, and to enable tenants to make deliberate changes in the size and layout of their holdings. Of course, not all tenants could command the resources required to purchase land or avoid the circumstances which obliged some to sell it, but whether they were buying land or selling it they could at least decide which piece of land was transacted. In this respect the effect of land transactions upon holding layout tended to be much more selective than that of inheritance custom. However, because the reasons and the circumstances under which tenants transacted land were so varied this effect was neither simple nor consistent. Thus, on the one hand the land market could promote the engrossing of holdings and consolidation of strips and on the other it could foster their fragmentation and subdivision. Moreover, since both these developments might be taking place simultaneously—some tenants accumulating land while others were disposing of it—it is necessary to distinguish between the net and gross effects of the land market. It was the former which determined whether the land market reinforced or counteracted inheritance custom.

Since the alienation of customary land required the sanction of, and represented a source of income to, the manorial authorities a full record of all land transactions was kept in the court rolls. Where such rolls survive, therefore, they provide an excellent guide to the dimensions and general trend of the land market. In this respect it is particularly unfortunate that Martham's series of court rolls is so fragmentary and that only a handful of rolls is extant from before the Black Death. Nevertheless, the latter do convey at least some impression of the size and frequency of land transactions round about the time that the 1292 extent was compiled. Indeed, in the last decade of the thirteenth century conveyances were one of the most frequent items of business dealt with by the manor court, the 12 courts of which records are extant sanctioning no less than 98 separate land transactions. This indicates an average turnover of about eight transactions per court and 30–40 transactions per year. Since only one piece of land was generally ever transacted at a time, the size of these transactions reflected the size of commonfield strips and was small. However, although the mean size of each transaction was only 1·6 rods their number was sufficient to ensure a turnover of about 12–16 acres a year, at which rate there could have been a complete turnover of land on this manor every fifty to seventy years. This observation is endorsed by a consideration of the much fuller series of court rolls which survives for the manor of Hakeford Hall in the township of Coltishall, to which reference has already been made. This series of rolls provides virtually continuous

[1] Campbell, thesis, pp. 189–219, 281–2.

although by no means complete coverage of the period 1275–1348 and reveals the existence of a land market which, allowing for the smaller size of this manor, was equally as active as that at Martham. Thus the 180 courts of which records are extant sanctioned a total of 900 conveyances at a mean rate of five conveyances per court. As at Martham, only one strip was generally ever transacted at a time, hence the mean size of each conveyance was small—only 1·8 rods—but again their number was sufficient to ensure a steady turnover of land, about 9–11 acres per year. What is perhaps most interesting of all, however, is the evidence which this series of rolls provides of a steady expansion in the activity of the land market during the first half of the fourteenth century. Between the 1290s and the 1340s the amount of land being conveyanced more than doubled. The possibility thus arises that the same may also have occurred at Martham with corresponding implications for the pattern of landownership. How great an impact the land market had already made by the close of the thirteenth century can be deduced from the 1292 extent.

As explained earlier, when the *tenementa* were first established each *tenementum* formed the holding of one tenant. By 1292, however, this neat coincidence between *tenementa* and holdings had disappeared, the former were divided between a number of different tenants and the latter were made up of land in more than one *tenementum*. This dispersal of landownership may to some extent have been the result of lateral inheritance, marriage settlements, and the like, but it was almost certainly also a product of the land market. Moreover, the wider the dispersal the greater the likelihood that it was the land market which was responsible. Thus out of a total of 376 holdings 48·6 per cent comprised land in at least two *tenementa*, 12·2 per cent comprised land in at least five *tenementa*, and 3·0 per cent comprised land in at least ten *tenementa*; all of which indicates a considerable redistribution of land. Nowhere is this redistribution more apparent than in the last group of holdings, those comprising land in at least ten *tenementa*. Although only 11 holdings fell within this category they included some of the largest holdings at Martham and together accounted for 14 per cent of the total area. In seven cases their owners belonged to families that had been established at Martham since the twelfth century, five of whom had actually succeeded in enlarging the size of their landholding. Here the implication would seem to be that these five families had used the land market to amass land, as can actually be traced in one case at the end of the thirteenth century.[1] Nevertheless, the importance of this process should not be exaggerated, for although a minority of tenants succeeded in expanding their holdings in this way the majority of tenants suffered a steady reduction in holding size. In fact, so pronounced was the latter development that it is tempting to conclude that the general fragmentation of landownership which took place during the thirteenth century was as much a product of the alienation of land as it was of the division of holdings between heirs. This was almost certainly the case at Coltishall where a systematic analysis of the individuals who transacted land shows that (with the exception of a few brief periods of economic hardship) buyers consistently outnumbered sellers and

[1] Thomas Godknape purchased 1 acre 1 rood in 1288, exchanged 1 acre 1 rood 23 perches for 1 acre 2 roods in 1290, and purchased 1 rood in 1298. In 1292 he held 12 acres 1 rood 14 perches distributed among 12 different *tenementa*.

that the amounts which individuals bought consequently rarely matched those which they sold.[1] Evidently not all tenants acquired their holdings by inheritance, some acquired theirs by purchase. Similarly, the parcellation of land may have been as much a product of the land market as of inheritance custom, the sale of small slips of land saving many peasants from starvation when harvests failed. Of course, there may also have been some consolidation of strips to compensate for this, but circumstances, namely the great and growing complexity of land-ownership, militated against this.

Table 4. *Martham: Composition of Holdings by* Tenementa *in 1292*

Number of tenants holding land in at least 10 *tenementa*	11
,, ,, ,, ,, ,, ,, ,, 9 ,,	14
,, ,, ,, ,, ,, ,, ,, 8 ,,	18
,, ,, ,, ,, ,, ,, ,, 7 ,,	24
,, ,, ,, ,, ,, ,, ,, 6 ,,	35
,, ,, ,, ,, ,, ,, ,, 5 ,,	46
,, ,, ,, ,, ,, ,, ,, 4 ,,	71
,, ,, ,, ,, ,, ,, ,, 3 ,,	105
,, ,, ,, ,, ,, ,, ,, 2 ,,	183
,, ,, ,, ,, ,, ,, ,, 1 *tenementum*	376

Until the middle of the fourteenth century, therefore, the land market complemented more than it countered the effect of inheritance custom: it lent an extra dimension to the fragmentation of holdings and parcellation of fields and further heightened the complexity of landownership. Subsequently, however, this situation was transformed. From 1349, and possibly even before,[2] population decline served to restrict the influence of partible inheritance on one hand, but to increase the relative supply of land on the other. The initial effect of the latter was to stimulate the leasing of land, but after about a generation, during which pressure upon land continued to slacken (witness the marked shortage of direct male heirs recorded by the obituaries)[3] and the supply of land consequently to increase, the duration of leases began to lengthen and eventually to give way to the permanent alienation of the land (see Fig. 1). Once this had taken place (the land market by now having become a buyer's rather than a seller's market), the engrossing of holdings and consolidation of strips were able to proceed apace. From the 1390s, as leasing began to wane, the land market consequently entered a period of unparalleled activity. The number of conveyances recorded in the court rolls was about the same as a century earlier (about 30–40 per year), but as each conveyance now frequently dealt with several strips at once and the strips themselves were becoming larger the total area transacted was over twice as great, averaging 25–35 acres per year. Moreover, this turnover was sustained by a tenant population which was significantly smaller than that which had existed at the end of the

[1] For further details see Campbell, in Smith and Wrigley, op. cit.

[2] There is nothing to indicate when population began to decline at Martham, although it is clear from circumstantial evidence that the Black Death exacted a heavy toll. This was the case almost everywhere in this thickly settled, corn-growing region (for an account of the Black Death in East Anglia see J. F. D. Shrewsbury, *A History of Bubonic Plague in the British Isles* (Cambridge, 1970), pp. 94–9). At Coltishall population decline does not appear to have begun before the 1340s.—Campbell, in Smith and Wrigley, op. cit.

[3] Of 142 tenants who died between 1350 and 1400 only 48·6 per cent were succeeded by sons.

Fig. 1. *Martham, Norfolk: Land Transactions Recorded in the Court Rolls*

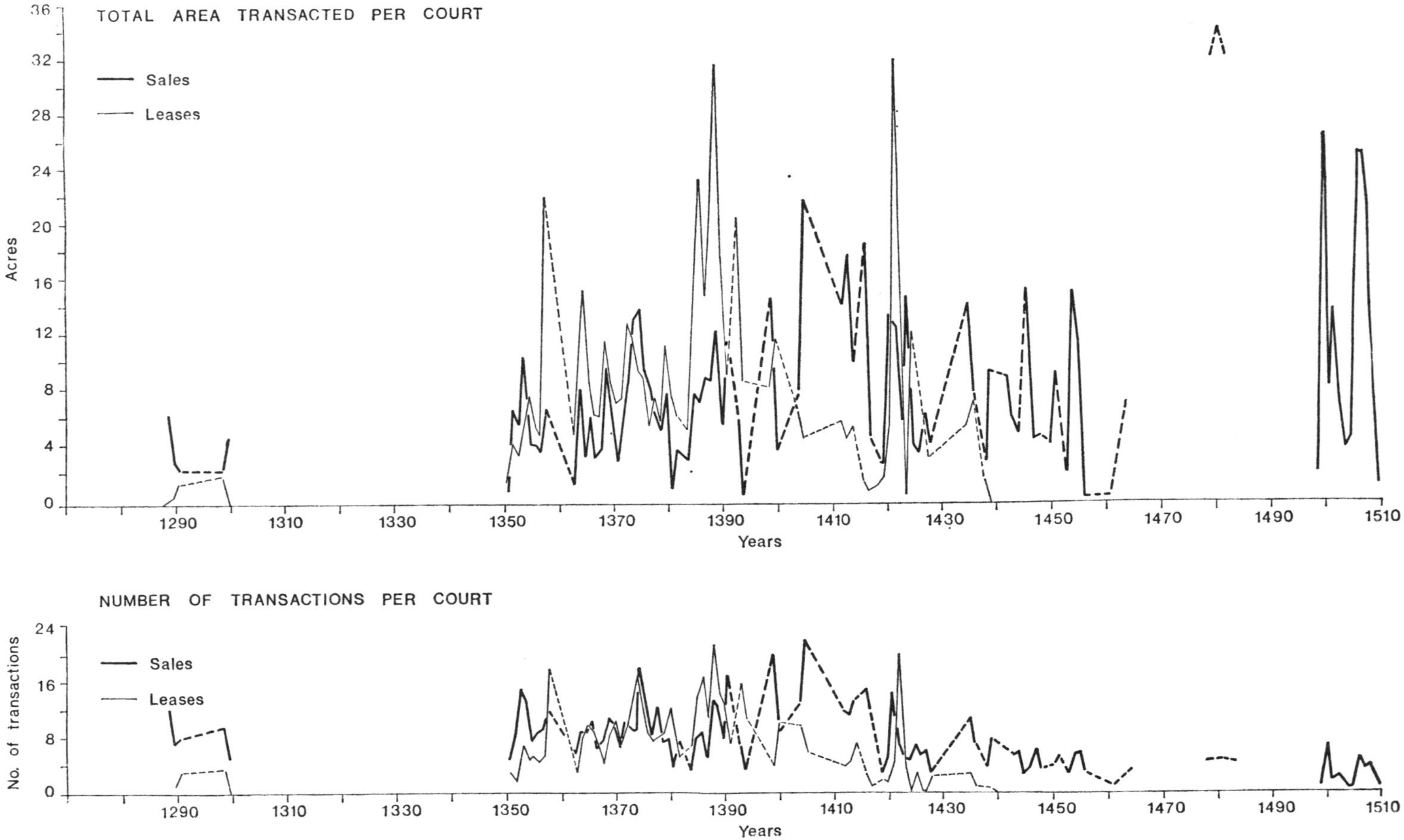

thirteenth century, a circumstance which lent still greater weight to the forces of agglomeration. The absolute turnover of land on the land market slackened somewhat during the middle years of the fifteenth century, but began to pick up again towards the beginning of the next, by which time a profound change had been effected in the pattern of landownership, as is revealed by a rental of 1497.[1]

The area encompassed by this rental is somewhat smaller than that covered by the 1292 extent, but allowing for this a comparison between the pattern of landownership at these two dates reveals the striking transformation which had taken place in the number and size of landholdings (see Table 5). Not only had

Table 5. *Martham, Norfolk: Number and Size of Landholdings in 1292 and 1497*

Size of holding (acres)	*1292* *No. of holdings*	*% of total no.*	*% of total area*	*1497* *No. of holdings*	*% of total no.*	*% of total area*
18 and over	0	0	0	11	14·3	42·8
17–under 18	1	0·3	2·0	1	1·3	2·5
16–under 17	0	0	0	1	1·3	2·2
15–under 16	0	0	0	1	1·3	2·1
14–under 15	1	0·3	1·7	2	2·6	4·1
13–under 14	0	0	0	4	5·2	7·5
12–under 13	3	0·8	4·4	2	2·6	3·4
11–under 12	2	0·5	2·7	4	5·2	6·3
10–under 11	3	0·8	3·7	2	2·6	3·0
9–under 10	1	0·3	1·1	2	2·6	2·5
8–under 9	6	1·6	6·0	2	2·6	2·3
7–under 8	4	1·1	3·5	5	6·5	5·2
6–under 7	8	2·1	6·2	4	5·2	3·6
5–under 6	19	5·1	12·3	5	6·5	3·6
4–under 5	18	4·8	9·6	4	5·2	2·3
3–under 4	25	6·7	10·3	4	5·2	1·8
2–under 3	47	12·4	14·1	8	10·4	2·6
1–under 2	85	22·6	13·8	9	11·7	1·8
Under 1	153	40·7	8·6	6	7·8	0·2
Total	376	100%	100%	77	100%	100%

the number of holdings declined but land had become increasingly concentrated into the hands of a few, less than a fifth of the total number of holdings now accounting for half the total area. That the polarization of landownership was latent in the action of the land market has been seen as early as the thirteenth century hence there can be little doubt that it was to this process that these much more substantial late fifteenth-century holdings owed their origin (having benefited from the freer market conditions and breakdown of traditional attitudes towards land which had followed in the wake of the demographic collapse of the mid-fourteenth century). It can be assumed that even if there had been no deliberate consolidation of strips some consolidation would have occurred as an inevitable concomitant of the concentration of strips into fewer hands, but how far this had modified field layout is a matter of conjecture. In other townships in this locality the first piecemeal enclosure was beginning to take place at this time and by the middle of the sixteenth century commonfields throughout eastern Norfolk were in active process of dissolution.[2] Nevertheless, the dissolution of

[1] N.R.O. Dean and Chapter Muniments, MS 2765. [2] Campbell, thesis, pp. 302–6.

these commonfields, like their formation, was an extremely protracted process and as late as the end of the eighteenth century small patches of commonfield land were still to be seen in this area.[1]

III

This consideration of the documentary evidence relating to the evolution of landholdings at Martham from the twelfth to the fifteenth century elucidates several aspects of the development of this township's commonfields. In particular, it demonstrates—as Thirsk has surmised—that the need to accommodate a growing population on a finite amount of land could lead to the progressive parcellation of that land and, ultimately, the emergence of subdivided fields. More than this, though, this study shows that the impact of population growth upon landholdings was tempered by the customs which governed and processes which influenced landownership, notably inheritance custom and the land market. Because of these processes population growth and the parcellation of land did not always proceed at the same rate. In fact, it would seem that although the initial breakup of consolidated holdings may have been achieved fairly quickly, the reduction of those holdings to a state of intense and relatively uniform subdivision could be very slow. The intense subdivision of arable land which prevailed at Martham in the late thirteenth century and which is revealed so clearly by the survey of 1292 represented the culmination of several centuries of development. That the evolution of subdivided fields was such a protracted process has not always been appreciated.

As important as the support which this evidence lends to that part of Thirsk's hypothesis which relates to the origin of a pattern of intermixed holdings, however, is its failure to bear out that part of her hypothesis which endeavours to account for the adoption of common rules of cultivation and grazing. At Martham the layout of holdings and fields reflected their haphazard origin and as such was positively inimical to the regularization and systematization of agricultural method. To have established a regular midland-type scheme of commonfield management would therefore have required a major reallotment and realignment of strips. According to Thirsk such a rationalization of landownership patterns may have been brought about gradually by means of piecemeal sales and exchanges on the part of individual cultivators,[2] but there is no evidence that the mounting complexity of holding layout ever induced this response at Martham. On the contrary, sales and exchanges tended to increase rather than reduce the confusion and appear to have been used to achieve a concentrated rather than an even scatter of strips. In the absence of a spontaneous reallotment of strips, of course, a planned redistribution of strips may have been used to achieve the same effect, but, again, there is no sign that this was ever contemplated at Martham. Despite the progressive diminution of holdings and proliferation of strips the inhabitants of this township showed no inclination to adopt a more communal system of agriculture. Indeed, the adoption of common rules of cultivation and

[1] W. Marshall, *The Rural Economy of Norfolk*, I (1975), 4.

[2] Thirsk, 'The Common Fields', 21–2. This is an essential element in Thirsk's thesis for it is by this means that she endeavours to account for the dearth of documentary references to the remodelling of field systems.

grazing may have been a retrograde step, for in their virtual absence eastern Norfolk supported higher standards of cultivation, higher levels of arable productivity, and a higher density of population than were to be found in localities with more closely co-ordinated and, therefore, supposedly more efficient commonfield systems. To assume that the mounting parcellation of arable fields needed to be offset by the progressive systematization of their management if diminishing returns were not to set in, would appear to be invalid. Some alternative explanation for the origin of common rights and regulations must consequently be sought. In this context the lack of interest in such rights and regulations in localities where individual peasants and the peasant community as a whole both enjoyed considerable freedom of action is perhaps significant. It suggests that the regularization of holding layout and systematization of cropping and grazing practices may not have emanated from within the peasant community but have been imposed upon it from without, possibly by the manorial lord (as in the case of the creation of the *eriungs* at Martham).[1] It is certainly intriguing that commonfields appear to have manifest greatest regularity of layout and management in precisely those areas where manor and vill were most often coincident.

[1] For continental evidence of seigneurial remodelling of field systems see S. Goransson, 'Field and Village on the Island of Oland', *Geog. Annaler*, XL B (1958), 101–58, and C. T. Smith, *An Historical Geography of Western Europe before 1800* (1967), pp. 236–41.

II

THE EXTENT AND LAYOUT OF COMMONFIELDS IN EASTERN NORFOLK

Eastern Norfolk is not generally thought of as a champion area. In 1611 Sir Henry Spelman wrote of the county as a whole 'the parts from *Thetford* to *Burneham*, and thence Westward, as also along the Coast, be counted *Champion*: the rest (as better furnished with woods) *Woodland'*;[1] whilst nearly two centuries later Nathaniel Kent commented of eastern Norfolk 'the eye seems ever on the edge of a forest, which is, as it were by enchantment, continually changing into hedgerows'.[2] In 1795 William Marshall gave a more prosaic description of the area: 'Some remnants of common fields still remain; but in general, they are not larger than well sized inclosures. Upon the whole, East Norfolk at large may be said to be A VERY OLD-INCLOSED COUNTRY.[3] These testimonies nothwithstanding, there is abundant evidence that commonfields were once extremely well developed in eastern Norfolk. This paper is concerned with their extent and layout and the changes which they underwent over time.[4]

*   *   *   *

A convenient starting point is provided by the parish maps and surveys which come into being in the late sixteenth century. For eastern Norfolk, there are manuscript maps of the 1580s relating to the parishes of Cawston and Lessingham (Figures 1 and 2),[5] whilst a third and somewhat later map, relating to the parish of Worstead, survives as an eighteenth century copy (Figure 3).[6] Written surveys are more numerous than the maps, and include those of Westwick (1547), Coltishall, Hempstead and Lessingham (all 1584), Blofield (1586), and Horstead-with-Stanninghall (also 1586).[7] Lessingham possesses both a written and a cartographic survey (the map of Horstead-with-Stanninghall – Figure 4 – is a reconstruction made by W. J. Corbett from the information contained in the Elizabethan survey book).[8] These eight parishes are widely, if somewhat unevenly, scattered through eastern Norfolk and are representative of a range of soil types.

The range of soil types in eastern Norfolk is greater than might be suspected from the subdued nature of the terrain. Soil fertility is most consistently high near the coast where the rich loam soils comprise some of the best arable land in the country. These loams occupy virtually the whole of Flegg and extend northwards along the coast as far as Trunch; they also occupy the low ridge of land

II

6

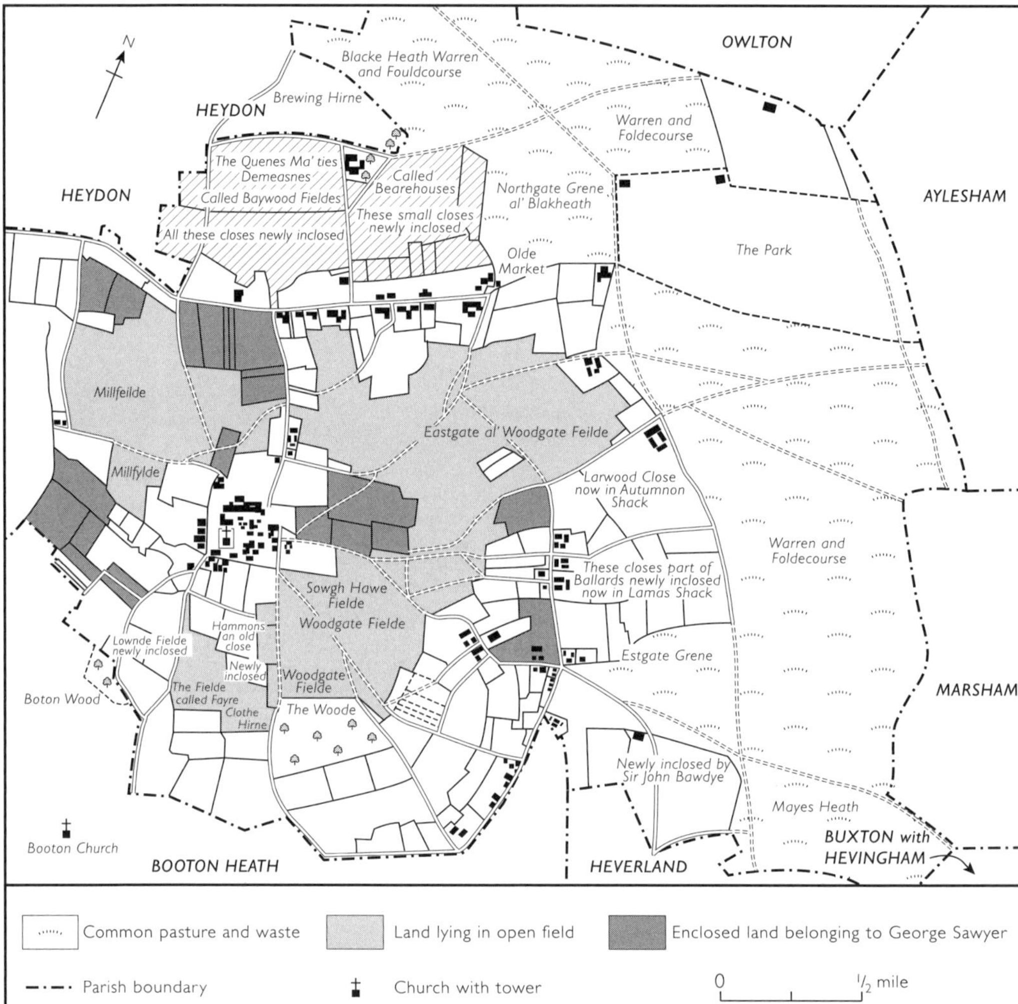

CAWSTON c.1580

Source: Norfolk Record Office, NRS 21404 A

Fig. 1

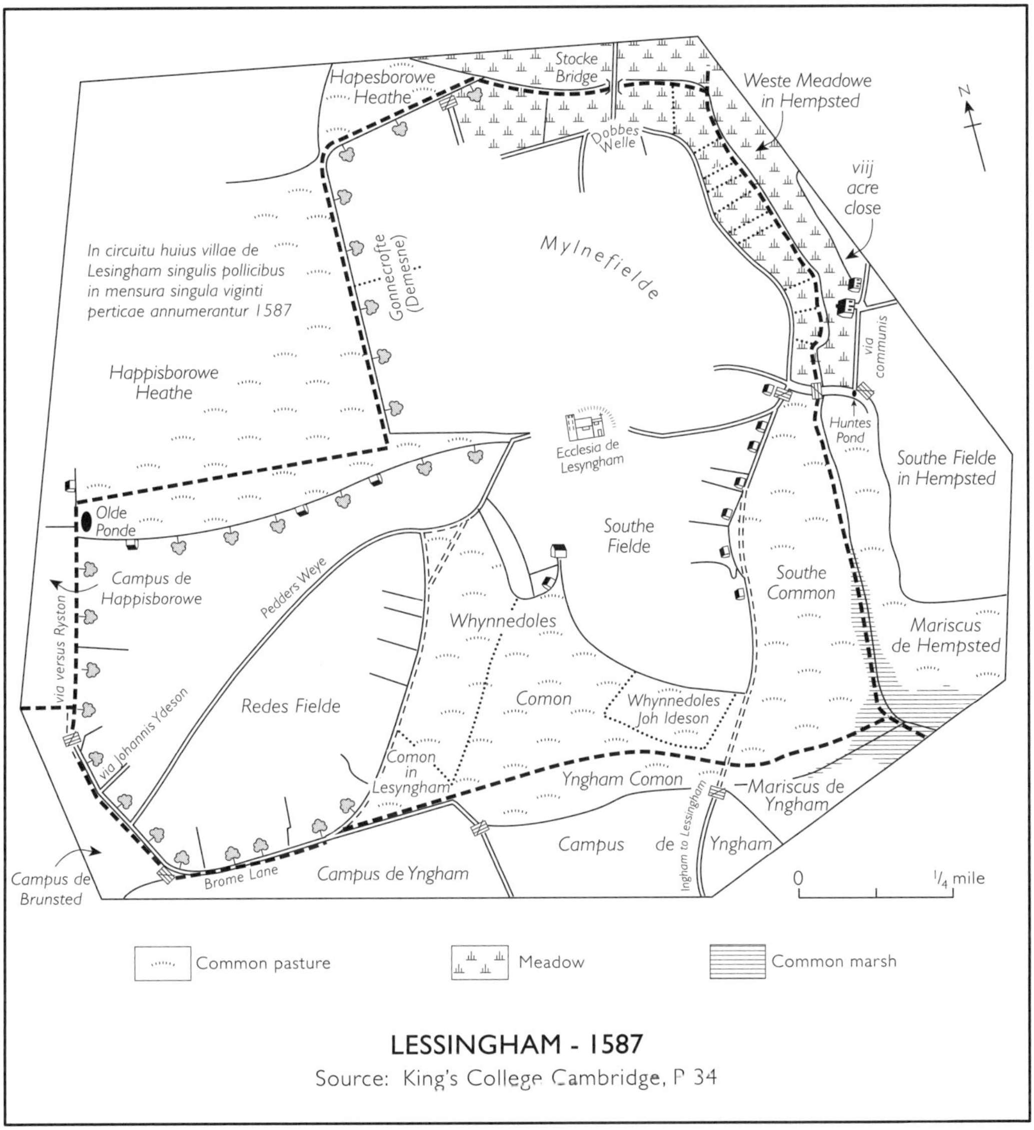
Hapesborowe Heathe
Stocke Bridge
Weste Meadowe in Hempsted
N
Dobbes Welle
viij acre close
In circuitu huius villae de Lesingham singulis pollicibus in mensura singula viginti perticae annumerantur 1587
Gonnecrofte (Demesne)
Mylnefielde
via communis
Happisborowe Heathe
Ecclesia de Lesyngham
Huntes Pond
Southe Fielde in Hempsted
Olde Ponde
Southe Fielde
Campus de Happisborowe
Pedders Weye
Southe Common
via versus Ryston
Whynnedoles
Mariscus de Hempsted
Redes Fielde
Comon
Whynnedoles Joh Ideson
via Johannis Ydeson
Comon in Lesyngham
Yngham Comon
Mariscus de Yngham
Campus de Yngham
Ingham to Lessingham
Campus de Brunsted
Brome Lane
Campus de Yngham
0
1/4 mile
Common pasture
Meadow
Common marsh
LESSINGHAM - 1587
Source: King's College Cambridge, P 34

Fig. 2

WORSTEAD

Source: Norfolk Record Office, Church Commissioner's Map 11905

Fig. 3

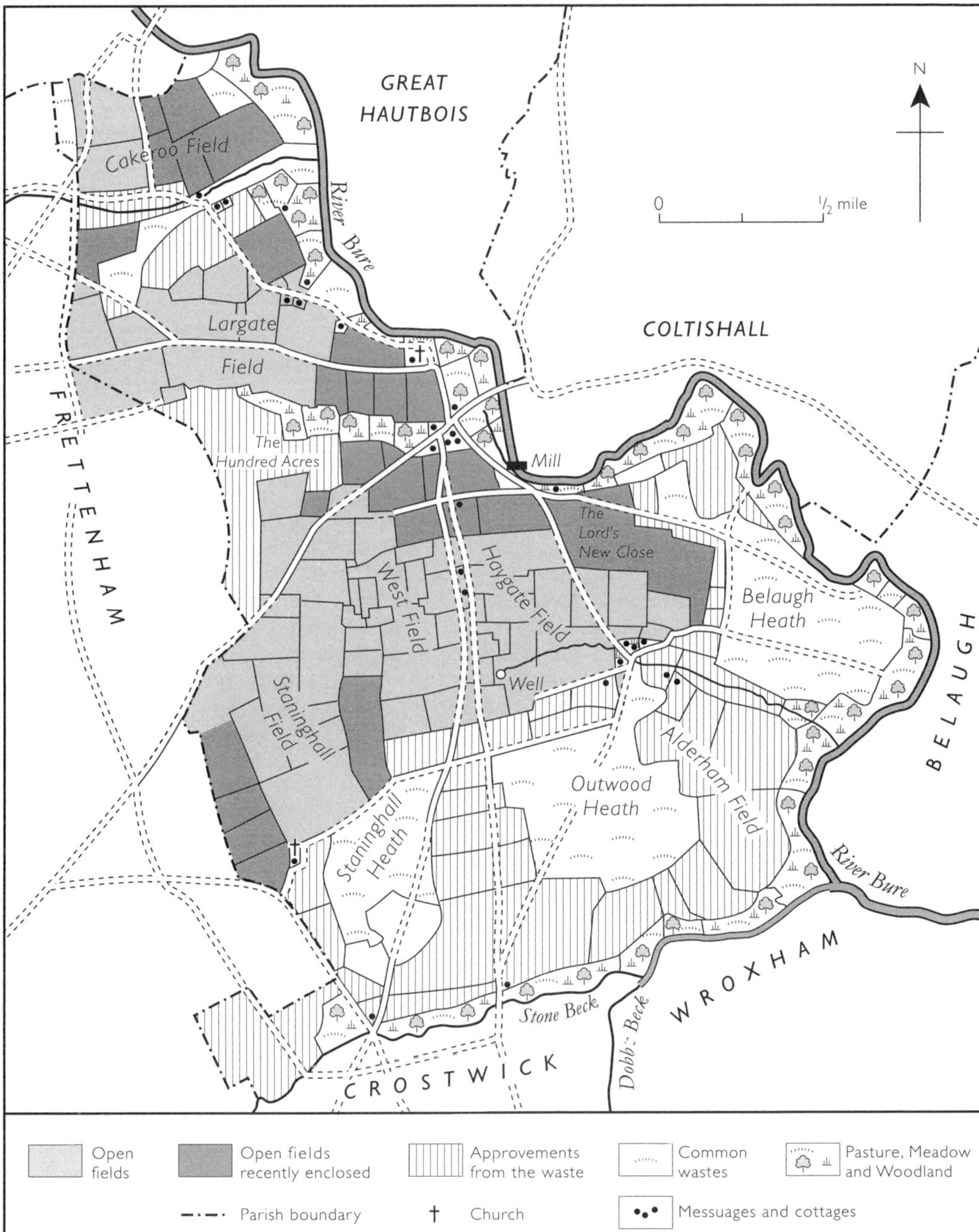

HORSTEAD - AND - STANINGHALL 1586

Source: Inset map from W.J. Corbett, "Elizabethan Village Surveys", *Transactions of the Royal Historical Society*, New Series, Vol. XI (1897)

Fig. 4

between the lower valleys of the Yare and Bure, and are found in patches throughout the area. Immediately to the north of Norwich, however, and thence extending in an arc curving west and north, the soils are lighter and more varied as the boulder clay drift gives way to a band of plateau and outwash sands and gravels. Loams still predominate, but north of the city sands and gravels have produced extensive infertile tracts for long occupied by heaths, of which Mousehold Heath is the best known but by no means the only example. But just as these soils are generally unsuitable for cultivation due to their lightness, so others have long remained uncultivated because of their wetness. The latter is particularly the case in the lower valleys of the rivers Ant, Bure and Yare, where the accumulation of alluvium and peat has led to the formation of this area's most distinctive physical feature, the extensive 'Broadland' marshes.

These diverse soil conditions are reflected in the relative importance of permanent, and more importantly of common, pasture in each of the surveyed parishes (Table 1). All the parishes contained at least some common pasture but the relative and absolute amounts varied considerably: where soils were poor, as at Cawston, or in part ill-drained, as at Hempstead, common pastures were usually very extensive, but where all or most of the land was cultivable, as at Coltishall, common pasture was minimal. The survey of Coltishall enumerates 1,126 acres of which common pasture comprised a mere 10 acres; three other pieces of common are mentioned, but as the total size of the parish is only 1,190 acres the most that they can have accounted for is a further 64 acres. Indeed, it is unlikely that there were more than 50 acres of common pasture in this parish in 1584: some additional grazing land, held in severalty, did exist but was of limited extent. The corollary of this situation is that the commonfields extended right up to the parish boundary, where they merged with those of the adjacent parishes of Great Hautbois, Sco Ruston, Tunstead, and Belaugh. This feature was repeated elsewhere in eastern Norfolk: the commonfields of Westwick linked with those of Worstead to the east, and the commonfields of Blofield linked with those of Hemblington, North Burlingham, Braydeston,

Township	Date of Survey	Total Acreage of Parish	Total Acreage Surveyed	Percentage Surveyed	Total Acreage Open Fields	Total Acreage Closes	Closes as percentage of total farmland
Westwick	1547	1,205.0	757.5	62.9	532.0	225.5	29.8
Coltishall	1584	1,190.0	1,126.5	94.7	861.0	255.5	22.9
Hempstead	1584	907.25	524.2	57.8	402.6	121.6	23.2
Lessingham	1584	641.0	446.9	69.7	267.8	179.1	40.1
Blofield	1586	2,321.0	1,938.6	83.4	1,223.5	715.1	36.9
Horstead-with-Stanninghall	1586	2,848.0	2,083.4	73.2	1,200.8	882.6	42.4

Eastern Norfolk: Extent of Commonfield in Six Townships in the Sixteenth Century

Table 1

Sources: N.R.O. Pet. MS 584; N.R.O. NRS 16646 37G; K.C.C. E28, N52, P34

and Strumpshaw. In fact, conditions at Blofield closely resembled those at Coltishall, for extensive commonfields filled by far the greater part of the parish: this was in marked contrast with a parish such as Horstead-with-Stanninghall where patches of infertile sand and gravel had led to the persistence of residual areas of common pasture amidst the arable fields. Evidently commonfields were most extensive where soils were most consistently good, notably in the hundreds of East and West Flegg, Walsham, and Blofield. In these particular districts little potentially cultivable land was left uncultivated and it was the adjacent marsh-lands, together with occasional heaths, which provided the principal reserves of common pasture.

The survey books nevertheless reveal that this simple distinction between commonfield and common pasture was already beginning to break down under the impact of enclosure. Closes accounted for at least a fifth of the farmland in most parishes, and twice this in several, notably Lessingham and Horstead-with-Stanninghall (Table 1). Some of these closes may represent land anciently held in severalty, particularly demesne land (as appears to have been the case at Lessingham), but in most cases they comprised former commonfield land. On the map of Cawston (Figure 1) several closes, including demesne land, are labelled 'newly inclosed' and similar references to piecemeal enclosures may be derived from many other parishes in this area. At Gimingham, for instance, it was agreed in 1583 that the inhabitants of the parish might keep all existing enclosures of commonfield land 'without any leavinge open at the time of shacke'.[9] At Bassingham, on the other hand, it was asserted that 'One who had purchased divers parcells together, in which the inhabitants have used to have shacke, and long time since have inclosed it . . . . . . the owner cannot exclude them of common there, notwithstanding that he will not common with them, but hold his own lands so inclosed.[10] However, by-laws such as this were rarely very effective and the enclosed area gradually encroached upon the unenclosed until a field layout was created, such as prevailed at Worstead in the seventeenth century (Figure 3), in which commonfields were reduced to a few isolated blocks.

That the enclosure movement was gaining momentum in the late sixteenth century was largely a function of the considerable progress which had already been made in the preliminary task of consolidating commonfield strips. This is clearly revealed by the parish surveys. Table 2 summarizes the size distribution of land parcels recorded in these surveys and demonstrates the extent to which consolidation had already preapred the way for enclosure. Strips containing less than ½ acre were still the most numerous but they only occupied a fraction of the total area, less than one per cent at both Lessingham and Horstead-with-Stanninghall. In spatial terms parcels of at least 3 acres in size, although much less numerous, were considerably more important. Such parcels contained at least 50 per cent of the total area at Westwick, Coltishall, Hempstead, and Blofield, 70 per cent of the total area at Lessingham, and 85 per cent of the total area at Horstead-with-Stanninghall (whose inhabitants were to agree to the enclosure of the common pastures and extinction of all common rights in 1599).[11] It is noticeable that parcels tended to be largest around the periphery of the commonfields and in the immediate vicinity of messuages, indicating, perhaps, the greater ease of consolidating strips in peripheral locations, and the greater desire to consolidate strips in central locations. Enclosure thus tended to spread outwards from the main settlement nuclei and inwards from the perimeter of the open fields, producing the sort of transitional landscape represented in the maps of Cawston and Worstead (Figures 1 and 3).

Size of Land Parcel	Westwick 1547	Hempstead 1584	Blofield 1586	Coltishall 1584	Lessingham 1584	Horstead-with-Stanninghall 1586
	Percentage of Total Number					
At least 15 Ac.	0.4	0.5	1.0	1.4	3.6	5.1
10 Ac. to less than 15 Ac.	1.2	0	0.4	1.7	1.8	3.6
5 Ac. " ". " 10 Ac.	4.2	4.1	4.9	4.5	7.2	15.8
4 Ac. " " " 5 Ac.	3.2	3.3	2.3	3.8	6.6	6.9
3 Ac. " " " 4 Ac.	2.0	4.4	4.8	6.3	8.4	9.9
2 Ac. " " " 3 Ac.	8.7	8.5	10.4	14.1	14.4	11.3
1 Ac. " " " 2 Ac.	25.2	22.1	32.4	29.6	33.5	24.4
½ Ac. " " " 1 Ac.	28.8	33.6	33.3	30.3	16.2	16.9
" " ½ Ac.	26.3	23.5	10.5	8.3	8.3	6.1
	Percentage of Total Area					
At least 15 Ac.	4.6	13.4	15.9	13.7	26.3	38.4
10 Ac. to less than 15 Ac.	9.5	0	2.7	10.0	8.7	9.5
5 Ac. " " " 10 Ac.	20.0	17.6	17.6	14.1	15.9	23.8
4 Ac. " " " 5 Ac.	8.8	9.5	5.6	7.9	10.2	6.4
3 Ac. " " " 4 Ac.	8.4	9.2	8.9	11.0	9.1	6.8
2 Ac. " " " 3 Ac.	12.8	12.5	13.3	14.7	11.5	5.5
1 Ac. " " " 2 Ac.	20.0	18.9	22.4	17.9	14.2	7.0
½ Ac. " " " 1 Ac.	11.9	14.7	11.8	9.4	3.3	2.3
" " ½ Ac.	4.0	4.2	1.8	1.3	0.8	0.3
Total No. of parcels	504	366	1,119	574	167	467
Mean size of parcel	1.5 ac.	1.4 ac.	1.7 ac.	2.0 ac.	2.7 ac.	4.5 ac.

Eastern Norfolk: Size Distribution of Land Parcels in Six Townships in the Sixteenth Century

Table 2

Sources: N.R.O. Pet. MS 584; N.R.O. NRS 16646 37G; K.C.C. E28, N52, P34

In spite of the differences which existed between these eight parishes, in the ratio of common pasture to commonfield and in the rates at which consolidation and enclosure were taking place, the layout of their fields shared one fundamental feature: a pervasive irregularity. There is no evidence that any of them ever possessed the sort of internal order that was the very essence of the regular commonfield systems of the midlands. In the latter, strips were aggregated into furlongs and fields and it was these which comprised the basic units of cultivation and grazing. In eastern Norfolk, in contrast, it was the individual parcels and holdings which provided the twin bases of commonfield management. As a result field divisions manifested an almost bewildering variation, reflecting the peculiar topography of each parish. To reduce this complexity to some sort of order amenable to written description, sixteenth century surveyors sometimes resorted to the device of dividing each parish into a number of 'precincts' each of which was further subdivided into a number of 'quarentines'. These, however, are abritrary divisions adopted for the sake of descriptive convenience and should not be mistaken for East Anglian equivalents of the 'fields' and 'furlongs' of the midlands.[12] Thus, at Coltishall the surveyors divided the parish into nine precincts, ranging in size from 43.5 to 248.675 acres: these nine precincts were then further subdivided into 83 quarentines, the largest of which was larger than both the first and fourth precincts and contained some 82.25 acres. Moreover, the constituent parcels of individual holdings were extremely unevenly divided between these precincts and there was a marked tendency for holdings to be concentrated within particular portions of the parish (see Table 3). Richard Ward, for example, held 55.25 acres at Coltishall in 1584 (he also held a substantial acreage in the neighbouring parish of Horstead-with-Stanninghall) concentrated in only three precincts – the sixth, seventh, and eighth – a third of which had been acquired by recent purchase. This pattern was repeated in other parishes in the area. Thus, at Lessingham the holdings of Nicholas Crowe, Clement Free, Agnes Joye, and William Bullock were confined to Redes Fielde, those of William Beare, John Dawson, and Elizabeth Jackson were concentrated in Mylnefielde, and those of Nicholas Curstewyck and Gregory Tompson were restricted to Southe Fielde. Similar examples could be cited from Westwick, Hempstead, Blofield, and Horstead-with-Stanninghall.

The disregard of field divisions by farm holdings is reflected in another aspect of their layout. Just as many commonfields merged across the parish boundary with those of adjacent parishes so, too, many holdings were made up of land held in a number of different parishes. This did not necessarily entail an excessive degree of dispersal for in most cases the parcels in question were immediately adjacent to the parish boundary. In this way at least 23 per cent of tenants at Blofield, 27 percent of tenants at Hempstead, 34 per cent at Coltishall, 37 per cent at Lessingham, and 37 per cent at Horstead-with-Stanninghall held land in more than one parish. This phenomenon may have been fostered by the general lack of coincidence between manor and parish which had long existed in this part of Norfolk. A single parish might contain any number of different manors

Name of Tenant	Area held in each precinct (acres)									Total Area Held
	1	2	3	4	5	6	7	8	9	
Walter Bayspole	26.75	35.75	20.75		36.25				31.125	150.625
George Sotherton					92.375	28.75	9.5			130.625
John Brend						34.75	17.375	58.625		110.75
Thomas Secker		2.0	4.0		34.75	57.125	1.25			99.125
Hugh Spendelove						10.0	45.625	2.0		57.625
Stephen Rose			11.25	31.75	14.0					57.0
Richard Ward						10.375	18.875	26.0		55.25
John Puttock					12.875	1.0	17.875	3.5	5.0	40.25
Thomas Puttock							7.75			7.75

Coltishall, Norfolk: Layout of Selected Land Holdings in 1584

Table 3

Source: K.C.C. E28

and, conversely, a single manor could include land in several different parishes. At Aylsham, for instance, in 1608/09, it was complained that 'The Limmits Butts and Bounds cannot be sett Foerth but by plott by reason other lords lands lye intermixed wth his Hignes lands their and doeth extend into other out townes neere adioyninge'[13] and a similar excuse was given at Gimingham in 1580 to account for the tenants' failure to deliver 'the names of all the Free-houlders & coppye houlders, belonginge to the mannor of Gymingham, w^thin the Soken'.[14] The pronounced tendency for holding layout to transcend parish boundaries was therefore part of the general lack of coincidence between different types of territorial unit – holdings, manors, and parishes – within this area at this time. Consequently, individual parcels and holdings formed the basic units of agrarian organisation; neither manors nor parishes provided an adequate basis for the communal organisation of agriculture. Likewise, communal rotations such as existed in the midlands, and irregular cropping shifts such as existed in western Norfolk, were both absent from east Norfolk and harvest shack remained the sole common right to which its commonfields were subject.[15]

The general freedom of action which this state of affairs conferred upon individual cultivators undoubtedly facilitated the consolidation of strips and fostered the growing preference for farming in severalty. These trends, in turn, were reinforced by the engrossing of holdings. The survey books reveal a mean holding size of 25-30 acres, which corresponds with a mean sown area of a little over 19 acres recorded by contemporary Probate Inventories. By this time only about a quarter of holdings were smaller than 5 acres and few holdings were larger than 160 acres: 90 per cent of holdings were in fact smaller than 80 acres in size and farms containing some 20-40 acres of land were most common (Table 4). As with everything else, however, conditions varied from township to township and Horstead-with-Stanninghall, precocious in so many other respects, was notable in containing fewer small holdings and more large holdings than any of the other surveyed townships.

Despite the modest size of most farms in eastern Norfolk it is clear from the goods and chattels recorded in Probate Inventories of the period that production, even on comparatively small holdings, was geared to the market.[16] Husbandry was for the most part mixed, with the arable and pastoral components mutually dependent. Wheat and barley were the most important grain crops, each accounting for over a quarter of the total area under crops although legumes (which accounted for 16 per cent of the cropped area) were more ubiquitous, being recorded on 82 per cent of farms. Legumes were undoubtedly grown in order to maintain soil fertility and to provide fodder for the small dairy herds which were kept on most farms. Dairy produce was plainly an important source of income and few farms were without a fully equipped dairy capable of producing cheese and butter in significant quantities. On some of the larger farms beef cattle replaced dairy cattle, young stock being bought in from breeders in Wales and the north of England and fattened up for the Norwich and Yarmouth markets. Working horses seem to have been present on most farms and this probably accounts for the acreage, about a tenth of the total, devoted to oats. Small amounts of rye and buckwheat were also grown, the latter providing feed for the poultry which were something of a local speciality. Farms in this area were well placed to take advantage of both the rapidly expanding Norwich and London markets; and the attendant commercialisation of agriculture almost certainly underlay the trends towards the engrossing of holdings and consolidation and enclosure of common-field land which have already been described.

Holding Size	Lessingham 1584	Hempstead 1584	Westwick 1547	Coltishall 1584	Blofield 1586	Horstead-with-Stanninghall 1586
	Percentage of Total Number					
More Than 160 acs.					3.2	17.2
80 ac. to Less Than 160 acs.		12.5	2.7	10.4	4.8	6.9
40 ac. " " " 80 ac.	16.7	4.2	21.6	6.3	14.3	24.0
20 ac. " " " 40 ac.	20.8	16.7	10.8	22.9	22.2	10.3
10 ac. " " " 20 ac.	16.7	16.7	13.5	18.8	17.5	20.7
5 ac. " " " 10 ac.	25.0	25.0	16.2	10.4	14.3	3.4
" " " 5 ac.	20.8	25.0	35.1	31.3	23.8	17.2
Total number of Holdings	24	24	37	48	63	29
Mean Size of Holding	19.4 ac.	22.8 ac.	24.5 ac.	26.1 ac.	29.9 ac.	72.3 ac.

Eastern Norfolk: Size Distribution of Land Holdings in Six Townships in the Sixteenth Century

Table 4

Sources: N.R.O. Pet. MS 584; N.R.O. NRS 16646 37G; K.C.C. E28, N52, P34

* * * *

Thus far attention has been focused upon the extent and layout of commonfields in the sixteenth century, for it is in this period that topographical evidence of them is most abundant. However, in order to ascertain field arrangements at the time when commonfield farming was in its prime it is necessary to turn to the evidence of earlier periods. During the middle ages sources invariably relate to individual manors rather than entire townships and rarely combine both the detail and comprehensiveness of the sixteenth century surveys. Nevertheless, a variety of types of evidence exist from which the extent and layout of commonfields may be reconstructed. Documentation is best for the neighbouring Norwich Cathedral Priory manors of Martham and Hemsby, both of which retain detailed extents which record landownership on a strip by strip basis: Martham retains a single extent which dates from 1292[17] whilst Hemsby retains two extents, one dating from 1422 and the other from circa 1500.[18] This evidence is of special significance for in the thirteenth and fourteenth centuries the district of Flegg, in which these two manors were located, was possibly the most populous and intensively farmed locality in the country.[19] Elsewhere in eastern Norfolk medieval field arrangements have to be reconstructed from more disparate sources of evidence: manorial court rolls, rentals, and account rolls.

The 1292 extent of Martham and 1422 extent of Hemsby record 1,057.9 and 999.1 acres of arable respectively, of which 211.05 and 188.1 acres respectively were held by the Prior of Norwich and the remainder – a combined total of 1,868.9 acres – by his tenants. In neither case is the entire township surveyed since the extents only record land belonging to Norwich Cathedral Priory which, although the principal, was not the sole landowner in these townships. The fields into which the arable was divided were substantial in number and varied in size. At Martham a majority of the surveyed area – 53.6 per cent of the total – was contained in fields with the suffix *feld*, of which *Estfeld* and *Westfeld* were relatively extensive but *Suthfeld* was quite small. In addition a substantial proportion of the area (35.8 per cent) was accounted for by much smaller fields. These fields have names with the suffixes *dele, grave, lond, mere, toft* and *wong*, as in *Sumertondele, Martinesgrave, Haverlond, Tomeres, Gunnestoft*, and *Westwong*. It is the equivalents of these fields which predominated at Hemsby. Here the name element *field* is wholly absent and the arable was divided into almost a hundred separate divisions, the largest of which – *Estonnesend* – contained only 29.6 acres. These fields vary considerably in size and bear names which, in addition to the elements present at Martham, include the suffixes *croft, furlong*, and *howe* (as in *Havercroft, Myddylfurlong*, and *Hungurhowe*).

With the probable exception of most of the demesne land held by the Prior of Norwich, these fields were characterised by an intense degree of sub-division. At Martham the land held by the peasantry was divided into 2,122 separate parcels with a mean size of 0.5 acres and at Hemsby the peasantry held 1,479 parcels with a mean size of 0.7 acres. The vast majority of these parcels were smaller than 3 acres and a very substantial majority were smaller than 1 acre (Figure 8). Such a high degree of parcellation seems to have been fairly characteristic of medieval commonfields in eastern Norfolk. At Coltishall, for instance, the size distribution of commonfield strips may be reconstructed from 1,300 land transactions recorded in the court rolls between 1275 and 1405.[20] Conveyances may be biased towards the over-representation of very small and fairly large pieces of land but provide the best available substitute for a detailed manorial extent. This observation is borne out by the fact that the resultant size distribution of land parcels at Coltishall, in the period 1275-1349, is closely comparable with that which prevailed at Martham in 1292: over 90 per cent of parcels transacted were smaller than 1 acre and 68.8 per cent were smaller than ½ acre. Conditions on the Bishop of Norwich's manor at Hevingham appear to have been broadly similar. Of 165 parcels transacted in the manor court during the last quarter of the thirteenth century, 86 per cent were smaller than 1 acre and 53.9 per cent smaller than ½ acre.[21] A century later the mean size of 95 parcels recorded in an extent of circa 1383 was 0.5 acres.[22]

Where parcellation was so extreme considerable importance obviously attached to the physical demarcation of strips. At Martham strips appear to have been separated from one another by slender balks of unploughed land. Thus the court rolls record the prosecution of Adam Goscelyn for ploughing up a *communam divisam in Westfield* measuring ½ foot by 20 perches,[23] and in 1506 it was stipulated that *unusquisque tenens permittit unam culturam in a forowe ex parte terre sue iacere non arratus pro bunda et meta inter tenentem et tenentem ac inter feodum et feodum sub pena cuiuslibet xijd.*[24] At Hevingham, on the other hand, mere stones seem to have been used to define ownership boundaries. This arrangement is recorded in a prosecution of the mid-fifteenth century: *Juratores ex officio presentant quod Nicholaus Plomere abradicavit unam bundam lapidis per capitale inter terra dicti Nicholai et terra Thome Kevyng in campo vocato Wellecroft . . .*[25] What is surprising, given that these commonfields comprised several thousand strips and that the divisions between strips were so insubstantial, is that prosecutions for the abuse of ownership boundaries do not occur with greater frequency. Likewise, there seems to have been very little confusion with regard to which strip was owned by whom; the fact that most tenants held only a small number of strips each must have made for ease of identification.

At Martham and Hemsby the amounts of land held in the commonfields were very small. The 2,122 parcels at Martham were held by no less than 376 different individuals, with the result that the mean holding was less than six parcels and 2.8 acres: no tenant held more than 18 acres and the holdings of two-thirds of all tenants were below average in size. Although it is unlikely that these quantities represent the full amounts owned by the peasantry (since many undoubtedly held additional land of other lords and in neighbouring townships) there is a clear implication that holding size was both relatively and absolutely small. Conditions at Hemsby were similar, if somewhat less extreme. In this case there were 173 tenants of whom 44 seem to have been outsiders from other townships. The mean holding thus comprised less than seven parcels and 4.7 acres if all tenants are included, and less than eight parcels and 5.1 acres if outsiders are excluded. Again, the holdings of a substantial majority of tenants were below average in size. At Coltishall information on holding size derived from the obituaries of 338 deceased tenants recorded in the court rolls reveal much the same picture. The mean size of all holdings recorded in the period 1275-1405 was a little below 2½ acres and medium size was 1 acre. Furthermore, holding size seems to have declined during the first half of the fourteenth century so that by mid century fully three-quarters of all tenants held less than 2 acres. Likewise at Hevingham, 72.5 per cent of tenants whose obituaries are recorded in the court rolls between

	Ac.	Rd.	Per.		Ac.	Rd.	Per.
Richard Anger				*Robert Gemere*			
Westwong	3	0	0	Estfeld	2	1	0
Hellecroft		1	0	Westfeld	1	1	0
Toft with Messuage		2	0	Suthfeld	1	0	30
	3	3	0	Tomeres		1	0
Robert Spite				Clovenhove			20
Westwong	3	2	20	Toft with messuage	1	2	0
Westfeld	1	2	0		6	2	10
Toft with messuage		1	20	*Nicholas Berte*			
	5	2	0	Estfeld	2	1	20
John Knight				Westfeld	1	0	20
Estfeld	3	0	0	Suthfeld	4	1	0
Westfeld	2	2	30	Morgrave		2	0
Fendovetoftes			20	Toft with messuage		3	0
Toft with messuage		1	0		9	0	0
	6	0	10				

Martham, Norfolk: Layout of selected land holdings in 1292

Table 5

Source: B. M. Stowe MS 936

1274 and 1299, 61.4 per cent of tenants recorded in an extent of circa 1284,[26] and 64.0 per cent of tenants recorded in an extent of circa 1383, held less than 2 acres. On other manors for which evidence of holding size is available, such as Worstead, Hautbois, Burgh and Thurne,[27] conditions were very much the same, thus leaving little doubt that in the middle ages the majority of holdings in eastern Norfolk were extremely small.

When the amounts held were so small and the number of field divisions was so large holding layout was, perforce, irregular. At Martham, Nicholas Berte held a relatively substantial amount of land – 9 acres – but this was far from equally divided among the commonfields: almost half his land lay in *Suthfeld* and more than twice as much lay in *Estfeld* as in *Westfeld*. As will be seen from Table 5, this situation was fairly typical. Moreover, even when a tenant such as Robert Gemere, held parcels relatively widely scattered through the commonfields their distribution was never anything other than erratic. To compound the situation, many of the Prior's tenants held parcels in the adjacent commonfields of Bastwick, Repps, Rollesby, Somerton, and Winterton. Much the same pattern may be identified at Hemsby where the layout of even the most substantial holdings appears to have been haphazard. Geoffrey Sporlle, for example, with a holding of 28.6 acres, was one of the most substantial tenants of the Prior, yet he held land in less than a third of the fields named in the extent. This apparent lack of any underlying rationale reflects the fact that cultivators did not have to comply with common rotations and cropping shifts; with the result that holding layout had no bearing upon the management of these commonfields, and vice versa.

A consideration of medieval evidence this reveals a basic pattern of extensive commonfields, intensively parcellated, and divided amongst a myriad of separate small-holders. This pattern presents a clear contrast with conditions as they existed in eastern Norfolk in the late sixteenth century, when commonfield farming was in decline. Yet medieval field systems were themselves far from static, and it is to the changes in field and holding layout which occurred during the middle ages that attention must now be turned.

* * * *

A variety of sources is available from which agrarian developments may be reconstructed during the two and a half centuries after 'Domesday'. This was a period of demographic and economic expansion in the country as a whole and eastern Norfolk was no exception.[28] As Domesday Book shows, it was already a closely settled district in 1086 and it was to become even more so during the ensuing centuries.

By the late eleventh century Norfolk was the most populous county in England whilst within the county population densities in the east were generally twice, and in places three times, as high as in the west.[29] Many of the peasant holdings in eastern Norfolk were already smaller than 10 acres, and Domesday Book records some which were smaller than 5 acres. In keeping with such a high density of population, the development of arable farming was advanced. This is revealed by the high density of plough-teams recorded (five teams per square mile in parts of Flegg), and by the fact that woodland was by this time very sparse. Indeed, it has been shown that many townships in this area were already dependent for fuel upon peat, dug from the Broadland marshes, and it has been estimated that 900 million cubic feet of peat were extracted during the three or four centuries that these peat diggings were operational.[30] These characteristics all became further

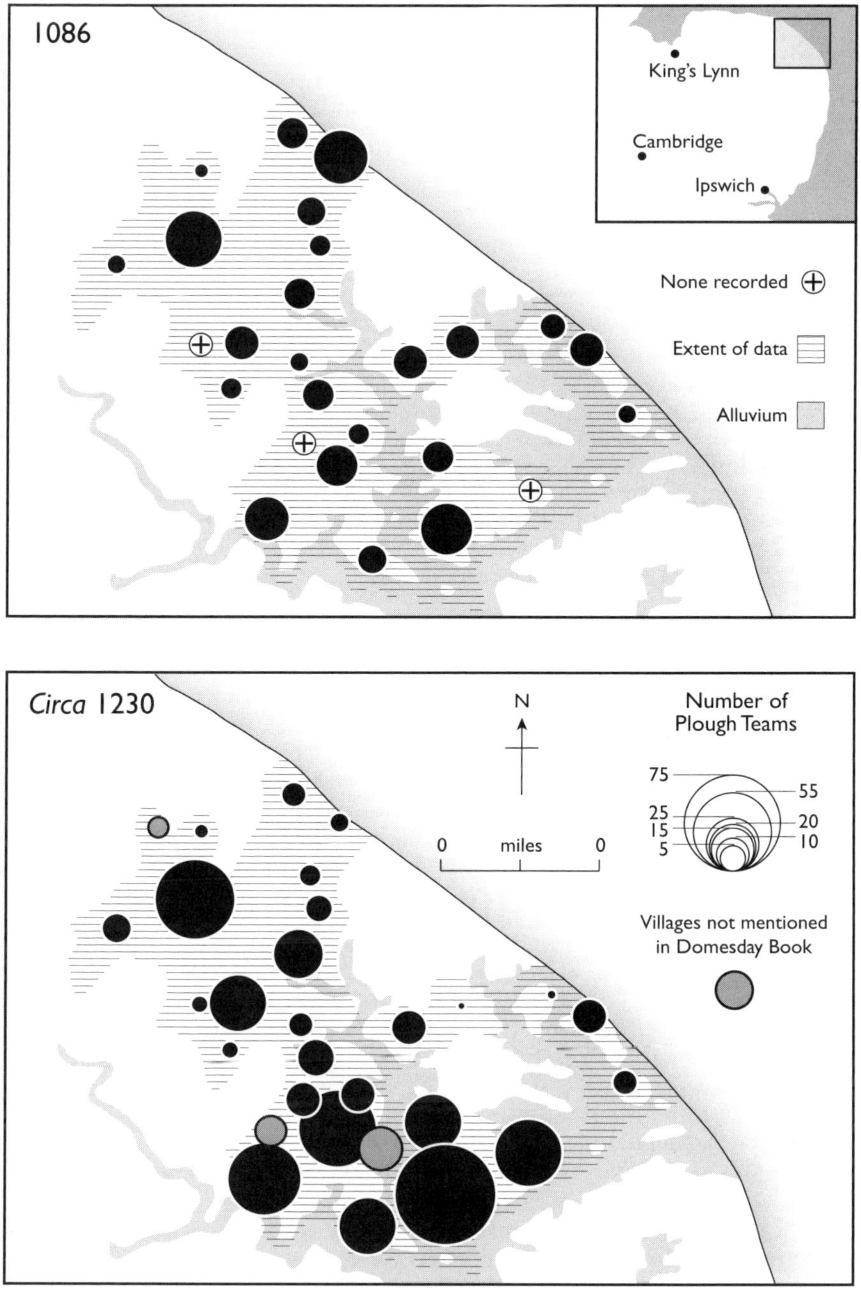

DEACONRY OF WAXHAM
Increase in the number of plough teams, 1086 - *c.*1230

Fig. 5

accentuated during the twelfth and thirteenth centuries, a trend which had important implications for the development of commonfields.

Evidence of the continued expansion of agriculture during the twelfth and thirteenth centuries is provided by the cartulary of the Abbey of St. Benet of Holme. This document contains a great deal of miscellaneous information relating to the St. Benet estates in eastern Norfolk, amongst which is a listing, made before 1234, of the number of plough-teams in 28 townships in the Deaconry of Waxham.[31] From a comparison between these data and equivalent information given in Domesday Book[32] it transpires that the density of plough-teams, already high in 1086, virtually doubled during the next century and a half (Figure 5). Increases in the number of recorded teams occurred at 20 of the 28 recorded townships and in some cases these were very pronounced, those at Neatishead and Ludham rising from 15 to 52 and 21 to 77 respectively. The increase in the general density of plough-teams was especially marked in the vicinity of these two townships and includes to vills which do not figure in Domesday Book, Ashmanaugh with 10½ plough-teams and Irstead with 14 plough-teams. Conversely, the most pronounced decreases in plough-teams were in certain coastal townships, notably Bacton, Ingham, Palling, and Waxham, which may reflect a change in the basis of enumeration or, perhaps, the loss of land to the sea.[33] With these few exceptions, however, this evidence indicates a significant increase in agricultural activity in this area during the twelfth and thirteenth centuries.

One form which this increase undoubtedly assumed was the extension of the cultivated area, as may be clearly seen at Horstead.[34] This manor was one of the few English possessions of the Abbey of La Trinité at Caen and a cartulary of the mid twelfth century reveals the extent to which assarting had been taking place. Thus, a total of 35 purprestures, or assarts, are recorded, ranging in size from ½ acre to 24 acres and together amounting to 170 acres. Significantly, the same document records a £40 reduction in the value of the manor's woodlands. Since houses had been built upon 19 of these purprestures it is likely that much of this assarting had been in response to population growth. A similar process may be identified on other manors in eastern Norfolk, as at Cawston where an extent of 1290/91 records 108.875 acres of assarts.[35] By the late twelfth century, however, the opportunities for assarting were rapidly being exhausted and from this date the cultivated area on many manors ceased to expand. The Prior of Norwich's manor of Martham provides the clearest instance of this. On both occasions when land tenure is recorded in the late twelfth and late thirteenth centuries,[36] the land area is the same, despite a 250 per cent increase in tenant numbers. An equivalent extent of the manors of Ripton Hall, Park Hall, and Crictot Hall in Hevingham records exactly the same phenomenon: no change in the cultivated area between the beginning of the thirteenth century and, in this case, the late fourteenth century, and yet a 275 per cent growth in tenant numbers.[37] The explanation for this is that when opportunities for colonisation were exhausted, methods of cultivation were intensified and holdings were subdivided.

The earliest evidence of husbandry methods employed in this area dates from the middle years of the thirteenth century and relates to eleven demesnes belonging to the Abbey of St. Benet at Holme.[38] From this it is clear that agriculture was already remarkably intensive. Rotations were complex and flexible and were distinguished by the relative unimportance of bare fallows and the prominence of a substantial course of legumes (legumes accounted for approximately one tenth of the total sown area). To a considerable extent legumes were a substitute for fallow since they served both to provide fodder and to improve the nitrogen content of the soil. Moreover, a heavy emphasis upon spring grown crops, particularly barley, allowed regular fallowing of the land on a half-year basis whilst the use of horses for ploughing derived maximum advantage from the changeover from natural to produced fodder. Thereafter, as later account rolls show, this system of cultivation was steadily intensified until a peak of development was attained in the late thirteenth century beyond which further progress does not appear to have been possible.[39] This was a period of veritable 'High Farming', with high inputs and high outputs. Thus, legumes came to occupy about a fifth of the total area on most demesnes, bare fallows were reduced in frequency to a point at which they were occasionally dispensed with altogether, great care was lavished on the maintenance of soil fertility by means of manuring, marling, and folding, and there was an increase in the number of livestock carried on the demesnes. The reward of these careful and intensive husbandry practices was a high and sustained level of productivity. This is to be seen in the very high proportion of the total arable area which was regularly under crops (grain crops generally occupied between two-thirds and three-quarters of the total arable area) and in high yields per acre achieved at the expense of unimpressive yields per seed (wheat yielded at an average of 15 bushels per acre and barley at an average of 14½ bushels per acre).[40] It was this capacity to raise the productivity of the land which undoubtedly allowed the progressive subdivision of holdings and the accommodation thereby of a steadily expanding population.

The decline in holding size which occurred during the thirteenth century as a result of the increase in tenant numbers at Martham and Hevingham is illustrated in Figure 6. On both manors small holdings proliferated at the expense of large, so much so that holdings in excess of 12 acres were virtually eliminated and the bulk of the peasantry were reduced to a state of near landlessness. Evidence of holding size at Lessingham[41] and Worstead at intermediate dates during the thirteenth century, and at Coltishall during the fourteenth century, amplifies this general chronology and indicates that holdings became smallest and most numerous in the early fourteenth century. At the root of this trend lay mounting population pressure. This is implicit in the growth of tenant numbers which occurred at Martham[42] and is corroborated by certain data of a quasi-demographic nature derived from the Coltishall court rolls, notably a succession of fourteenth century nominative listings and information concerning heirship recorded in the obituaries of 350 deceased tenants.[43] From this evidence it appears that population growth persisted without serious set-back until well into the second quarter of the fourteenth century. However, it was not population growth *per se* which was responsible for this progressive subdivision of holdings; it was the operation of customary inheritance and an active market in peasant land (both free and unfree) in conjunction with population growth which was the determining factor.

Partible inheritance was the custom on many, but by no means all, of the manors in eastern Norfolk in the middle ages. It was certainly the custom at Martham and Coltishall and it was also practised at Aylsham, Belaugh, and Burgh, together, very probably, with Cawston, Lessingham, Thurne, and Worstead.[44] At Martham the operation of this custom is manifest in the size and layout of many of the holdings recorded in the 1292 extent. For example, the separate holdings of Robert de Hil senior, Robert de Hil junior, and John de Hil were clearly the product of the tri-partite division of a single holding between them. Similarly, when William Herberd died in 1290 the court rolls record that his holding of a messuage and 6 acres 3 rods 30 perches was inherited by his two

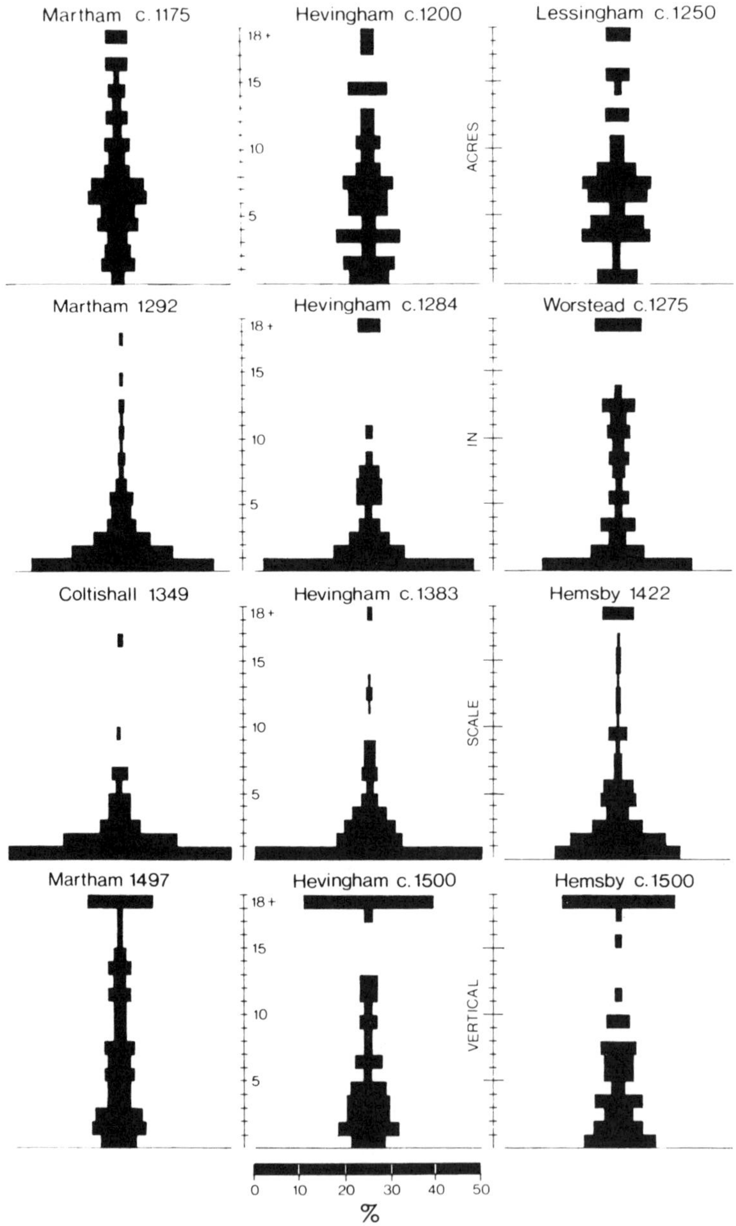

Fig. 6
Eastern Norfolk: Size Distribution of Land Holdings c1175-c1500
(Percentage of total number)

Sources: British Museum, Stowe MS 936; Add. MS 24,316 ff. 51-6; K.C.C., E34; N.R.O. Dean and Chapter Muniments, MS 2765 and Register V ff. 132-5; N.R.O., NRS 14761 29 D 4; NRS 19280 33 F 9; NRS 14479 29 C 1; Middleton, Killin, and Bruce, 19.11.68 (Supp. Catalogue), 1 and 2

sons, John and William.[45] Subsequently, in 1292, John was recorded in possession of 3 acres 3 rods 10 perches and William in possession of 3 acres 30 perches (a combined total of 7 acres), thus showing that their father's holding had been divided between them. At Coltishall the obituaries recorded in the court rolls allow some evaluation of the incidence of partible inheritance and this shows that during the period 1275-1348, 39.3 per cent of holdings were divided between more than one heir.[46] However, partible inheritance was not ubiquitous in this locality and at Hevingham, Horstead, and Marsham various forms of impartible inheritance were practised.[47] Nevertheless, impartible inheritance did not exclude all possibility of subdivision as may be seen in the case of the Bishop of Norwich's manor at Hevingham.[48] Ultimo-geniture (Borough English) was the custom on this manor and accordingly it was the youngest son who inherited on the death of his father (this custom is considered by some to represent a modified form of partible inheritance);[49] yet on at least one occasion this custom was interpreted in such a way that, although the younger son received the lion's share of his father's holding, a small share was reserved for the elder brother. A much more common cause of division, however, was the failure of sons altogether and the survival instead of several daughters. When this occurred each daughter received an equal share of the holding, as in 22 out of a total of 157 cases of inheritance recorded at Hevingham between 1275 and 1509.[50] In this context it is interesting to note that at Coltishall a quarter of all cases of divided inheritance recorded in the period 1275-1348 occurred due to the absence of sons. All forms of inheritance, therefore, led to some degree of subdivision: on the other hand, not all subdivision was the result of inheritance and by the late thirteenth century the land market provided an equally important means of obtaining land.

Evidence of the importance of the land market in this locality at this time is abundant and there can be little doubt that sales and purchases exercised a profound influence on the pattern of landownership. At Martham 98 separate transactions, dealing with 37 acres are recorded in the proceedings of 12 manorial courts between 1288 and 1299:[51] at Hevingham 302 separate transactions, dealing with 110 acres are recorded in the proceedings of 95 manorial courts between 1274 and 1343:[52] and at Coltishall 1,025 separate transactions, dealing with 404 acres are recorded in the proceedings of 180 manorial courts between 1275 and 1349.[53] The effect of this very active market upon landownership is revealed by a closer examination of the pattern of buying and selling at Coltishall.[54] Between 1275 and 1349 the land market at Coltishall was congested due to the participation of a very large number of separate buyers and sellers: during the peak period of activity between 1317 and 1341, 356 separate individuals can be identified, 43.8 per cent of whom bought land, 33.2 per cent of whom sold land, and 23.0 per cent of whom both bought and sold land. These figures illustrate the tendency for buyers of land to outnumber sellers which was a persistent feature of the land market in this period, except during years of adverse economic conditions such as prevailed in the 1320s. The net effect of this market, therefore, tended to be to promote the relative dispersal, rather than the concentration, of landownership. This is borne out by the fact that a majority of those who participated in the market (55.1 per cent of the total) gained land. At Coltishall, therefore, the land market reinforced the trend towards a greater number of smaller holdings and at the same time led to a very high turnover of land. The scale of the latter may be seen at Martham where 48.7 per cent of holdings recorded in 1292 were made up of land formerly held by at least five different individuals, and 3.0 per cent of holdings comprised land formerly held by at least ten different individuals.[55] These figures imply a substantial redistribution of land, probably as a result of the land market, and at the same time demonstrate the relative unimportance of a process of engrossing.

For a number of reasons, therefore, and in a variety of ways, holdings became progressively greater in number and smaller in size during the thirteenth and early fourteenth centuries. As a corollary of this commonfields became increasingly parcellated until they too reached a peak of subdivision in the first half of the fourteenth century. There are several reasons why this should have been so, not least the fact that the processes which fostered the fragmentation of holdings also tended to promote the multiplication of unenclosed strips. Thus, at Martham the tripartite division which led to the creation of the separate holdings of Robert de Hil senior, Robert de Hil junior, and John de Hil was achieved by means of the subdivision of each of nine separate pieces of land: in this way 18 additional land parcels were established. Similarly, the division of William Herberd's holding between his two sons involved the subdivision of several of the land parcels which it comprised. These examples show that the operation of inheritance custom was a major cause of parcellation. A contributory cause was the land market. Until the middle of the fourteenth century the volume and pattern of conveyancing at Coltishall was heavily influenced by harvests; in so far as when harvests failed – as in 1293-94, 1297, 1314-17, 1321-22, 1330-31 and 1346-47 – many peasants had no alternative but to sell land to buy food. Under these circumstances individuals naturally endeavoured to keep their losses to a minimum and hence there was a pronounced tendency to dispose of small slips of land, many of which almost certainly represented portions of larger parcels.[56] Of course the land market was also used to consolidate strips; but given the large number of different owners, the even greater number of separate strips, and the general competitiveness of the land market, this was achieved rarely and with considerable difficulty.

The net effect of these trends may be seen at Martham, where it is possible to undertake a partial reconstruction of holding layout in the late twelfth century from information contained in the 1292 extent. This exercise indicates that the 250 per cent increase in holding numbers which took place during the thirteenth century was accompanied by an increase in parcellation of approximately 50 per cent.[57] Not surprisingly, parcellation was greatest where there was most scope for it, as in the case of several peripheral blocks of land newly won from the waste. For instance, a 6 acre piece of land adjacent to the common pasture of Martham and originally held by a single tenant – Syware Blakeman – had by 1292 been divided into ten separate parcels held by as many different tenants (Figure 7). The name of this piece of land, *Blakesmannestoft*, implies that it originated as a late encroachment on the waste but by the end of the thirteenth century it had become assimilated into, and indistinguishable from, the rest of the commonfields. Other pieces of land – *Hardyngestoft, Hiltoftes, Morgrimestoft,* and *Godknapestoft* – appear to have shared a similar fate, originating as the consolidated holdings of individual tenants but then becoming subdivided and incorporated into the commonfields. The thirteenth century, therefore, emerges as a period of growth in the development of Martham's commonfields, with both an extension of area and an intensification of parcellation. Nevertheless, the fact that the parcellation of fields progressed at a lower rate than the proliferation of holdings, and was more characteristic of certain portions of this township's fields than others, suggests that once parcel size had been reduced to a certain minimum, further subdivision rarely occurred. In other words, parcellation was approaching a maximum in the late thirteenth century and accordingly commonfield layout was beginning to stabilise.

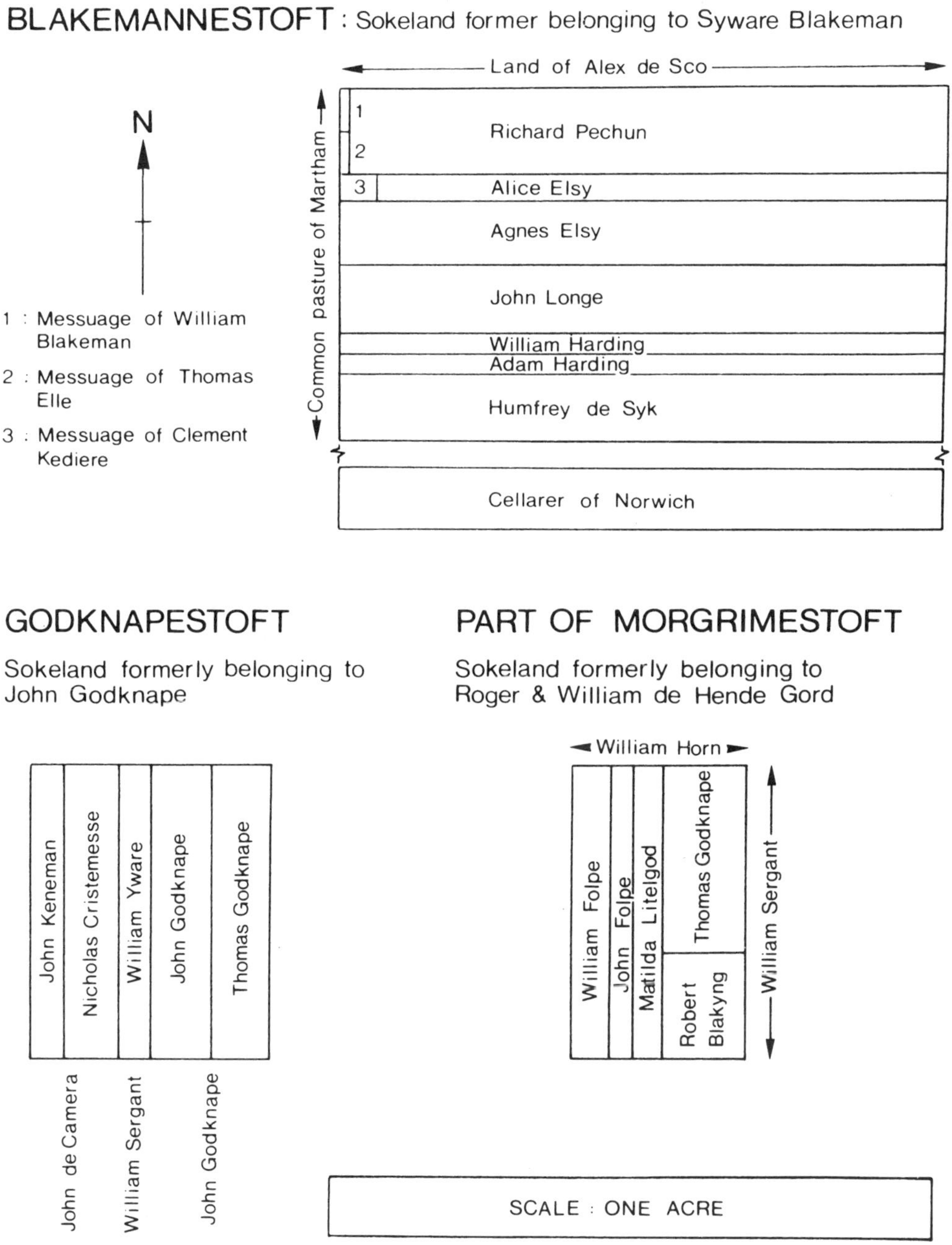
BLAKEMANNESTOFT : Sokeland former belonging to Syware Blakeman
Land of Alex de Sco
N
1 : Messuage of William Blakeman
2 : Messuage of Thomas Elle
3 : Messuage of Clement Kediere
Common pasture of Martham
1
2
Richard Pechun
3
Alice Elsy
Agnes Elsy
John Longe
William Harding
Adam Harding
Humfrey de Syk
Land of William Tange
Cellarer of Norwich
GODKNAPESTOFT
Sokeland formerly belonging to John Godknape
John Keneman
Nicholas Cristemesse
William Yware
John Godknape
Thomas Godknape
John de Camera
William Sergant
John Godknape
PART OF MORGRIMESTOFT
Sokeland formerly belonging to Roger & William de Hende Gord
William Horn
William Folpe
John Folpe
Matilda Litelgod
Thomas Godknape
Robert Blakyng
William Sergant
SCALE : ONE ACRE

Fig. 7

From this account it follows that the twelfth and thirteenth centuries witnessed a growth in the commonfields of eastern Norfolk. The operation of inheritance custom and the land market led to the proliferation of holdings and this in turn fostered the parcellation of land. Commonfields thus grew in both extent and subdivision. Moreover, this was all part of a larger movement which involved the growth of population and changes in agricultural technique.

* * *

The growth in population which had exerted such a profound influence upon the development of eastern Norfolk's commonfields in the twelfth and thirteenth centuries was finally terminated and reversed in the fourteenth.[58] To judge from the experience of Coltishall, the first half of the fourteenth century appears to have been a period of latent demographic crisis as the peasantry experienced mounting economic difficulty. Nevertheless, crisis did not actually materialise until 1349, when the outbreak of plague precipitated an immediate and substantial reduction in numbers.[59] Plague mortality was at least 55 per cent at Coltishall and it was probably much the same elsewhere.[60] As late as 1440 a 76 year old resident of Hevingham, born 15 years *after* the Black Death, recalled that 'after the grete pestelence . . . all thyngge was out of mende'.[61] Subsequent outbreaks of plague undoubtedly occurred and must have been partly responsible for the continued decline in population during the remainder of the fourteenth century and greater part of the fifteenth century. This is apparent in the reduced numbers enumerated in a succession of nominative listings entered in the Coltishall court rolls, and in the 60 per cent decline in tenant numbers which occurred on the Prior of Norwich's manor at Hemsby between 1422 and the end of the century (although the number of messuages, houses, and cottages remained virtually unchanged). Most telling, however, is the evidence of obituaries recorded in the Coltishall and Martham court rolls.[62] These show a marked shortage of surviving children. Sons survived in less than half of all cases and co-heirship was greatly reduced in frequency: at Coltishall it occurred in only a fifth of all cases recorded between 1349 and 1405, and at Martham in a quarter of all cases recorded between 1350 and 1450. These figures indicate an adverse replacement rate and a downward trend in population. The latter exerted a profound influence upon farming arrangements in this locality.

The most immediate and enduring effect of the reversal in population trends was a reduction in the incidence of subdivision, both of holdings and of land parcels. As has been observed, the incidence of co-heirship was reduced by half and as successive generations became progressively smaller so agglomeration displaced subdivision as the net product of inheritance custom. Furthermore, customary inheritance itself began to decline. The most conspicuous sign of this is a growth of conveyances made to selected beneficiaries in anticipation of death. This represents a significant departure from established practice and became increasingly prevalent from the second half of the fourteenth century. At Martham conveyances made under these circumstances are usually distinguished by the accompanying phrase *languens in extremis*. Such conveyances usually dealt with larger amounts of land than other forms of conveyance and accounted for a quarter of all the land conveyanced between 1351 and 1509.[63] The decline of customary inheritance was further encouraged by the practice of disposing of land by will which grew up during the second half of the fifteenth century. By this date the principle was firmly established that tenants were free to dispose of their property as they wished and, accordingly, customary inheritance was only invoked when an individual died intestate. This was to have important long-term implications for it meant that subdivision and parcellation did not recur when population growth was finally resumed in the sixteenth century.[64]

The relative demise in the importance of customary inheritance was matched by a corresponding growth in the importance of the land market, to the extent that it became by far the single most important process determining the ownership of land. In the case of Martham, where the amounts of land transferred by inheritance and conveyancing were roughly equal during the fourteenth century, the respective proportions were one-quarter and three-quarters by the end of the fifteenth century. Behind this shift lay an absolute increase in the turnover of land on the market: at Martham the turnover of land doubled during the course of the fifteenth century, and at Coltishall and Hevingham it increased approximately five-fold.[65] The principal component of this change was a marked growth in the size of individual conveyances. Land was no longer bought and sold a parcel at a time: instead, entire holdings, or parts of them, now changed hands. The transition was gradual and becomes increasingly apparent during the second half of the fifteenth century. In 1463, for example, Robert Rovy bought a messuage and 10.5 acres in Martham,[66] and 20 years later Robert Bysshop, Alice his wife, and Robert Nicholas paid £43-6s-8d for a messuage and 41.75 acres in Hevingham.[67] The sale, in both cases, of a messuage with the land becomes increasingly characteristic at this time and betrays the fact that these were almost certainly entire holdings. The latter example also shows that the amounts of cash involved in some of these transactions were very considerable, which indicates the transformation which the market had undergone since the impecunious days of the early fourteenth century when harvest failure had obliged many individuals to sell land in order to buy food. Nevertheless, the story of the land market during this period is not the same for all manors nor is it one of steady growth. Variations in the rate of conveyancing existed between manors and over time, the reasons for which are not immediately apparent. Of one thing, however, it is possible to be certain: from the late fourteenth century the land market served to promote the agglomeration rather than the subdivision of land. This is implicit in the greater scale of individual conveyances and explicit in the growing size of holdings and of strips.

Figures 6 and 8 illustrate the changes in holding size and parcel size which occurred in eastern Norfolk during the middle ages. As will be noted, evidence is particularly good for the Prior of Norwich's two manors of Martham and Hemsby and the Bishop of Norwich's manor of Hevingham. On all three of these manors there was a marked increase in holding size from the end of the fourteenth century. Thus, mean holding size grew from 2.8 acres at Martham in 1292, to 4.7 acres at Hemsby in 1422, 9.3 acres at Martham in 1497,[68] and 12.2 acres at Hemsby circa 1500.[69] Behind this progressive expansion in mean holding size lay a decline in the number of small holdings and a growth in the importance of large. The proportion of the total area occupied by holdings of less than 5 acres declined from 56.4 per cent at Martham in 1292, to 20.5 per cent at Hemsby in 1422, 8.7 per cent at Martham in 1497, and 5.6 per cent at Hemsby circa 1500. Conversely, the proportion of the total area occupied by holdings of 12 acres and more grew from 8.1 per cent at Martham in 1292, to 57.7 per cent at Hemsby in 1422, 64.6 per cent at Martham in 1497, and 82.1 per cent at Hemsby circa 1500. These figures show quite plainly that by the late fifteenth century the peasant small holder was rapidly being displaced by a new and much smaller group of relatively substantial tenant farmers. At Hemsby, for instance, 16

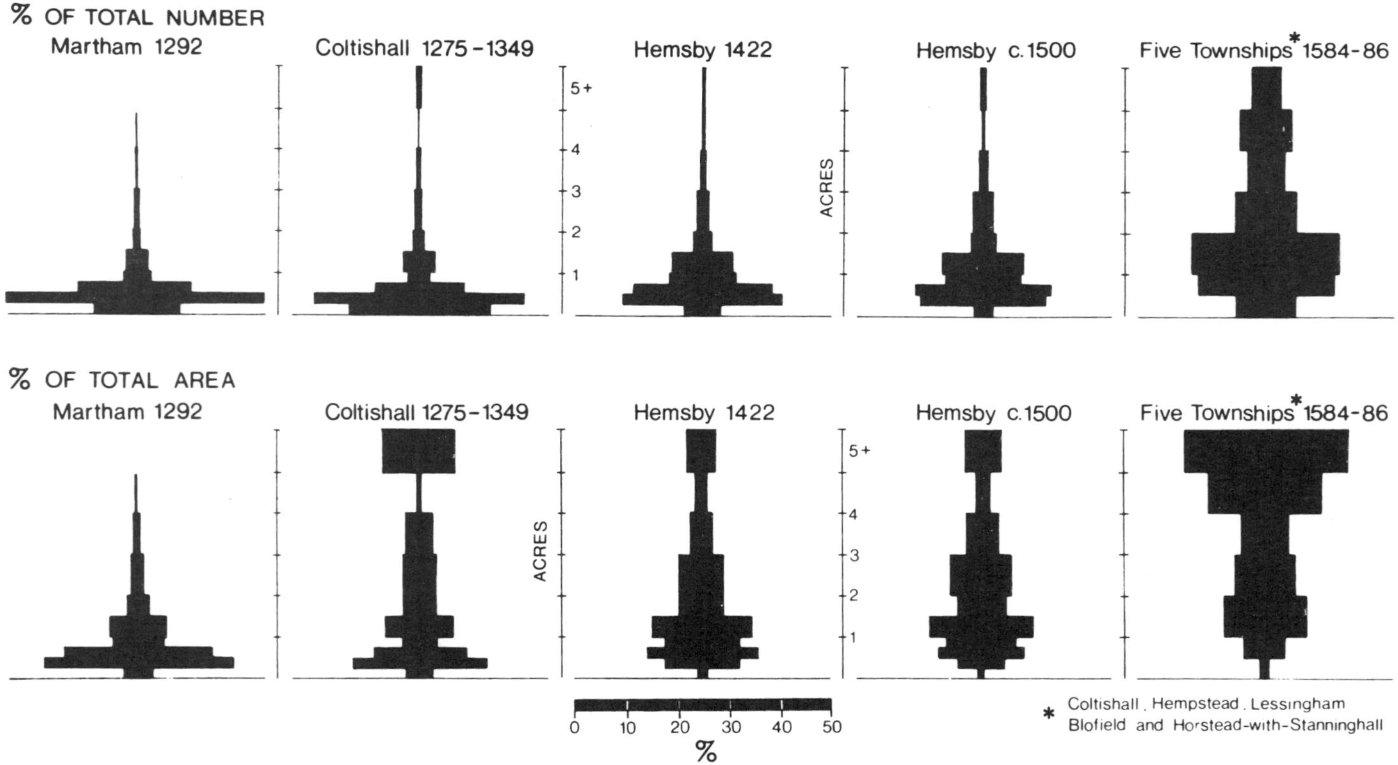

Fig. 8
Eastern Norfolk: Size distribution of land parcels 1292-1586

Sources: British Museum, Stowe MS 936; K.C.C., E 29-34; N.R.O., Middleton, Killin and Bruce, 19.11.68 (Supp. Catalogue), 1 and 2; K.C.C., E 28, N 52, P 34; N.R.O., NRS 16646 37 G

individuals held 78.7 per cent of all the land at the beginning of the sixteenth century.

The emergence of this incipient yeoman class may be traced in the court rolls of the period. At Martham the largest holding recorded in the 1292 extent contained some 17.1 acres yet by 1373 the obituaries recorded in the court rolls reveal the existence of a holding of 18 acres and, not long afterwards, another of 27.75 acres.[70] In 1434 a 30 acre holding is recorded[71] and by 1497 there were four holdings of more than 30 acres, the largest of which comprised 38 acres. A long and complex history of land purchase lay behind the creation of many of these holdings, the most impressive example of which is provided by the Bysshop family of Hevingham. In 1382/83 a certain John Bysshop held just ½ acre from the manor of Crictots Hall, yet by the beginning of the sixteenth century successive members of the family had built this holding up to 49.75 acres.[72] The Bysshop family chose to remain at Hevingham, and they prospered, but many of their contemporaries sold up and left. John Doules is a case in point; until 1382/83 he had held a messuage in Hevingham but that very year he left the manor and paid chevage for the right to live 10 miles away at Cringleford.[73] John Doules is no isolated case and between 1381 and 1451 at least 52 others sold up and left the manor. Their loss was the gain of families such as the Bysshops.

A concomitant of the engrossing of holdings was the consolidation of strips (see Figure 8), but although this made some progress the transformation was less dramatic. Mean parcel size grew from 0.5 acres at Martham in 1292, to 0.7 acres at Hemsby in 1422, and 0.85 acres at Hemsby circa 1500. Likewise at Hevingham mean parcel size grew from 0.65 acres in 1382/83 to 0.9 acres circa 1500-10. In the case of both Hemsby and Hevingham change consisted principally of a decline in the number of very small parcels rather than a growth in the number of large. At Hemsby, where the information is fullest, parcels smaller than 0.5 acres registered the greatest decline and parcels of 2-3 acres the greatest growth. Just as in the thirteenth century the parcellation of land proceeded more slowly than the subdivision of holdings so too, in the fifteenth century, the consolidation of strips seems to have lagged behind the engrossing of holdings. This suggests that, at least initially, the consolidation of strips may have been incidental to the engrossing of holdings, it being the latter to which tenants attached the greater priority. It also suggests that effective consolidation could only take place once the number of land holdings had been reduced and the size of land holdings had been increased. Consequently, consolidation may not have become an objective in its own right until relatively late and it is perhaps for this reason that the most conspicuous changes in the number and size of land parcels were postponed until the sixteenth century.

The processes which were eventually to lead to the dissolution of eastern Norfolk's commonfields were therefore already beginning to take effect from the second half of the fourteenth century. From this time field layout began to undergo gradual modification. Nevertheless, the momentum of change was initially relatively slow and it was only in the late fifteenth century that the cumulative effect of these processes began to become pronounced.

* * *

When Sir Henry Spelman described that part of Norfolk 'from *Thetford* to *Burneham*, and thence Westward, as also along the Coast' as *'Champion'* and 'the rest' as '*Woodland*', he was describing a situation which was comparatively new. The hedgerows which Nathaniel Kent subsequently regarded as such a feature of eastern Norfolk were largely a product of the sixteenth century and after. Before then the landscape had worn a very different aspect, that of a champion area *par excellence*, residual traces of which lingered on into the late eighteenth century to catch the attention of William Marshal. Extensive commonfields were already established at Martham by the late twelfth century and they persisted in most parishes in eastern Norfolk for at least a further four centuries. They were most fully developed in the fourteenth century at which time it is possible that in terms of sheer fragmentation of land they were without parallel in the whole country. The high degree of parcellation reached in the fourteenth century did not always prevail and during the long period for which these commonfields were in existence they were in a constant state of change. The rate and direction of change were determined by social and economic factors which, in turn, were partly a function of the prevailing demographic trend. But the relationship was not a simple one; for population growth in the twelfth and thirteenth centuries promoted the growth of commonfields, while population growth in the sixteenth century reinforced their decline.

I wish to acknowledge the assistance of Miss J. Orr in the preparation of some of the material upon which this paper is based.

[1] J. Arlott (ed.), *John Speed's England: A Coloured Facsimile of the Maps and Text from the Theatre of the Empire of Great Britain First Edition, 1611* (Part II, 1953), f. 35.

[2] Nathaniel Kent, *A General View of the Agriculture of Norfolk* (1796).

[3] William Marshall, *The Rural Economy of Norfolk*, I (1795), 4.

[4] For a discussion of the nature of common rights and regulations in eastern Norfolk see B.M.S. Campbell, 'The Regional Uniqueness of English Field Systems? Some Evidence from Eastern Norfolk', *Agricultural History Review* 1, 29 (1981).

[5] Norfolk Record Office (hereafter N.R.O.), NRS 21404 A; King's College Cambridge (hereafter K.C.C.), P 34.

[6] N.R.O., Church Commissioners' Map 11905.

[7] N.R.O., Pet. MS 584; K.C.C., E 28 and P 34; N.R.O., NRS 16646 37 G; K.C.C., N 52.

[8] W. J. Corbett, 'Elizabethan Village Surveys', *Trans. Roy. Hist. Soc.* new ser. XI (1897), 67-87.

[9] C. M. Hoare, *The History of an East Anglian Soke: Studies in Original Documents* (1918), 332.

[10] Water Rye, *Some Rough Materials for a History of the Hundred of North Erpingham in the County of Norfolk*, I (1883), 28.

[11] K.C.C., N 10.

[12] M. R. Postgate, 'Field Systems of East Anglia', Chapter 7 in A. R. H. Baker and R. A. Butlin (eds.), *Studies of Field Systems in the British Isles* (1973), 290-1.

[13] Public Record Office, London (hereafter P.R.O.), E 315 Vol. 360 f. 39v. For a medieval counterpart see B. M. S. Campbell, 'Field Systems in Eastern Norfolk during the Middle Ages: A study with Reference to the Demographic and Agrarian Changes of the Fourteenth Century' (University of Cambridge, unpublished Ph.D. thesis, 1975), 260-2.

[14] P.R.O., DL 44/295.

[15] Campbell (1981). For a classification of different types of field system see B. M. S. Campbell, 'Commonfield Origins: The Regional Dimension', in T. Rowley (ed.), *The Origins of the Commonfield System* (1981)

[16] This observation is based upon the evidence of 230 inventories relating to the Hundreds of North and South Erpingham, Tunstead, Happing, East and West Flegg, Blofield, and Walsham, in the period 1580-1600 (N.R.O., Norwich Diocesan Inventories). See also E. Kerridge, *The Agricultural Revolution* (1967), 87-9.

[17] British Library, Stowe MS 936. A perch of 18½ feet is used throughout the extent, however, in this article all measurements have been converted into statute acres.

[18] N.R.O., Middleton, Killin, and Bruce 19.11.68 (Supp. Catalogue), 1 and 2.
[19] Campbell (1975), 17-23 and 337-57.
[20] K.C.C., E 29-38.
[21] N.R.O., NRS 13714 28 D 6; NRS 14634 29 D 2; NRS 14473 29 C 1.
[22] N.R.O., NRS 19280 33 F 9.
[23] N.R.O., Dean and Chapter Muniments, Martham Box 1, 30.
[24] N.R.O., NNAS 5952 20 D 5.
[25] N.R.O., NRS 19559 42 D 2.
[26] N.R.O., NRS 14761 29 D 4.
[27] N.R.O., Dean and Chapter Muniments, Register V ff. 132-5; P.R.O., SC 11 Portf. 22 no. 10; W. Hudson, 'The Abbot of St. Benet and his Tenants after the Peasant Revolt of 1381' *Antiquary* 29 (1894), 256.
[28] E. Miller and J. Hatcher, *Medieval England: Rural Society and Economic Change 1086-1348* (1978).
[29] H. C. Darby, *The Domesday Geography of Eastern England* (3rd edition, 1971), 97-152.
[30] J. M. Lambert, J. N. Jennings, C. T. Smith, C. Green, J. N. Hutchinson, 'The Making of the Broads', *Royal Geographical Society Research Series* 3 (1960), 53.
[31] British Library, Cott. MS Galba E ii f. 203 – the listing may be dated from the death, in 1234, of William de Glanville, whose name appears in it.
[32] Figures of Domesday plough-teams are provided by Mr. R. Versey of the Department of Geography, Cambridge: see Campbell (1975), 376.
[33] 40.83 acres were lost to the sea at Hemsby between 1422 and circa 1500.
[34] Bibliothèque Nationale, Paris, MS Latin 5650 ff. 50-2.
[35] P.R.O., SC 11 Roll 471.
[36] For a full discussion of this evidence see B. M. S. Campbell, 'Population Change and the Genesis of Commonfields on a Norfolk Manor', *Economic History Review* 2nd ser. 2, XXXIII (1980).
[37] N.R.O., NRS 19280 33 F 9 – this evidence is discussed in Campbell (1975), 275-9.
[38] N.R.O., Norwich Diocesan Records, Est/1 and 2.
[39] Campbell (1975), 79-95 and 337-57. Also B.M.S. Campbell, 'Technological Progress and High Productivity in Medieval Agriculture: The Evidence of Eastern Norfolk' (unpublished paper presented at a seminar in Peterhouse College, Cambridge, December 1979).
[40] For comparative yield figures in the sixteenth and seventeenth centuries see M. Overton, 'Estimating Crop Yields from Probate Inventories: An Example from East Anglia, 1585-1735', *Journal of Economic History* XXXIX (1979), 363-78.
[41] British Library, Add. MS 24, 316 ff. 51-6.
[42] Campbell (1980).
[43] B.M.S. Campbell, 'Population Pressure, Inheritance and the Land Market in a Fourteenth Century Peasant Community', Chapter 2 in R.M. Smith (ed.), *Land, Kinship and Lifecycle* (forthcoming).
[44] P.R.O., E 315 Vol. 360 f. 38; K.C.C., E 34; P.R.O., SC 12 Portf. 22 no. 10; P.R.O., SC 11 Roll 471; British Museum, Add. MS 24,316 ff. 51-6; Hudson, 256; N.R.O., Dean and Chapter Muniments, Register V ff. 132-5.
[45] N.R.O., Dean and Chapter Muniments, Martham Box I, 6.
[46] Campbell (1980).
[47] P.R.O., NRS 14473 29 C 1; K.C.C., N 52 f. 329.
[48] Campbell (1975), 279-80.
[49] R. J. Faith, 'Peasant Families and Inheritance Customs in Medieval England', *Agricultural History Review* 14 (1966), 82-4.
[50] 28 per cent of tenants at Martham in 1292 and 7.3 per cent of tenants at Hemsby in 1422 were women.
[51] N.R.O., Dean and Chapter Muniments, Martham Court Rolls 1288-99.
[52] N.R.O., NRS 13714 28 D 6; NRS 14634 29 D 2; NRS 14473 29 C 1; NRS 13684 28 D 3.
[53] K.C.C., E 29-34.
[54] A full discussion of the dynamics of the land market at Coltishall is given in Campbell, 'Population Pressure' (forthcoming).
[55] Campbell (1980).
[56] Campbell, 'Population Pressure' (forthcoming).
[57] Campbell (1980).
[58] J. Hatcher, *Plague, Population and the English Economy 1348-1530* (1977).
[59] Campbell, 'Population Pressure' (forthcoming).
[60] For a description of the impact of the plague outbreak of 1348/49 upon East Anglia see J.F.D. Shrewsbury, *A History of Bubonic Plague in the British Isles* (1970), 94-9.
[61] N.R.O., NRS 19559 42 D 2.
[62] Campbell (1980).
[63] Campbell (1975), 133-5.
[64] R.M. Smith, 'Population and its Geography in England 1500-1730', Chapter 8 in R. A. Dodgshon and R.A. Butlin (eds.), *An Historical Geography of England and Wales* (1978).

[65]N.R.O., NRS 11312 26 B 3; NNAS 5924-29 20 D 3; NNAS 5930-50 20 D 4; NNAS 5951-52 20 D 5; K.C.C., E 34-39; N.R.O., NRS 19558 42 D 2; NRS 14772 29 D 4; NRS 14486 29 C 1; NRS 14637 29 D 2; NRS 19559 42 D 2; NRS 14763 29 D 4; NRS 19560 42 D 2. This material is tabulated in Campbell (1975), 377-80.

[66]N.R.O., NNAS 5951 20 D 5.

[67]N.R.O., NRS 14763 29 D 4.

[68]N.R.O., Dean and Chapter Muniments, MS 2765.

[69]N.R.O., Middleton, Killin, and Bruce 19.11.68 (Supp. Catalogue), 2.

[70]N.R.O., Dean and Chapter Muniments, Martham Box II.

[71]N.R.O., NNAS 5940 20 D 4.

[72]N.R.O., NRS 19280 33 F 9; NRS 14479 29 C 1, NRS 13714 28 D 6.

[73]N.R.O., NRS 19558 42 D 2.

III

The Regional Uniqueness of English Field Systems? Some Evidence from Eastern Norfolk

THE assumption that England possessed a number of regionally distinct field systems, differentiated from one another by certain unique attributes, has been implicit in much writing on English field systems. Yet in the present state of knowledge, with the full geographical extent and precise mode of operation of the common-field system imperfectly known, the possibility remains that the different systems which existed transcended regional boundaries and thus were not exclusively regional in character. Local and regional idiosyncracies of terminology and tenure certainly existed, but by themselves these do not constitute evidence of unique local or regional field systems. In fact, an examination of field systems on strictly functional grounds may well demonstrate the contrary, as in the case of Kent, where A R H Baker has shown that there was little peculiarly Kentish about the 'so-called "Kentish system"'.[1]

This issue of the regional uniqueness, or otherwise, of English field systems is not only important in its own right but also bears upon our understanding of the origin and development of the common-field system. Any explanation of the origin of the system must account for the fact that field systems became co-ordinated and systematized in different ways, and to a different extent in different parts of the country. As yet no convincing reasons for this have been advanced. H L Gray thought that spatial variations in field systems could be attributed to colonization by different ethnic groups but this view has now been largely discredited.[2] Even Joan Thirsk's more recent hypothesis, that regional variations in field systems reflect regional variations in population density, in the relative importance of pastoral and arable farming, and in soil and terrain, leaves certain facts unexplained.[3] There are several exceptions to her observations that 'the classic common-field system represented an intensive system of farming for corn that was characteristic of all well-populated villages in plains and valleys in all parts of the kingdom', and that 'field-systems and the rigour of their rules and regulations varied according to the type of farming practised, and perhaps according to the size of populations'.[4] Conspicuous exceptions are the greater part of East Anglia and the extreme south-east of England. Any further advance in our understanding of the genesis of the common-field system in England will therefore partly depend upon a fuller knowledge of the distribution, mode of operation, and course of development of each of its variant forms.

Among the most interesting areas for the study of field systems are areas which were characterized by intensive arable farming and

[1] A R H Baker, 'Some Fields and Farms in Medieval Kent', *Archaeologia Cantiana*, LXXX, 1965, pp 152–74: B M S Campbell, 'Commonfield Origins: The Regional Dimension', in T Rowley (ed), *The Origins of Open Field Agriculture* (forthcoming).

[2] H L Gray, *English Field Systems*, Cambridge, Mass, 1915. See also A R H Baker, 'Howard Levi Gray and English Field Systems: An Evaluation', *Agricultural History*, 39, 2, 1965, pp 86–91.

[3] J Thirsk, 'The Common Fields', *Past & Present*, 29, 1964, pp 3–25; J Thirsk, 'The Origin of the Common Fields', *Past & Present*, 33, 1966, pp 142–7; J Thirsk, 'Preface to the Third Edition', pp v–xv in C S and C S Orwin, *The Open Fields*, Oxford, 3rd edn, 1967.

[4] Ibid, p xi.

high population densities. In such areas subdivided fields were often especially well developed and the problems of reconciling the mutually dependent but conflicting demands of pastoral and arable husbandry were particularly acute. One such area was East Anglia. During the middle ages parts of this large and diverse region (notably eastern and south-eastern Norfolk) supported higher densities of population, and were characterized by higher levels of assessed lay wealth and more intensive methods of farming than any other part of the country.[5] At present, however, specific knowledge of the field systems which operated in East Anglia is confined to western Norfolk and adjacent portions of Suffolk and Cambridgeshire, away from the most economically advanced localities.[6] Moreover, the excellence of sixteenth and seventeenth century sources has attracted attention away from earlier periods so that even in western Norfolk little is known of field systems before the major agrarian changes of the later middle ages. It therefore remains to be proven that the field system which is known to have existed in western Norfolk in the post-medieval period had at one time prevailed throughout East Anglia. Since this system possessed certain highly individual characteristics this consideration is of some importance to the wider issue of regional uniqueness.

The individuality of field systems in western Norfolk largely derived from a unique fusion of the two opposing elements of flexibility and control; rights of common grazing, on the aftermath of the harvest (harvest shack) and on strips lying fallow throughout the year, applied to fields characterized by the utmost irregularity of layout and holdings which employed a highly flexible system of cropping. The irregularity of field layout and flexibility of cropping posed no great obstacle to the institution of harvest shack but presented serious problems to common grazing of the arable at other times of the year. These problems were resolved by means of an institution known as the foldcourse. Foldcourses comprised two essential elements; on the one hand the imposition of irregular cropping shifts to rationalize the distribution of unsown strips (including the provision of compensation for cultivators disadvantaged by possessing a disproportionate amount of land in the fallow shift), and on the other, the supervised grazing of communal flocks upon the fallow. Difficulties of access to the fallow strips and of control and manoeuvrability of livestock meant that rights of fallow grazing were confined to sheep. Accordingly, as soon as spring lambing was past, sheep were collected into communal flocks which were fed upon the heaths and sheepwalks by day and folded upon the fallow arable by night, whose soil they tathed with their treading, dung and urine. Within the commonfields the sheep were controlled by means of moveable folds, which permitted the grazing of relatively small blocks of fallow and also facilitated a more systematic pattern of grazing and dunging than would otherwise have occurred. In fact, the fertilization of the arable appears to have been the principal objective of the system, for the tathe of the sheep fold seems to have been of greater benefit to the arable fields than was

[5] For the distribution of population and wealth in medieval England see the maps in H C Darby (ed), *A New Historical Geography of England*, Cambridge, 1973, pp 46, 139, 191. For East Anglian agriculture see B M S Campbell, 'Field Systems in Eastern Norfolk during the Middle Ages: A Study with Particular Reference to the Demographic and Agrarian Changes of the Fourteenth Century' (University of Cambridge, unpublished PhD thesis, 1975), pp 83–95, 105–17, 337–54.

[6] Studies which have concentrated upon the field systems of this area are Gray, op cit, pp 305–54; J Saltmarsh and H C Darby, 'The Infield-Outfield System on a Norfolk Manor', *Econ Hist Rev*, III, 1935, pp 30–44; K J Allison, 'The Sheep-Corn Husbandry of Norfolk in the Sixteenth and Seventeenth Centuries', *Ag Hist Rev*, V, 1, 1957, pp 12–30; M R Postgate, 'The Field Systems of Breckland', *Ag Hist Rev*, X, 2, 1962, pp 80–101; M R Postgate, 'The Open Fields of Cambridgeshire' (University of Cambridge, unpublished PhD thesis, 1964); M Spufford, 'A Cambridgeshire Community, Chippenham from Settlement to Enclosure', *Occasional Paper, Dept of English Local Hist, Univ of Leics*, XX, 1965. For the most recent account of East Anglian field systems see M R Postgate, 'Field Systems of East Anglia', being Chapter 7 in A R H Baker and R A Butlin (eds), *Studies of Field Systems in the British Isles*, Cambridge, 1973.

the meagre pasturage available on the fallow to the sheep.

The use of sheep as walking dung machines, to transfer nutrients from the permanent pasture (which remained an essential adjunct to the system) to the arable, was a principle common to most areas of sheep-corn husbandry, but only in East Anglia does it appear to have been codified and integrated into the common-field system.[7] Even more unusual is the fact that responsibility for this component of the field system was vested in the manorial lord rather than the corporate authority of the entire commonfield community (especially as in East Anglia there was generally a total lack of coincidence between manor and vill). In common-field villages elsewhere all rights of common grazing belonged to all the cultivators. It may be that in East Anglia the irregularity of holding layout and flexibility of cropping arrangements were such that the organization of fallow grazing demanded the superior authority of the manorial lord, but the subordinate position in which this placed the majority of cultivators was a potential weakness of the system. An unscrupulous lord could fold his tenants' sheep on the demesne to the neglect of their own land, thereby appropriating dung to his own use: or, if he was more interested in his wool clip than his corn yields, he could overstock the fields and pastures and expand his own flock at the expense of his tenants'.[8] Ultimately, abuses such as these brought the system into disrepute and led to its decline and dissolution. This, however, was a development of the post-medieval period. What follows is an attempt to reconstruct field arrangements in the more fertile and thickly settled locality of eastern Norfolk at their medieval zenith.

I

In the late thirteenth and early fourteenth centuries manorial records show that extensive common fields dominated the landscape of eastern Norfolk, and that the subdivision of land was intense. Two and a half centuries later, by which time the common fields had already been much modified by consolidation and enclosure, the first topographical descriptions of entire townships become available and it is possible to assess the relative disposition of arable and pasture and reconstruct the original cadastra of the commonfields. This evidence shows that the ratio of pasture to arable was generally low, significantly lower than in western Norfolk, although subject to wide variation from township to township according to prevailing soil conditions. It also shows that, as was usual in East Anglia, the layout of the commonfields was highly irregular.[9]

The first of these points requires little elaboration. Eastern Norfolk was a closely-settled locality and arable land was consequently at a premium with the result that common pastures tended to remain only where soils were too light or poorly drained for cultivation. Given the uneven distribution of such soils some townships (particularly those in the vicinity of the Broadland marshes and the sandy heaths to the north of Norwich) were quite generously endowed whereas others, often in close proximity,

[7] E Kerridge, 'The Sheepfold in Wiltshire and the Floating of the Water Meadows', *Econ Hist Rev*, 2nd ser, VI, 1954, pp 282–9; J Thirsk (ed), *The Agrarian History of England and Wales*, IV, Cambridge, 1967, pp 33, 51, 56, 70, 92, 136, 188, 250; E Kerridge, *The Farmers of Old England*, 1973, pp 20–1, 77–84.

[8] For examples see K J Allison, 'The Lost Villages of Norfolk', *Norfolk Archaeology*, XXXI, 1957, pp 116–62; K J Allison, 'Flock Management in the Sixteenth and Seventeenth Centuries', *Econ Hist Rev*, 2nd ser, XI, 1958, pp 98–112.

[9] The maps and surveys upon which these and the following observations are based include those of Westwick (1547), Norfolk Record Office (hereafter NRO) Pet MS 584; Cawston (*c*1580), NRO NRS 21404 A; Coltishall (1584), King's College, Cambridge (hereafter KCC) E 28; Hempstead with Lessingham (1584), KCC P 34; Blofield (1586), NRO NRS 16646 37 G; Horstead-with-Stanninghall (1586), KCC N 52; Horsford (1613), NRO Aylsham Collection 54; Worstead (undated 17th century), NRO Church Commissioners' Map 11905. For a fuller discussion see B M S Campbell, 'The Extent and Layout of Commonfields in Eastern Norfolk', *Norfolk Archaeology* (forthcoming).

were not. At Horsford, Hevingham and Cawston, for instance, the open fields formed large but isolated islands of arable in a sea of pasture, whereas the converse applied on the rich loams of Flegg and adjacent areas, where the arable fields frequently stretched without interruption from township to township. At Coltishall, according to a survey of 1584, the common fields merged across the parish boundary with those of each of the neighbouring townships of Belaugh, Tunstead, Sco Ruston, and Hautbois and common pasture consisted of a mere 80 acres out of a total acreage of 1190. As will be seen later, the near elimination of common pasture in many townships in eastern Norfolk had important implications for the numbers of livestock that were kept and, more especially, for methods of feeding them.

As variable as the distribution of pasture and arable, in fact possibly even more so, was the internal layout of the common-fields. Confronted with this topographical complexity many sixteenth-century surveyors resorted to the purely descriptive device of dividing townships into a number of sectors or precincts, each of which was further subdivided into a number of quarantines or furlongs, the constituent strips of which were then itemized. The divisions which they adopted were usually determined by physical features such as roads, streams, field boundaries, and the orientation of strips, and their descriptions therefore tend to reflect the peculiar topography of each parish. Thus at Coltishall there were nine precincts, ranging in size from 43½ acres to 248⅝ acres, while at Lessingham a contemporary survey describes only four precincts although, as at Coltishall, they show a complete lack of uniformity in size.[10] This variation in the number and size of the precincts described by the surveyors, as of the other units which they employed, illustrates the all-pervading irregularity which characterized the number and size of common-fields in this locality. Such field names as are given relate to indeterminate areas, and it is plain that, with the exception of the strip, none of the areal units described possessed more than topographical significance. The individual strip was the fundamental unit of cultivation and to gauge a reliable impression of the original size and number of the strips, before they were affected by consolidation and enclosure, it is necessary to turn to earlier sources of evidence.

By the first half of the fourteenth century land holdings in eastern Norfolk were both attenuated in size and fragmented in layout. Statistics derived from data relating to a single manor almost certainly understate holding size but nevertheless convey the distinct impression that the majority of peasant holdings were extremely small. On the Prior of Norwich's manor at Martham, for example, an extent of 1292 shows mean holding size to have been as low as 2¼ acres (and it probably declined further during the next fifty years), whilst of 2122 arable strips 94 per cent were smaller than 1 acre, 66 per cent were smaller than ½ acre and 16 per cent were smaller than ¼ acre.[11] Evidence from other manors in this locality reveals conditions which were much the same. At Coltishall mean holding size, as indicated by the obituaries of 400 deceased tenants recorded in the court rolls between 1280 and 1400, was less than 3 acres, whilst of 900 land parcels transacted in the court rolls between 1275 and 1349, 91 per cent were smaller than 1 acre, 69 per cent were smaller than ½ acre, and 32 per cent were smaller than ¼ acre.[12] At Hevingham similar evidence reveals an almost identical state of affairs at the close of the thirteenth century, with few holdings larger than 3 acres and 86 per cent of land parcels smaller than 1 acre.[13] Moreover, as the

[10] For a detailed analysis of field layout at Coltishall see Campbell, op cit, 1975, pp 147–55.

[11] BM: Stowe MS 936.

[12] KCC: E 29–38.

[13] NRO: NRS 14761 29 D 4, NRS 14634 29 D 2, NRS 14473 29 C 1. Similar conditions prevailed at Lessingham (BM: Add MS 24,316, ff 51–6), Worstead (NRO: Dean and Chapter Muniments, Register V, ff 132–5), Hautbois (PRO: SC 11 Roll 475), Burgh (PRO: SC 12 Portf 22 no 10), Hemsby (NRO: Middleton, Killin and Bruce,

detailed extent of Martham confirms, these holdings were not only diminutive in size but were also highly irregular in layout, a characteristic which was exacerbated by the combined action of an active peasant land market and, on most manors, a custom of partible inheritance. Such small and irregular holdings were fundamental to the way in which these extensive, minutely subdivided arable fields were worked, promoting the adoption of certain common rights but inhibiting the development of others. Of the former the most fundamental was arguably the right of shack feed on the aftermath of the harvest.

Harvest shack was a practical, commonsense response to the opportunities for pasturing livestock on the aftermath of the harvest where permanent pasture was scarce and arable fields were subdivided. It therefore tended to be found wherever holdings were heavily fragmented and strips were small, as was indeed the case in eastern Norfolk. At Horstead-with-Stanninghall an enclosure award of 1599 records the dissolution of 'libertie of shack in the tyme of shacke', and passing reference is made to either *shack* or *tempore aperte* at Antingham, Bassingham, Cawston, Lessingham, Martham, North Walsham, and South Walsham.[14] At Gimingham there is reference to 'the shack tyme of winter' whilst at Catton and Hellesdon rights of common pasture applied to the arable fields *a tempore quo blada in eisdem terris crescentia leventur et adventet usque ad tempus quod dicte terre reseminentur.*[15] Explicit testimony of the duration of this right is, however, available only for three townships in this locality; at Postwick and Reedham rights of common pasturage applied to the arable fields for a period of six months (1 August — 2 February and 29 September — 25 March respectively), whilst at Ingham this period was slightly longer, lasting for seven months from 29 September to 3 May.[16] On the other hand, once grass and plant growth ceased in mid-November the forage available on the arable became so meagre as to be virtually valueless. This no doubt underlies the fact that villein tenants were obliged to pay *bossagium* and *faldagium* and place their sheep in the lord's fold only up until 10 November, and extra cowherds and shepherds employed on the demesnes to supervise the stubble grazing of livestock were rarely retained for more than three months in the autumn. At Martham, for instance, in 1380 two shepherds were employed during the last three weeks in September whilst the harvest was being gathered in, of whom one was retained for a further ten weeks until the middle of December, when he too was laid off.[17]

But just as the fragmentation of holdings encouraged the institution of harvest shack so it also hindered the institution of collective grazing rights at other times of the year. An essential precondition for common grazing of the fallow was the segregation of sown from unsown strips. In western Norfolk in the sixteenth and seventeenth centuries this was achieved by the imposition of a system of irregular cropping shifts, but in this case neither the subdivision of fields nor the complexity of holding layout was as great as in eastern Norfolk in the middle ages. Where holdings were so small, so fragmented, and so irregular, and moreover subject to constant change, the problems presented to the operation of even the most ingenious communal rotational scheme would have been virtually insurmountable. These problems were further compounded by the complexity and intensity of cropping practices in this area.

19.11.68), and Thurne (W Hudson, 'The Abbot of St. Benet and his Tenants after the Peasant Revolt of 1381', *Antiquary*, 29, 1894, p 256).

[14] KCC: N 10; Walter Rye, *Some Rough Materials for a History of the Hundred of North Erpingham in the County of Norfolk*, I, Norwich, 1883, pp 14, 28; NRO: NRS 21404 A, NNAS 5930 20 D 4; PRO: DL 44/295, E 142 no. 83(4); KCC: P 34.

[15] PRO: DL 44/295; NRO: Dean and Chapter Muniments, Register I, ff 252 and 254v.

[16] PRO: C134 File 104 (1) and File 95 (14); C135 File 74 (8).

[17] NRO: NNAS 5896 20 D 1.

Detailed evidence of rotations, and thus the incidence of fallowing, is available only for demesnes (most of which comprised at least some open-field land) and may be deceptive as the tiny holdings of the peasantry were probably cultivated with greater intensity. Even so, by the second half of the fourteenth century, when most account rolls begin to record the area of fallow on a regular basis, no demesne in this locality was fallowing land more frequently than once every five or six years, and several were fallowing it as infrequently as once a decade (see Table 1). Furthermore, fallowing had been even less frequent during the first half of the fourteenth century when 'high-farming' was at its peak, often less than once every twelve years. There were even occasions on some demesnes when fallows were dispensed with altogether and the entire arable area brought into cultivation, although this policy was rarely pursued for more than two or three years in succession. Notwithstanding the intensity of cropping which was the consequence of this near elimination of fallows, none of these demesnes appears to have suffered from a deterioration of productivity.[18] Far from it, most of them sustained a level of output per acre which was exceptional by the standards of the day. Wheat, for instance, the most demanding cereal crop, yielded an average of 15 bushels per acre on almost all demesnes in eastern Norfolk, and on the most productive, such as the neighbouring demesnes of Martham and Hemsby (both of which belonged to Norwich Cathedral Priory), averaged well over 20 bushels, and rose to over 30 bushels in a good year. Barley, the principal grain grown in this area, yielded at a broadly similar if somewhat lower rate, yields of 25 bushels being by no means unknown.

This seeming paradox, of the near elimination of fallows coupled with a high and sustained level of productivity, was the product of a progressive and carefully

TABLE 1

Proportion of total arable annually fallowed on various demesnes in eastern Norfolk, 1269–1428

Demesne	Time-span (years)	Number of A/C Rolls	% total arable area fallowed min	mean	max
Pre-1350					
Halvergate	1268–74	4	0	0	0
Flegg	1340–41	1		0	
Heigham By Norwich	1302–06	2	2.5	3.15	3.8
Acle	1268–80	7	1.3	5.2	11.4
South Walsham	1270–97	9	0	6.8	13.4
Martham	1294–1350	19	*0*	*8.1*	*25.3*
Hemsby	1294–1342	13	*1.8*	*9.4*	*14.2*
Suffield	1272–1300	9	*2.1*	*11.7*	*21.2*
Hanworth	1272–1306	19	*0*	*12.4*	*25.5*
Knapton	1345–48	2	*19.15*	*20.3*	*21.5*
Post-1350					
North Walsham	1367–1427	5	3.7	7.8	9.9
Potter Heigham	1389–90	1		9.4	
Thwaite	*c*1386–87	1		10.4	
Scottow	*c*1364–65	1		10.6	
Ashby	1378–92	2	11.1	13.15	15.2
Flegg	1351–1428	14	0	13.3	25.2
Ludham	*c*1354–55	1		13.3	
Martham	1355–1420	19	2.9	13.4	27.9
Hoveton	1392–1422	2	10.8	13.55	16.3
Heigham By Norwich	1380–81	1		15.9	
Plumstead	1359–1420	15	11.0	16.0	27.7
Hevingham	1357–58	1		17.3	
Shotesham	1352–53 or 1368–69	1		17.8	

Figures in italics are estimates.

Sources: PRO SC 6/929/1–7, SC 6/936/2–8 and 18–32, SC 6/937/1–10, SC 6/944/1–9 and 23–31; NRO Dean and Chapter Muniments MS 4652–65, 4945–64, 5127–43; NRO Diocesan Est/2, 9, 11, 12; NRO Church Commissioners' 101426 2/13 and 11/13; NRO NNAS 5892–5903 20 D 1, NNAS 5904–16 20 D 2; NRO NRS 13996 28 F 3; Windsor, St George's Chapel XV 53 98–9.

[18] For a converse situation on less intensively cultivated demesnes see J Z Titow, *Winchester Yields: A Study in Medieval Agricultural Productivity*, Cambridge, 1972.

balanced system of husbandry. This is illustrated by the pains that were taken to conserve and improve soil fertility. Animal manure, for example, the prime source of fertilizer, was put to maximum use by being carefully collected and then systematically spread upon the land and ploughed in, an operation which was essential if losses from the twin processes of leaching and oxidization were to be kept to a minimum. So great was this concern to make maximum use of available supplies of livestock manure that some demesnes in the vicinity of the Broadland marshes even went to the length of gathering up the manure from the sheep and cattle pens on the marshes and transporting it back to the demesnes to be spread on the fields. Elsewhere supplies of farmyard manure were supplemented by marl, or even, in one instance, night soil purchased from Norwich, and the whole procedure was reinforced by repeated ploughings which stirred up the nutrients within the soil and improved its texture. As well as these direct measures of improving fertility an important part was also played by the choice of crops and the way in which they were rotated with one-another. Thus, a pronounced emphasis upon spring-sown crops — barley, oats and legumes — ensured that about two-thirds of the arable was annually fallowed on a half-yearly basis, whilst the cultivation of legumes on a large scale both restored the nitrogen content of the soil and supplied valuable fodder to the livestock, whose manure was returned to the soil.[19] Also, since crop rotations were themselves extremely flexible, allowance could readily be made for local and annual variations in soil conditions. In fact, the only purpose for which the occasional bare fallow appears to have been retained was to cleanse the land of weed growth, a function for which the employment of heavy seeding rates and use of oats as a smother crop were but partial substitutes.[20]

If it is assumed that the peasantry cultivated their small holdings at least as intensively as the manorial lords cultivated their demesnes, then, with less than 10 per cent of all common-field strips lying fallow each year, the difficulties of instituting a system of common rotation would have been as great as its desirability would have been small. In other words, under this system of cultivation the amount of land available for temporary pasturage would have been too meagre to have merited the concessions and compromises involved in arranging for it to be grazed in common. Indeed, by interfering with a highly flexible and effective system of cropping such an action might even have been counter-productive. The weight of circumstantial evidence, therefore, suggests that rights of common grazing did not apply to fallow strips in these common fields. The sole exception to this rule appears to have been the township of Little Plumstead, located on the edge of Mousehold Heath just five miles north-east of Norwich, where there is a solitary reference in an *Inquisition Post Mortem* to fallow arable remaining subject to rights of common pasturage for the whole year.[21] Elsewhere, however, cultivators made their own arrangements for grazing fallow strips, probably using a system of tethering, just as they enjoyed complete freedom in matters of cultivation. Only in shack time, after the harvest, were the arable fields grazed in common.

It follows from the conclusion that fallow strips were grazed in severalty rather than in common that rights of foldcourse, as they were known in western Norfolk, did not apply to these common fields. This is borne out by a limited amount of direct evidence.

[19] Legumes accounted for between a fifth and a seventh of a sown area.

[20] Fallow land was generally ploughed at least four, and occasionally as many as six times before being returned to cultivation. The most common seeding rates were 4 bushels per acre for wheat and legumes, 6 bushels per acre for barley and 8 bushels per acre for oats.

[21] PRO: C135 File 64 (2).

For instance, on several manors, as at Blickling in 1410–11, tenants paid to have their land tathed by the lord's fold, a transaction which would not have been necessary if the manorial flock was entitled to common grazing on the fallow.[22] So highly valued was the tathe of the fold — it was valued at 2*s* per acre at Antingham and 2*s* 6*d* per acre at both Blickling and Saxthorpe — that it was evidently reserved to the exclusive use of the demesne. This is specified as having been the case at Gimingham and Hevingham, an account roll of the former referring to *agistamento de xiiij*^{xx} *bidentes de collecto in falda domini per messorem pro dominica terra compostand hoc anno* (the term *bidentes de collecto* probably refers to the cullet sheep which villein and other tenants were obliged to place in the lord's custody).[23] That sheep folds and flocks were confined to the demesnes is also implicit in certain more general statements. At Horstead-and-Stanninghall the fellows of King's College, Cambridge were entitled, as lords of the manor, to 'feede and depasture their sheepe in and uppon the . . . Commons and heathe ground and in and upon the . . . inclossuer called the hundred Acres and upon other demeasnes of the said Manor wth libertie of shacke for the same sheepe yeerlie in shacke tyme', and at Catton there was 'A Foldcourse for 300 sheep being only on the Shack of the Lands belonging to the s^{d}. manor'.[24]

Whilst reinforcing the impression that there was no attempt to co-ordinate the distribution of fallow strips and subject them to rights of foldcourse or any other collective grazing right, these references nonetheless testify to seigneurial intervention in certain aspects of animal husbandry. In particular, manorial lords enjoyed certain privileges with regard to sheep folding. On many manors, such as Gimingham and Hevingham, the lord was entitled to the tathe of his customary tenants' sheep, a right which is specified in a number of manorial extents. Thus, at Heigham-by-Norwich in 1275 it was recorded of Simon Bele, villein, that *debet habere omnes bidentes suas in falda domini a Pentecosta usque ad festum Sancti Martini.*[25] In addition, at Horstead-and-Stanninghall and Catton, as also at Gimingham and Hevingham and several other townships, the manorial lord was entitled to feed his flock upon the common pastures and upon the common fields in shack time after the harvest. Such rights went by the collective name of 'liberty of the fold' *(libertatibus faldagii)* and in this form are recorded on many, although not all, manors in the area.[26] Nevertheless, there is nothing to suggest that manorial lords enjoyed superior rights of pasturage over their tenants' land when it lay fallow: in this respect there was a fundamental difference between 'liberty of the fold' and the right of foldcourse.

This fundamental difference between the right of foldcourse and its diminutive, 'liberty of the fold', reflected the varying importance of sheep rearing within the rural economy. In sandy western Norfolk sheep rearing was at least as important as the grain production with which it was so closely integrated, and

[22] *Et de vijs. vjd. receptis de diversis hominibus pro iij acris terre cum falda domini compostand hoc anno pro acra ijs. vjd.* (NRO: NRS 10196 25 A 1). Similar payments are recorded at Antingham (NRO: MS 6031 16 B 8) and Saxthorpe (NRO: 19677 42 E 3).

[23] NRO: MS 6001 16 A 6. The entry relating to Hevingham is much the same (NRO: NRS 14747 29 D 4). Other references to 'cullet sheep' occur at Ashby (NRO Diocesan Est/9) and Martham (NRO: NNAS 5900 20 D 1). See also N Davis, 'Sheep Farming Terms in Medieval Norfolk', *Notes & Queries*, 16, 1969, pp 404–5, and Allison, op cit, 1957, p 21.

[24] KCC: N 10; NRO Dean and Chapter Muniments MS 2669. Also an inquisition on behalf of the Abbot of St. Benet at Holme found that he had *faldam . . . in solo ipsius Abbatis apud Antyngham* (NRO NRS 3102 13 B 2) whilst at South Walsham Lady Margery Folyet possessed *cursus unius falde tempore operto* (PRO: E 142 no 83(4)).

[25] NRO: Diocesan Est/2 (2/2). An almost identical arrangement prevailed at Taverham (BM: Stowe MS 936, f 30).

[26] Rye, op cit, pp 16, 27, 37, 77; NRO: Dean and Chapter Muniments, Leases 1st Ledger Book, f 128. References to 'foldcourses' at Cawston (NRO MS 12538 30 D 5), Roughton (Rye, op cit, pp 165–7) and Westwick (NRO Pet MS 584/2/16) are to sheep-walks 'where never plough or fould for tashing doth com' (BM: Add MS 27403, cited in Postgate, op cit, 1973, p 317).

most manors carried large flocks, but in eastern Norfolk, where soils were heavier and more fertile, corn production was pre-eminent and sheep rearing was relegated to a much less prominent position.[27]

Demesne flocks, according to manorial stock accounts, were generally either small or non-existent. This is partially explained by the fact that on large estates, such as those of Norwich Cathedral Priory and St Benet's Abbey, it was the practice to manage flocks on an inter-manorial basis and account for them separately. In 1343, for instance, St Benet's Abbey had 1900 sheep on the marshes attached to its granges of Hoveton, Ashmanhaugh, Worstead, Barton, *Hardele* and *Kybald*, whilst in 1420/21 a flock of 610 was inter-manorial between the three lay manors of Blickling, Gunton and Erpingham.[28] However, sheep rearing on this scale was possible only in the immediate vicinity of extensive marshes and heaths, and consequently on most manors the virtual absence of sheep from the stock account is probably a genuine indication of the unimportance of sheep in the rural economy. When this was the case, manorial lords appear to have relied upon their tenants' sheep to fold the demesnes, although to judge from the areas folded each year even these sheep were few in number. Folding was generally unrelenting from Lamas-tide until Martinmas (3 May — 10 November), a period of twenty-seven weeks,[29] yet on no demesne for which account rolls survive did the area folded ever exceed 35 acres, which implies that flocks rarely comprised more than 200 animals.[30] At Martham, one of the most productive demesnes in eastern Norfolk, situated in the immediate vicinity of extensive marshland pastures, the acreage folded averaged only twenty-four acres between 1363 and 1400, whilst at Plumstead, a manor with an even more favourable location, it averaged twenty-six acres and never exceeded thirty-two acres.[31] These were both intensively cultivated demesnes and it is possible that the smallness of the area available for folding rather than the small size of sheep flocks underlies these low figures, but, on the other hand, there is nothing to suggest that these demesnes failed to take up their full option on their customary tenants' sheep. On the contrary, there are cases in the court rolls of Martham of tenants who were prosecuted for evading their obligation to place their sheep in the lord's fold and at Plumstead the tathe of the fold was supplemented with additional manure bought in from outside.[32] On other demesnes the area folded was frequently smaller still, as at Flegg where it averaged only nine acres between 1355 and 1427, and there were some demesnes, such as Suffield, where folding does not appear to have taken place at all.[33] In the few rare instances when the number of cullet sheep is recorded the small size of these flocks is confirmed. At Gimingham the number of cullet sheep folding the demesne ranged from 143 in 1358/59 to 280 in 1367/68, but was generally below 200 in number.[34] With flocks of this size it is no surprise to find that the acreages folded were never large and that the tathe of

[27] At Sedgeford in north-western Norfolk — Norwich Cathedral Priory's principal sheep manor — the profits of the fold regularly yielded over £15 a year in the late thirteenth century, whereas at Taverham, close to Norwich and in the vicinity of extensive sandy heaths, they rarely yielded more than £5, and on the important corn manor of Plumstead, usually less than £1 (NRO: Dean and Chapter Muniments, *Proficuum Maneriorum*).

[28] NRO: Diocesan Est/2 (2/11), NRS 10535 25 B 5.

[29] This is the period most commonly specified in manorial extents during which tenants paid *faldagium* and were obliged to place their sheep in the lord's fold: NRO Diocesan Est/2 (2/2); PRO SC 11 Roll 471, DL 29 289/4747.

[30] 'A thousand sheep would fold an acre of common-field land in a night', Thirsk, *The Agrarian History of England and Wales*, p 188. Similarly, Kerridge has estimated that it took 200 sheep a week to tathe one statute acre (E Kerridge, *The Agricultural Revolution*, 1967, pp 74–5).

[31] NRO: NNAS 5894–5903 20 D 1, NNAS 5904–05 20 D 2, Dean and Chapter Muniments MS 5127–38.

[32] NRO: Dean and Chapter Muniments MS 4998–99.

[33] NRO: Diocesan Est/9; PRO SC 6/944/1–6.

[34] PRO: DL 29 288/4719–20; NRO: MS 6001 A 6, NRS 11331–32 26 B 6, NRS 11058–60 25 E 2, NRS 11069 25 E 3.

the fold was jealously reserved to the demesne.

Under these circumstances manorial lords therefore had neither the need nor the desire to annex their tenants' land as well as their sheep and 'liberty of the fold' remained a right by which dung was appropriated to the demesne.[35] Indeed, several references in extents and account rolls imply quite strongly that villein tenants alone paid *faldagium* and were subject to the lord's liberty of the fold.[35] It would thus appear that this right was less an instrument of common-field management than of seigneurial exploitation and, as such, cannot be classed as part of the field system of this area. In fact, 'liberty of the fold' detracted from, rather than contributed to, the operation of these common fields. The right of the lord's flock to harvest shack meant that he had a vested interest in preventing enclosure and gave him the power to obstruct it, whilst his own right to the tathe of his customary tenants' sheep reduced the productivity of their land. However, these were essentially indirect influences upon the operation of these common fields; they did not seriously alter the pattern of remarkable freedom from all but the most fundamental common rules and regulations.

II

As the foregoing discussion has revealed, eastern Norfolk's common fields were more remarkable for their extent and degree of subdivision than for any superimposed field organization. By the close of the thirteenth century the only concession which appears to have been made to the fragmentation and intermixture of holdings was the institution of common grazing rights on the aftermath of the harvest. With this exception, their form found little expression in their function. As these common fields were so singularly lacking in system, therefore, it follows that there could have been little that was peculiarly East Anglian about them; not even, in a negative sense, the absence of so many of the rules and regulations usually associated with common-field systems (for which Kentish field systems provide a direct parallel).[36] Some individuality certainly derived from the lords' superior position in matters of sheep folding; this, too, was related to analagous rights elsewhere, both within and outside East Anglia, and was more a manifestation of seigneurial privilege than an essential and integral part of the field system.[37] Moreover, the lords' control over their tenants' sheep was not absolute: it was restricted to customary tenants. The area's substantial freeholding population remained exempt.[38] The latter, even more than the former, therefore, enjoyed almost complete autonomy in the management of their land, an autonomy which, if the husbandry of the demesnes is at all representative, they undoubtedly exploited to the full.

Such conclusions clearly have a number of wider implications at both a regional and a more general level, not least in the modification which they make to established models of East Anglian field systems. Much still remains to be known about the function of field systems elsewhere in East Anglia, but meanwhile the evidence of field systems in eastern Norfolk does at least demonstrate that the peculiar pastoral practices long deemed to have been the special hall-mark of this region

[35] BM: Stowe MS 936, f 30; NRO: Diocesan Est/2 (2/2) and (2/8); PRO: SC 11 Roll 471.

[36] Cambell, op cit, forthcoming, Baker, op cit, 1965. Also A R H Baker, 'Field Systems of Southeast England', in Baker and Butlin, op cit, pp 393–419. Subdivided fields devoid of common rules of cultivation and grazing have also been identified in the Lincolnshire fenland, see H E Hallam, *Settlement and Society: A Study of the Early Agrarian History of South Lincolnshire*, Cambridge, 1965, pp 137–61.

[37] H S Bennett, *Life on the English Manor*, Cambridge, 1937, p 77; P D A Harvey, *A Medieval Oxfordshire Village: Cuxham, 1240–1400*, Oxford, 1965, p 62; M M Postan (ed), *The Cambridge Economic History of Europe*, I, *The Agrarian Life of the Middle Ages*, Cambridge, 2nd edn, 1966, p 601.

[38] For the number and distribution of freemen see B Dodwell, 'The Free Peasantry of East Anglia in Domesday', *Norfolk Archaeology*, XXVII, 1939, pp 145–57; H C Darby, *The Domesday Geography of Eastern England*, Cambridge, 1952, pp 361–2.

were, in fact, absent from large parts of it. Indeed, when more is known about field systems in Suffolk and Essex (as yet a serious gap in our knowledge) it may even transpire that the sort of loosely controlled common-field system identified in eastern Norfolk was the predominant form of East Anglian field system.[39] On currently available evidence it does seem that foldcourses, and the cropping shifts which were such an essential part of them, were confined to areas where soils were light and readily exhausted, pastures were extensive, and sheep rearing was the predominant pastoral activity. The light sand and good sand regions of western Norfolk and north-western Suffolk fulfil these criteria and, for that matter, also provide abundant evidence of foldcourses; but this is less true of much else of East Anglia, particularly on the boulder clays of high Suffolk and south-central Norfolk. In these areas, although there may have been a seigneurial monopoly of sheep folds, the existence of foldcourses remains to be proven.

Nevertheless, although East Anglian field systems are now revealed as less uniform than has hitherto been supposed, the differences between them should not be exaggerated; certain characteristics were shared by all of them, including an irregularity of field layout, haphazard inter-mixture of holdings, flexibility of cropping practice, common grazing of the aftermath of the harvest, and, perhaps most significantly of all, a seigneurial monopoly of sheep folds. This last feature, contrary to earlier belief, was subject to considerable variation within the region and did not always play a direct role in the management of the common fields. Nonetheless, that seigneurial control of sheep folding prevailed in principle, even where foldcourses were non-existent, is important, for it suggests that the East Anglian foldcourse may indeed represent 'the survival and special development of a seigneurial monopoly which was once widespread'.[40] In other words, what in eastern Norfolk remained a mere manorial imposition, in western Norfolk, under different physical and social circumstances, was transformed into an integral and essential part of the field system. Proof that the foldcourse system actually originated in this way has yet to be found and requires a careful examination of medieval rather than later evidence, but such an interpretation does render this institution somewhat more explicable, and could explain why the right of foldcourse inhered in the manorial lord and not in the whole common-field community. Moreover, if foldcourses were imposed from above rather than evolved from below, then their failure to develop beyond a rather rudimentary stage in eastern Norfolk, even where the necessary physical conditions prevailed (as in the vicinity of the Broadland marshes and sandy heaths to the north of Norwich), may have been as much a reflection of the weakness of the manorial nexus in this locality as of any fundamental difference of geographical environment.

Placed in perspective, therefore, the foldcourse system emerges, like the infield-outfield system of Breckland, as a sub-regional response to particular environmental and social conditions.[41] As such it was uniquely East Anglian, but whether the same was true of field systems in the remainder of the region is another matter. Certainly, field systems in eastern Norfolk showed little essential difference from loosely organized field systems elsewhere. This point is important, for earlier writers, preoccupied with uniqueness, have tended to stress institutions such as the foldcourse and aspects of land tenure peculiar to this region and have played down the very real similarities which existed between these and other field systems. In this context there is a particularly close resem-

[39] For a case study of Suffolk field systems see D P Dymond, 'The Parish of Walsham-le-Willows: Two Elizabethan Surveys and their Medieval Background', *Proc of the Suffolk Inst of Archaeology*, XXXIII, 2, 1974, pp 195–211.

[40] A Simpson, 'The East Anglian Foldcourse; Some Queries', *Ag Hist Rev*, VI, 2, 1958, pp 87–96.

[41] Postgate, op cit, 1962.

blance between the field systems of eastern Norfolk and those of northern Kent.

With the reservation that there is yet some doubt whether Kentish subdivided fields were ever grazed in common after the harvest, there is little in terms of system to distinguish them from the common fields of eastern Norfolk.[42] In neither case did common rotations exist, and in neither case were fallow strips pastured in common, with the result that both field systems gave scope to almost unlimited individualism in matters of husbandry. Moreover, the freedom which individual land holders enjoyed in the cultivation of their land was paralleled by their freedom to divide their holdings between heirs and even dispose of land *inter vivos*. As a result small holdings predominated, population densities were exceptionally high, and husbandry tended to be both intensive and productive. Yet underlying the high level of economic development attained by both eastern Norfolk and northern Kent by the high middle ages, and the close affinity displayed by their field systems, were settlement histories which could scarcely have been more different. Northern Kent was settled relatively early by Romans and Jutes, whereas eastern Norfolk was settled relatively late by Angles, Frisians and Danes. Such contrasting settlement histories, and yet such similar field systems, conflict with the hypothesis that regional differences in field systems were the outcome of the introduction to this country of different common-field systems by colonists coming from different parts of the continent.[43] Ethnic differences may well have underlaid many of the detailed variations in tenure and custom but they do not seem to have given rise to functionally distinct and regionally unique field systems.

Rather than supporting a theory of regional uniqueness and an ethnic origin of field systems, the field systems of eastern Norfolk and northern Kent (with their intense subdivision of land and haphazard inter-mixture of holdings) accord better with a concept of gradual organic growth. In fact, both localities furnish examples of townships (Martham in Norfolk and Gillingham in Kent) in which subdivided fields have been shown to have evolved spontaneously during the early middle ages.[44] Nevertheless, although supporting a theory of gradual growth, the development of these field systems differed in one all-important respect from that postulated by Thirsk; neither in eastern Norfolk nor in northern Kent did it culminate in the regularization of holding layout and the co-ordination of farming practices. Yet, according to Thirsk:

> '. . . as the parcels of each cultivator became more and more scattered, regulations had to be introduced to ensure that all had access to their own land and to water, and that meadows and ploughland were protected from damage by stock. The community was drawn together by sheer necessity to cooperate in the control of farming practices.'[15]

Undoubtedly the systematization of field systems conferred many real advantages but, on the other hand, as the experience of eastern Norfolk clearly demonstrates, failure to systematize was not necessarily inimical to agriculture and may even have bestowed certain advantages of its own. Thus, in eastern Norfolk the absence of communal controls meant that cultivators were free to innovate and thereby raise the intensity and productivity of agriculture. By the early fourteenth century the standard of cultivation in these irregular common fields had attained an exceptionally high level. This would suggest that the intensification of agricultural method was an alternative response to the

[42] See above, n 36.

[43] This was the interpretation put forward by H L Gray in 1915. See Baker, op cit, 'Howard Levi Gray and English Field Systems'.

[44] B M S Campbell, 'Population Change and the Genesis of Commonfields on a Norfolk Manor', *Econ Hist Rev*, 2nd ser, XXXIII, 2, 1980, pp 174–92, A R H Baker, 'Open Fields and Partible Inheritance on a Kent Manor', *Econ Hist Rev*, 2nd ser, XVII, 1964, pp 1–23.

[45] Thirsk, op cit, 1964, reprinted in R H Hilton (ed), *Peasants, Knights and Heretics: Studies in Medieval English Social History*, Cambridge, 1976, p 16.

regularization of existing holdings and farming practice, as populations grew and land became more and more subdivided. These two responses were to a large extent mutually exclusive, for the methods of husbandry employed in eastern Norfolk by the early fourteenth century would have been incompatible with field systems which were any more closely controlled (ie which employed common rotations and enforced communal grazing of the fallow). It is also arguable that an absence of common rules left field systems more adaptable, better able to accommodate further increases in population.

The question thus arises, what caused some townships to regularize the layout of holdings and fields and adopt common rules of cultivation and grazing, and others not? Obviously there is no simple answer, but the experience of eastern Norfolk does suggest that one decisive factory may have been the balance of power between the manorial authorities and the peasant community. A conspicuous feature of East Anglian field systems is the part evidently played by the manorial lord in the institution of certain rights, notably the right of foldcourse, and in attempts, mostly abortive, to standardize the size of customary holdings.[46] The effectiveness of the manorial lords in making any more radical changes was, however, severely curtailed by the weakness of the manorial nexus on the one hand, and the existence of a well developed peasant proprietorship on the other. In eastern Norfolk not only was there virtually no coincidence between manor and vill but the peasantry, many of whom were freemen and sokemen, were accustomed to divide, alienate, and cultivate their holdings as they pleased, and were patently unwilling to surrender any of these rights in favour of the introduction of a communal system of husbandry. So it would therefore seem that, far from having been evolved from below, the systematization of field systems was imposed from above; hence the generally close association between areas of strong manorialism and regular, highly systematized, field systems. How much more credible, therefore, is a theory which attributes the regularization of field systems, not to the corporate action of the peasant community, but to the intervention of some superior authority such as the manorial lord. The manorial lord is more likely than the peasant community to have been successful in reconciling the host of individual interests involved, and he possessed the authority to carry through and enforce the major reallocation of land which the regularization of holding and field layout would have required. It is to be hoped that further research into medieval field systems will cast additional light on this particular relationship.

[46] Manors where customary holdings had at one time been standardized include Martham (BM: Stowe MS 936), Lessingham (BM: Add MS 24, 316, ff 51–6), Cawston (PRO: SC 11 Roll 471), and Hevingham, Crictots Hall (NRO: NRS 19280 33 F 9).

IV

COMMONFIELD ORIGINS – THE REGIONAL DIMENSION

Medieval England possessed a plurality of commonfield systems: yet why this was so, like the related question of commonfield origins, awaits a satisfactory explanation. H.L. Gray was the first to identify and describe different commonfield systems, and he made their existence the keystone of his ethnic explanation of commonfield origins (Gray, 1915; Baker, 1965, a). However, although his regional classification of commonfield systems is still largely accepted, his views on their origin are now discredited. In contrast C.S. and C.S. Orwin (Orwin and Orwin, 1938), in their subsequent hypothesis of commonfield origins, paid little attention to regional variations in field systems and concentrated upon the regular commonfield system of the Midlands. In their view 'wherever you find evidence of open-field farming and at whatever date, it is sufficient to assume that you have got the three-field system at one stage or another' (Orwin, 1938, 127). This preoccupation with the Midland system is echoed in Joan Thirsk's more recent explanation of commonfield origins (Thirsk, 1964 and 1966, 142-7; Titow, 1965; Hilton, 1976, a). In the Thirsk model the Midland system represents the ultimate stage in a long process of evolution, other English field systems reflecting the effects of local and regional peculiarities of environment, settlement history, population density, and agrarian economy, upon the evolutionary process. But is it right to regard the Midland system as the 'norm' from which other field systems deviated?

The Midland system was certainly the most enduring of commonfield systems and for this reason is the system most fully represented in post-medieval sources. But in the fourteenth century, when commonfield farming was most widespread, its pre-eminence was less well marked. A rough estimate on the basis of the 1377 Poll Tax indicates that at most half, and possibly no more than a third, of England's rural population lived in townships whose commonfields were operated according to the Midland system. At least half the remainder lived in townships whose field systems conformed to alternative regimes. Moreover, thirteenth and fourteenth century manorial records leave no doubt that in their medieval heyday the physical development of these alternative field systems often matched, and sometimes even exceeded, that of their Midland counterparts: waste was virtually eliminated, holdings were highly fragmented, and fields were intensely parcellated. (Campbell, 1975). Most significantly, in the medieval period it was the *non*-Midland field systems which coincided with the areas of greatest population density, highest levels of assessed lay wealth, and most advanced and productive agriculture (Brandon, 1972; Campbell, 1975; Darby, 1973; Gray, 1915; Raftis, 1957; Searle, 1974; Saunders, 1930; Smith, 1943).

That being said, the precise nature and exact geographical spread of the various English commonfield systems remain inadequately known. Indeed, no generally accepted set of criteria exists for the identification and definition of different commonfield systems. Some studies have described them in strictly functional terms, but others have utilised a range of functional, morphological, tenurial, and even terminological characteristics (Baker and Butlin, 1973). The resultant lack of a common basis for comparison is a serious defect and it is difficult to resist the suspicion that some of the differences between systems are less real than has been supposed. This suspicion will linger as long as the criteria by which field systems are identified remain ill-defined. Gray's original claim that field systems were unique to different regions is thus still tacitly accepted, and English field systems continue to be classified and described accordingly (Baker, 1965, b).

Consideration of different commonfield systems reveals that they could comprise up to six basic elements: communal ownership of the waste, arable and meadow divided into unenclosed strips, individual holdings made up of a scatter of strips, fallow grazing by the stock of all the cultivators, the disposition of fallow strips controlled by the regulation of cropping, and communal regulation of all these activities. Closer examination of these six main elements allows their refinement, and specification of the following 14 functional attributes:

1. communal ownership of the waste	THE WASTE
2. arable and meadow characterised by a combination of closes and unenclosed strips	FIELD LAYOUT
3. arable and meadow characterised by a predominance of unenclosed strips	
4. holdings made up of an irregular distribution of strips	HOLDING LAYOUT
5. holdings made up of a regular distribution of strips	

6. full rights of common pasturage on the harvest shack	FALLOW GRAZING
7. limited rights of common pasturage on half-year fallows	
8. limited rights of common pasturage on full-year fallows	
9. full rights of common pasturage on half-year fallows	
10. full rights of common pasturage on full-year fallows	
11. imposition of flexible cropping shifts	REGULATION OF CROPPING
12. imposition of a regular crop rotation	
13. seignorial regulation of certain collective activities	MODE OF REGULATION
14. communal regulation of all collective activities	

These 14 items relate to arable field systems, that is to say, field systems where there was a relative shortage of pasture, and provide the criteria upon which a revised functional classification of such field systems can be based. A different set of attributes would need to be specified for pastoral field systems (Dodgshon, 1973; McCourt, 1954-5).[1]

Consideration of the various ways in which these 14 functional elements could be combined (see Table 5.1) allows the identification of five principal types of field system plus several sub-types. These may be characterised as follows:

A. *Non-common subdivided fields* – where common rights are confined to the waste and arable strips are cropped and grazed in severalty. Examples of this type include the subdivided fields of the Lincolnshire Fens and, possibly, those of Kent (Hallam, 1965; Baker, 1965, b).[2]

B. *Irregular commonfield systems with non-regulated cropping* – where fallow strips are subject to rights of common grazing for part, or the whole, of a year, but where cropping takes place in severalty. Here a two-fold distinction can be drawn between:
   (i) those field systems where common grazing rights were confined to the harvest shack (as in eastern Norfolk and parts of eastern Suffolk (Campbell, 1981; Gray, 1915).
   (ii) those field systems where common grazing rights applied both to harvest shack and to land lying fallow at other times of the year

(as in south and east Devon (Fox, 1972 and 1975).

C. *Irregular commonfield systems with partially regulated cropping* – where fallow strips are subject to rights of common grazing, the disposition of fallow strips is partially controlled by the imposition of flexible cropping shifts, but where holding layout remains irregular. These field systems assume three main forms:
   (i) where control of a system of flexible cropping shifts is vested in the seignorial authorities and limited rights of fallow grazing are instituted, either by restricting fallow grazing to certain types of livestock (e.g. sheep) or by confining it to certain strips (e.g. those lying within a particular cropping shift), or by a combination of the two. The principal example of this type of commonfield system is the foldcourse system of western Norfolk and adjacent portions of Suffolk and Cambridgeshire (Allison, 1957; Postgate, 1973).
   (ii) where limited rights of common grazing are instituted, together with associated cropping shifts, and communal control is established of all aspects of the system (e.g. the field system of western Cambridgeshire (Postgate, 1964, 1973).
   (iii) where full rights of common grazing are instituted, together with associated cropping shifts, and communal control is established of the whole system (e.g. the field systems of the Chilterns, parts of Essex, and the Thames Valley (Roden, 1973).

D. *Irregular commonfield systems with fully regulated cropping* – where fallow strips are subject to full rights of common grazing, the disposition of fallow strips is controlled by the imposition of common rotations, but where, since holdings are made up of a combination of closes and unenclosed strips, holding layout remains irregular. Examples of this type of field system include the so-called woodland systems of the Midlands (Roberts, 1973).

E. *Regular commonfield systems* – where fallow strips are subject to full rights of common grazing, the disposition of fallow strips is controlled by the imposition of common rotations, and where (since unenclosed strips predominate) holding layout is perforce regular (e.g. the two-, three- and four-field systems of the Midlands (Gray, 1915).

This functional gradation of field systems poses an obvious problem of explanation and, indeed, is open to several different interpretations. Foremost among these is the Thirsk model, according to which common rights and regulations, like the subdivided fields to which they related

A FUNCTIONAL CLASSIFICATION OF ENGLISH MEDIEVAL FIELD SYSTEMS (A)			1 Communal ownership of the waste	2 Arable & meadow characterised by both closes and unenclosed strips	3 Arable & meadow characterised by a predominance of unenclosed strips	4 Holdings made up of an irregular distribution of strips	5 Holdings made up of a regular distribution of strips	6 Full rights of common pasturage on the harvest shack	7 Limited rights of common pasturage on half-year fallows	8 Limited rights of common pasturage on full-year fallows	9 Full rights of common pasturage on half-year fallows	10 Full rights of common pasturage on full-year fallows	11 Imposition of flexible cropping shifts	12 Imposition of a regular crop rotation	13 Seignorial regulation of certain collective activities	14 Communal regulation of all collective activities
NON-COMMON SUBDIVIDED FIELDS			• •	• •	• •	• • • •										
IRREGULAR COMMONFIELD SYSTEMS	Non-regulated cropping (B)	variant i	• •	•	•	• •		• •								• •
		variant ii	• • • • • •	• • • •	• • • •	• • • • • •		• • • • • •	• • • •	• •	• •	• •				• • • • • •
	Partially regulated cropping (C)	variant i	• •	•	•	• •		• •	• •	• •			• •		• •	
		variant ii	• •	•	•	• •		• •	• •	• •			• •			• •
		variant iii	• •	•	•	• •		• •			• •	• •	• •			• •
	Fully regulated cropping (D)		•	•		•		•			•	•		•		•
REGULAR COMMONFIELD SYSTEMS (E)			•		•		•	•			•	•	•	•		•

PRINCIPAL COMPONENTS OF FIELD SYSTEMS

Table 5.1

Editorial Note: Table 5.1 was originally printed in two parts on pp. 116–117. Table 5.1 is now supplied here in one part on p. 116, thus leaving p. 117 intentionally blank.

evolved under the impetus of population growth (Thirsk, 1964 and 1966; Baker and Butlin, 1973, 619-56). This led, on the one hand, to an expansion of arable at the expense of pasture, and, on the other, to the proliferation of holdings and strips, thereby simultaneously placing a mounting premium on the temporary forage available on the fallow arable, and creating a need to regularise field and holding layout, so that access by all cultivators to their land and to water was ensured, and meadow and ploughland were protected from stock. To these ends, therefore, holding layout was regularised, communal rotations were introduced to rationalise the disposition of fallow strips, and common grazing rights were established on the latter. As a result better provision was secured for the livestock upon which arable production depended for its traction and manure, without any depletion in the cultivated area. However, this change from non-common subdivided fields to a regular commonfield system was not achieved at a stroke, it proceeded by stages. The first step may have been the establishment of informal agreements between parcenars and groups of cultivators, which in time may have become steadily extended and developed until they became transmuted by custom into formal binding rights. Common grazing of the stubble and aftermath of the harvest may have been the first of these rights to develop for it offered obvious advantages of convenience and was easily instituted, requiring no physical changes in the layout of fields and holdings. Subsequently, the attraction of fallow grazing available at other times of the year coupled, perhaps, with some rationalisation of holding layout, may have led to the adoption of more extended fallow grazing rights. The adoption of such rights may, in turn, have prompted the co-ordination of cropping patterns, so that blocks of contiguous strips remained fallow in the same year and grazing was thereby facilitated. In this way the process of systematisation may have progressed through increasingly developed forms of irregular commonfield system until the ultimate stage, the establishment of a regular commonfield system, was reached. A precondition for attainment of this final stage was a partial or total reallotment and realignment of strips, which could have been undertaken by a village assembly acting in the common interest.[3] However, because systematisation evolved slowly and under different environmental and social conditions, it did not develop everywhere in the same way or to the same extent. In particular, a basic distinction existed between the commonfield systems of populous grain-producing districts, in fertile valleys and plains, and those of less populous districts where there was a greater emphasis upon pastoralism, in upland areas and in the vicinity of extensive forests, marshes and fen. In this way the Thirsk model accommodates regional variations in field systems.

This thesis was first propounded in 1964, since when much evidence has been presented to verify that part of it which deals with the formation of subdivided fields and inter-mixed holdings (Baker, 1964; Spufford, 1965; Sheppard, 1966; Campbell, 1980; Bishop 1935). In contrast, evidence to support the part relating to the genesis of common rules and regulations has been less forthcoming (Campbell, 1981). Indeed, the assumption that increased demand for food led to the progressive rationalisation of the layout of fields and holdings and co-ordination of cropping and grazing practices, can be questioned on two fronts. First there are grounds for questioning the feasibility and advisability of such a course of development. Second, the existence of an alternative and possibly more plausible response to population growth can be demonstrated.

On grounds of practicality, there are several reasons for doubting whether the co-ordination and systematisation of commonfields progressed quite as smoothly, and were quite so directly related to population growth, as the Thirsk model postulates. To begin with, structural innovation (the collective reorganisation of existing fields, holdings and husbandry practices) required a consensus if it was to proceed. Common rights and regulations affected all cultivators and consequently required their unanimous consent before they could be instituted; this applied in particular to the reallotment of land which would have accompanied the regularisation of holding layout and imposition of common rotations. Structural innovation could only take place, therefore, when all the affected parties perceived that it was in their mutual interest. Moreover, to perceive the desirability of structural innovation was one thing: to effect such a change was another, and required a capacity for collective action. Unless communities possessed such a capacity, economic need alone could not have led to structural innovation. But to presuppose that regular commonfield systems were the creation of organised peasant communities begs the essential question whether such communities existed prior to the creation of the commonfield system. They are as likely to have been the effect of the system, as they are to have been the cause.

Even granting the existence of communities capable of making and executing the relevant collective decisions, it remains questionable whether they would have taken the decision to institute a regular commonfield system when peasant numbers were increasing. Such a decision was not to be taken lightly; once executed structural innovation was not readily reversed. This was because it entailed more than the mere physical reorganisation of holdings and fields, complex and controversial though this would have been: at its root lay the institution of a code of inviolable rights and regulations which, once created, could not easily be dissolved without recourse to a higher authority (as in the case of enclosure by Act of Parliament). Furthermore, repeal of these rights and regulations was, like their creation, subject to individual veto. The innate conservatism of peasant cultivators, which was an obstacle to the dissolution of the commonfield system, must surely also have militated against its creation. It is unlikely that individual cultivators would have been willing to agree to an irreversible change of unproven advantage: and they would have been particularly loath to do so when population growth was placing a mounting premium upon proprietorship and leading to a progressive narrowing of the margin of subsistence. Moreover, the larger the number of affected parties, the more remote would have been the prospect of getting them all to agree to structural innovation.

To relate adoption of the regular commonfield system to mounting population pressure presents a further difficulty: the former was innately static, thriving on the maintenance of the *status quo,* whereas the latter was essentially dynamic, and promoted change. Once formed, the regular commonfield system would consequently have been incompatible with continued population growth. If the advantages which the system offered were not to be eroded, therefore, it would have been requisite that the integrity of the new arrangements be preserved by making corresponding changes in the prevailing social and legal code. This accounts for the fact that impartible inheritance and an insistence upon the inalienability of land were invariable concomitants of the regular commonfield system (Homans, 1941; Faith, 1966, Howell, 1976). These, and related customs, served a dual purpose: they not only insulated the system from the divisive effects of population growth, but they also deterred population growth itself. However, the imposition of these customs would probably have met with considerable opposition if it had been attempted whilst peasant numbers were increasing, for, when this was the case, existing proprietory rights and inheritance practices tended to be jealously preserved. Indeed, population growth and customs which favoured the proliferation of holdings tended to be mutually reinforcing and, once established, the momentum which they generated was almost impossible to restrain (Faith, 1966).

More important than these *a priori* objections to the Thirsk model is the fact that it is possible to envisage an alternative, and arguably more plausible, response to the need to raise food production in a situation where holdings were intermixed and reserves of colonisable land had been exhausted. New agricultural methods could have been adopted and known agricultural methods intensified (agricultural innovation and involution). This is an intrinsically natural and practical solution to the need to increase agricultural productivity and several writers have demonstrated that it has been a characteristic agricultural response to population growth in subsistence societies (Chayanov, 1925; Boserup, 1965; de Vries, 1974; Grigg, 1976). Most technological changes could have been introduced within the existing framework of holdings and fields, and would not have required a major reallotment of land or change in regulations. Nor would individual cultivators have needed to obtain the consent of the rest of the commonfield community before they would adopt an innovation; the decision to innovate was an individual, not a collective one. In the absence of common rights and regulations individuals would have been under no obligation to make their husbandry practices conform with those of other cultivators, and the intensity, and the techniques of agricultural production could therefore have been adapted to individual circumstance. Experimentation was possible, in most cases changes were reversible, and technological innovation could thereby progress by relatively safe and easy stages. Since change could take place gradually, across a broad front, the risk attached to adopting any one innovation was minimised: the maintenance of an artificial equilibrium was not at stake.

Of course, technological innovation would not have been possible unless the necessary technology was available, and this has been the subject of considerable debate (White, 1962; Fussell, 1968; Titow, 1969). Some writers have represented medieval agriculture as technologically inert and incapable of increasing arable production without jeopardising soil fertility and precipitating falling yields (Titow, 1972). Yet there is a growing body of evidence to show that the necessary technology, by which agricultural production could be intensified, did exist, and it seems likely that this was employed with considerable success in several localities in eastern and south-eastern England by the close of the thirteenth century (Richard, 1892; Slicher van Bath, 1960). The technological means by which sustained increases in output per unit area were attained in these localities were manifold, but at their core lay the substitution of fodder crops for bare fallows and natural pasture. In the thirteenth century the principal fodder crops were peas and beans, although vetches were also known: oats, too, were grown as a fodder crop. To derive maximum advantage from this change to feed produced from natural fodder, horses were substituted for oxen in ploughing on account of their greater speed and capacity to work longer hours. Horse ploughing and harrowing also allowed more thorough preparation of the seed-bed, which was further improved by ploughing in farmyard manure from stall-fed livestock. Although considerably more labour-intensive than the pasturing of livestock upon the fallow fields, the manual spreading of manure was far more effective, as it reduced the twin losses from leaching and oxidisation and ensured much more even coverage. The importance attached to the maintenance of soil fertility is further reflected in the assiduous use made of all available supplies of fertiliser, including the folding of sheep upon isolated fallow strips, and the spreading of marl and urban refuse where available. The cultivation of leguminous fodder crops also enhanced soil fertility by raising its nitrogen content, whilst a heavy emphasis upon spring-sown crops allowed regular fallowing on a half-year basis. Flexible rotations, thick sowings, and careful weeding and harvesting also contributed to the productivity of this remarkably intensive system of husbandry. Its rewards were the near elimination of bare fallows, an expansion of the cultivated area to the maximum extent possible, and the attainment of high and sustained yields per acre (although often at the expense of only moderate yields per seed).[4] Where these technological innovations were made, as in eastern Norfolk, irregular commonfield systems with a minimum of common rights and regulations successfully supported exceptionally high population densities (Campbell, forthcoming).

By the thirteenth century, therefore, technological innovation and involution had become an extremely viable and effective alternative to structural innovation. The association of very progressive agricultural methods with highly irregular field systems also shows that chaos and inefficiency were not inevitable corollaries of a failure to rationalise the layout of holdings and fields and co-ordinate their cultivation and grazing. In fact, there were a number of localities in the thirteenth century which possessed all the preconditions which should supposedly have led to the adoption of a regular commonfield system; yet such a system was not adopted because the alternative response of technological innovation had been followed (Campbell, 1980; Baker, 1965, b; Gray, 1915, 302-3, 331-2). Indeed, to a considerable extent these two methods of raising food production were mutually exclusive. Innovations such as the substitution of fodder crops for bare fallows, flexible rotations, and the stall-feeding of livestock, would have been incompatible with a fully regularised commonfield system.

Although technological innovation was arguably the more viable response to population growth, structural innovation was of course adopted in some measure in many localities. This, however, is less likely to have been derived from population growth than from population stagnation and decline. Research into population trends prior to the demographic transition of the nineteenth century has now established that numerous short-term fluctuations in population were superimposed upon a long-term sequence of demographic 'cycles' (Helleiner, 1967; Wrigley, 1969). After a long initial phase of sustained population growth, each of these 'cycles' comprised a period of prolonged demographic contraction, heralded by a brief intermediate phase of mounting demographic crisis. Thus, sustained population growth during the twelfth and thirteenth centuries culminated in a period of acute demographic crisis during the first half of the fourteenth century: approximately a century and a half of population decline and stagnation then ensued (Hatcher, 1977; Miller and Hatcher, 1978). The next demographic 'cycle' commenced in the first half of the sixteenth century. Rapid population growth during the sixteenth and first part of the seventeenth centuries was terminated by renewed demographic crisis in the second quarter of the latter century; this ushered in a further century of demographic malaise, comprising stagnation in some localities, contraction in others (Smith, 1978). Finally, a last wave of sustained population growth during the second half of the eighteenth century culminated, not in demographic crisis, but the demographic transition, and the establishment of a new demographic pattern (Flinn, 1970). This 'cyclic' pattern was the product of important temporal shifts in mortality and fertility rates brought about by the complex interaction of biological and economic forces (Wrigley, 1966, 1968, 1969; Chambers, 1972). Given the demonstrable existence of this pattern from the mid-twelfth to the late eighteenth centuries, therefore, it is *a priori* likely that earlier periods also experienced successive waves of population growth separated by long intervals of demographic recession. As yet little is known of the chronology of population trends prior to the twelfth century, but the evidence of the Irish Annals, the Anglo-Saxon Chronicle, and other contemporary accounts does point to a number of periods of heightened mortality, and consequently, perhaps, of demographic decline, notably in the late seventh century, the closing decades of the eighth century, and again in the late eleventh century (Creighton, 1894; Bonser, 1963; Howe, 1972).

Such variations in the size of populations have obvious implications for the development of field systems and other agrarian institutions. But whereas the importance of periods of rising population has long been appreciated, especially with regard to the fragmentation of holdings and parcellation of fields, the influence of periods of stagnant and declining population has yet to receive serious consideration (except Fox, 1972, 1975). Such periods may, however, have made an equally important, if different, contribution to the development of field systems. It is arguable that they were peculiarly conducive to the rationalisation and systematisation of existing fields, holdings and husbandry practices. A principal reason for this was that as populations fell so it would have become advantageous to pool scarce labour resources in order to use them more efficiently. Establishment of a regular commonfield system enabled this; substantial labour savings derived from conducting activities in common which had previously been undertaken in severalty. The common grazing of fallow strips is a case in point. By combining common rotations with common grazing rights it was possible to pasture entire furlongs and fields with a minimum of supervision. The alternative, under a system of farming in severalty, was to fold or tether livestock on individual, scattered fallow strips, a practice which required the fencing of adjacent strips against the depredations of the livestock, should they have got loose. Further economies in labour were afforded by the greater convenience of a regular layout of holdings and fields.

Circumstances of falling rather than rising population would also have presented fewer practical problems to the rationalisation and systematisation of commonfields. Where the population was falling there would have been no pressure to intensify methods of cultivation and hence fewer objections to adopting a permanent rotational scheme which provided for the fallowing of blocks of strips on a regular basis. On the contrary, there would have been much to recommend a device which helped to maintain existing levels of productivity. Moreover, where the demand for land was slack it would have been fairly easy to effect changes in the layout of fields and holdings by the exchange and consolidation of strips, and, since there was little pressure upon land and all cultivators stood to derive advantages from this exercise, it can be presumed that few would have objected to it. In fact, the more the number of cultivators declined the more likely would have been the prospect of securing the consensus necessary for structural innovation. Circumstances would also have favoured the social changes consequent upon the institution of a regular commonfield system. Where younger sons were few, and there was no shortage of holdings, there would have been little hardship in making the changeover from partible to impartible inheritance and an insistence upon the inalienability of land (Campbell, forthcoming).

These general considerations still leave unexplained the marked regional variations in the character of structural innovation. Obviously, regional variations in population may have some bearing on this matter. Areas of relatively high or relatively low population density, for instance, may have been less conducive to structural innovation than areas of moderate population density. This is because if the population density was too high the obstacles to structural innovation may have been insurmountable, its advantages may have been of little relevance, and technological innovation may have been more appropriate. Conversely, if the population density was too slight there may have been little need for rationalisation and systematisation: rather than a trend towards more developed forms of commonfield system there may have been reversion to consolidated holdings and enclosed fields held in severalty, accompanied by the adoption of a more extensive form of agriculture. This would account for the fact that the regular commonfield system was absent from the most and the least populous areas. However, although the distribution of population undoubtedly exerted a general influence upon the distribution of regular and irregular commonfield systems, it cannot account for the detailed distribution of the different functional types of field system. Other conditioning factors almost certainly existed.

In this connection, and granted the doubts which have been expressed about the capacity of peasant communities to undertake structural change, considerable interest attaches to the institutional factor of lordship. Lordship was subject to important geographical variations, and three main ways may be envisaged in which variations in its authority, structure and continuity could have determined regional differences in field systems. First, the existence of peasant communities with a capacity for acting collectively may have been a function of strong and undivided lordship. Second, although peasant communities may have been the prime movers, lords may have been the instruments of structural innovation: in other words, once a peasant community had taken the relevant decision it may have referred the matter to its lord, on account of his superior authority, for implementation. Third, the creation of a regular layout of holdings and fields and the institution of common rights and regulations may have been a direct, seignorial imposition. In each of these three cases, it should be noted, strong and undivided lordship would have been most favourable to the functional development of the commonfield system.

Little is as yet known about the development of peasant communities as corporate entities, but it is plain that a capacity for collective action could have been of primary significance to the creation of centrally organised, peasant farming systems. This capacity could have originated in a number of ways, not least as a response to seignorial authority. Thus, peasants may have derived an initial sense of shared identity from the protection afforded by their lord, an identity which would have been reinforced by the legal bonds which tied tenant to lord. Progressive subordination of peasants by their lords would subsequently have prepared the way for collective action by the removal of social differences between peasants and the creation of a homogeneous class of customary tenants. The potential for collective action may then have been lent cohesion by, and found expression in, peasant resistance to seignorial exploitation and oppression. In this way peasant communities may have crystallised as corporate entities. Given that strong and undivided lordship was fundamental to this process, it follows that there may have been an important indirect relationship between spatial variations in the nature of lordship, and spatial variations in the organisation of peasant farming systems.

The impact of lordship upon the development of field systems may, however, have been more direct than this. It is possible that commonfield systems were the product of co-operation between lords and their tenants. The complex and controversial tasks of rationalising the layout of fields and holdings and instituting common rights and regulations may have been referred to lords by their tenants. After all, the lord would have been better placed to make an impartial *remembrement.* He had the authority necessary to execute this task and he probably had readier access to the requisite skills of valuation and surveying, and his court provided an established framework for the subsequent enforcement and regulation of the system. For lords to have been capable of acting in this way, however, it was necessary that their jurisdiction extend to the entire vill and its inhabitants. Again, therefore, circumstances would have been most favourable to the functional development of field systems in vills of strong and undivided lordship.

It is entirely, plausible, however, that lords intervened in the agricultural practices of their tenants for less altruistic reasons. Exploitation may have been their motive. Under the feudal mode of production peasants were personally unfree and legally subordinate to their lords, who were empowered to appropriate their surplus labour product (i.e. the labour and goods which were surplus to the satisfaction of the immediate subsistence requirements of the peasant and his family (Dobb, 1946; Duby, 1974; Hilton, 1976a). This surplus labour product was appropriated either in the form of direct labour services on the lord's demesne or in various forms of rent. Lords thus had a vested interest in the agricultural organisation of their tenants. Where labour was relatively scarce, therefore, they stood to gain from a more efficient deployment of peasant labour. As already observed, imposition of the regular commonfield system had precisely this effect; by pooling labour resources and arranging for certain activities to be carried out in common it facilitated the release of labour to work on the lord's demesne. Division of the demesne into strips located in the commonfields furthered this arrangement. The whole scheme may therefore have originated with the lord, tenants having no alternative but to comply. On the other hand, where labour was relatively abundant it would have been unnecessary to go to these lengths to secure the necessary labour to work the demesne, and recourse may have been made to alternative methods of appropriating the peasants' surplus labour product. The juridical authority vested in the lords allowed them to act in this autocratic manner, whilst the creation of standardised customary holdings furnished them with a suitable opportunity. Since the standardisation of holding size and regularisation of holding layout would both have required a substantial reallotment of land, both could have been instituted together. Indeed, the fact that virgates and bovates were agrarian as well as tenurial units implies that this was so. If the systematisation of commonfields was a function of the development of the feudal system, it again follows that it would have progressed furthest where there was a tradition of strong and undivided lordship and where there was a high propotion of customary tenants.

It thus transpires that there was a variety of ways in which differences in the authority, structure and continuity of lordship may have exerted a determining influence upon the development of field systems (Goränsson, 1958; Smith, 1967; Sheppard, 1976). In all cases this influence appears to have been most formative where a lord's territorial jurisdiction encompassed an entire vill, and where a substantial majority of the inhabitants of that vill were subordinate to the authority of the lord. Circumstances would have been less propitious for the rationalisation and systematisation of commonfields where only one, or neither, of these conditions prevailed. In such vills the capacity of the lord to influence the entire community, or to impose his will on the entire vill, would have been correspondingly reduced. Other things being equal, a distinction should consequently have existed between the field systems of areas where lordship was strong and those of areas where it was weak.

The evidence by which this supposition may be tested is limited because knowledge of the distribution of lordship and distribution of field systems is defective. Nevertheless, the evidence of Domesday Book, the Hundred Rolls, and the *Inquisitiones Post Mortem* does suggest that within lowland England lordship was consistently stronger in some areas than others (Kosminsky, 1956; Darby, 1977; Miller and Hatcher, 1978). In particular, the central and southern counties (the heartland of the old Anglo-Saxon state) appear to have been characterised by a higher incidence of vills of undivided lordship, and a lower incidence of freemen, than the eastern and south-eastern counties (where the continuity of lordship was disrupted by the Scandinavian incursions of the ninth and tenth centuries). Significantly, it was in the central and southern counties that regular commonfield systems were most fully developed, and in the eastern and south-eastern counties that various forms of irregular commonfield system prevailed. This general coincidence between areas of strong lordship and regular commonfield systems, and areas of weak lordship and irregular commonfield systems, also shows up at a more local level. Thus, in the heart of the Midlands, a fundamental contrast existed between the two adjacent Warwickshire hundreds of Kineton and Stoneleigh (Harley, 1958; Roberts, 1973). In the late thirteenth century the former was an area of relatively old settlement, strong lordship, and regular commonfield systems: its population evinced remarkably little increase during the twelfth and thirteenth centuries and contained very few freemen. The latter, on the other hand, was an area of more recent settlement; lordship was less firmly established, freemen were better represented, and many townships recorded a substantial increase in population during the twelfth and thirteenth centuries. In this hundred an irregular commonfield system with fully regulated cropping prevailed. A similar dichotomy may be seen in Norfolk. In the west of the county an irregular commonfield system with partially regulated cropping was the norm, whereas in the easternmost hundreds field systems were even less systematised and an irregular commonfield system with non-regulated cropping existed. This reflects the fact that the seignorial nexus was even weaker in the east than it was in the west; lordship was characterised by a higher degree of fragmentation and the freeholding element in the population was more substantial (Douglas, 1927; Darby, 1952; Allison, 1957, 1958; Rainbird Clarke, 1960; Campbell, 1981).

These examples suggest that if the functional gradation of field systems is explicable in terms of a single factor, that factor may be the structure of lordship: the greater the authority and continuity of lordship, the more fully systematised the commonfield system.

Nevertheless, it has not been the purpose of this paper to present a new monocausal explanation of commonfield origins. Rather, the intention has been to clarify definitions, remove preconceptions and pose questions. In this context the plurality of commonfield systems has been of central concern, for it is upon this issue that current explanations of commonfield origins founder. Only if the commonfield system is studied in its entirety and in its full socio-economic context will it be possible to account for the temporal processes which gave rise to this spatial complexity. As has been argued, such diverse variables as technological innovation in agriculture, the course of population change, and the institutional framework of society may have been crucial to the functional development of field systems. Yet current knowledge of each of these variables, particularly during the pre-Conquest period, is sadly deficient. So too is our knowledge of regional variations in field systems. Greater knowledge of the variant forms of commonfield system which existed, of their precise formation, their distribution, and their development, is fundamental to the explanation of commonfield origins.

Notes

1. Pastoral field systems were characterised by a shortage of arable land and a relative abundance of pasture and were principally distinguished by periodic cultivation of the common waste.

2. Although as yet unproven, it remains possible that the subdivided fields of north and east Kent were subject to rights of common grazing on the aftermath of the harvest.

3. The Thirsk model also postulates that the piecemeal buying, selling, exchanging and leasing of land may have been alternative means by which holding layout became regularised.

4. In the first half of the fourteenth century the mean yield of wheat and barley on the most productive demesnes might be as high as 20 bushels per acre, rising to over 30 bushels per acre in a very good year. The equivalent mean yield ratios of these two crops were 6 fold and 3½ fold respectively.

BIBLIOGRAPHY CHAPTER IV

Allison, K.J., 1957, 'The sheep-corn husbandry of Norfolk in the sixteenth and seventeenth centuries', *Agricultural History Review* 5, pp. 12–30.

Allison, K.J., 1958, 'Flock management in the sixteenth and seventeenth centuries', *Economic History Review*, 2nd series, 11, pp. 98–112.

Baker, A.R.H., 1964, 'Open fields and partible inheritance on a Kent manor', *Economic History Review*, 2nd series, 17, pp. 1–23.

Baker, A.R.H., 1965a, 'Howard Levi Gray and *English field systems*: an evaluation', *Agricultural History* 39, pp. 86–91.

Baker, A.R.H., 1965b, 'Some fields and farms in medieval Kent', *Archaeologia Cantiana* 80, pp. 152–74.

Baker, A.R.H., and Butlin, R.A., eds., 1973, *Studies of field systems in the British Isles*, Cambridge.

Bishop, T.A.M., 1935, 'Assarting and the growth of the open fields', *Economic History Review*, 1st series, 6, pp. 13–29.

Bonser, W., 1963, *The medical background of Anglo-Saxon England: a study in history, psychology and folklore*, London.

Boserup, E., 1965, *The conditions of agricultural growth: the economics of agrarian change under population pressure*, London.

Brandon, P.F., 1972, 'Cereal yields on the Sussex estates of Battle Abbey during the later Middle Ages', *Economic History Review*, 2nd series, 25, pp. 403–20.

Campbell, B.M.S., 1975, 'Field systems in eastern Norfolk during the Middle Ages: a study with particular reference to the demographic and agrarian changes of the fourteenth century', unpublished PhD thesis, University of Cambridge.

Campbell, B.M.S., 1980, 'Population change and the genesis of commonfields on a Norfolk manor', *Economic History Review*, 2nd series, 33, pp. 174–92.

Campbell, B.M.S., 1981, 'The regional uniqueness of English field systems? Some evidence from eastern Norfolk', *Agricultural History Review* 29, pp. 16–28.

Campbell, B.M.S., forthcoming (1984), 'Population pressure, inheritance and the land market in a fourteenth-century peasant community', pp. 87–134 in R.M. Smith, ed., *Land, kinship and life-cycle*, Cambridge.

Chambers, J.D., 1972, *Population, economy and society in pre-industrial England*, Oxford.

Chayanov, A.V., 1925 (English edition 1966), trans. R.E.F. Smith, ed., D. Thorner and B. Kerblay, *The theory of peasant economy*, American Economic Association, Homewood, IL.

Clarke, R.R., 1960, *East Anglia*, London.

Creighton, C., 1894, *A history of epidemics in Britain, Volume 1, A.D. 664 to the Great Plague*, Cambridge.

Darby, H.C., 1952, *The Domesday geography of eastern England*, Cambridge.

Darby, H.C., ed., 1973, *A new historical geography of England*, Cambridge.

Darby, H.C., 1977, *Domesday England*, Cambridge.

Dobb, M., 1946, *Studies in the development of capitalism*, London.

Dodgshon, R.A., 1973, 'The nature and development of infield-outfield in Scotland', *Transactions of the Institute of British Geographers* 59, pp. 1–23.

Douglas, D.C., 1927, *The social structure of medieval East Anglia*, Oxford Studies in Social and Legal History 9, Oxford.

Duby, G., 1974 (English edition), trans. Clarke, H.B., *The early growth of the European economy: warriors and peasants from the seventh to the twelfth century*, London.

Faith, R.J., 1966, 'Peasant families and inheritance customs in medieval England', *Agricultural History Review* 14, pp. 77–95.

Flinn, M.W., 1970, *British population growth 1700–1850*, London and Basingstoke.

Fox, H.S.A., 1972, 'Field systems of east and south Devon. Part 1: east Devon', *Transactions of the Devonshire Association* 104, pp. 81–135.

Fox, H.S.A., 1975, 'The chronology of enclosure and economic development in medieval Devon', *Economic History Review*, 2nd series, 28, pp. 181–202.

Fussell, G.E., 1968, 'Social change but static technology: rural England in the fourteenth century', *History Studies*, 1, pp. 23–32.

Goränsson, S., 1958, 'Field and village on the Island of Öland', *Geografiska Annaler* 40B, pp. 101–58.

Gray, H.L., 1915, *English field systems*, Cambridge, MA.

Grigg, D.B., 1976, 'Population pressure and agricultural change', *Progress in Geography* 8, pp. 133–76.

Hallam, H.E., 1965, *Settlement and society: a study of the early agrarian history of south Lincolnshire*, Cambridge.

Harley, J.B., 1958, 'Population trends and agricultural developments from the Warwickshire Hundred Rolls of 1279', *Economic History Review*, 2nd series, 11, pp. 8–18.

Hatcher, J., 1977, *Plague, population and the English economy 1348–1530*, London and Basingstoke.

Helleiner, K., 1967, 'The population of Europe from the Black Death to the eve of the vital revolution', pp. 1–95 in E.E. Rich and C.H. Wilson, eds., *Cambridge economic history of Europe, Volume 4, The economy of expanding Europe in the sixteenth and seventeenth centuries*, Cambridge.

Hilton, R.H., ed., 1976a, *Peasants, knights and heretics: studies in medieval English social history*, Cambridge.

Hilton, R.H., intro., 1976b, *The transition from feudalism to capitalism*, London.

Homans, G.C., 1941, *English villagers of the thirteenth-century*, Cambridge, MA.

Howe, G.M., 1972, *Man, environment and disease in Britain: a medical geography through the ages*, Harmondsworth.

Howell, C., 1976, 'Peasant inheritance customs in the midlands, 1280–1700', pp. 112–55 in J. Goody, J. Thirsk, and E. P. Thompson, eds., *Family and inheritance: rural society in western Europe 1200–1800*, Cambridge.

Kosminsky, E.A., 1956 (English edition), trans. R. Kisch, ed. R.H. Hilton, *Studies in the agrarian history of England in the thirteenth century*, Oxford.

McCourt, D., 1954–5, 'The infield-outfield system in Ireland', *Economic History Review*, 2nd series, 7, pp. 369–76.

Miller, E., and Hatcher, J., 1978, *Medieval England: rural society and economic change 1086–1348*, London.

Orwin, C.S., and Orwin, C.S., 1938, *The open fields*, 3rd edition, Oxford, 1967.

Postgate, M.R., 1964, 'The open fields of Cambridgeshire', unpublished PhD thesis, University of Cambridge.

Postgate, M.R., 1973, 'Field systems of East Anglia', pp. 281–324 in Baker and Butlin, eds., *Studies of field systems.*

Raftis, J.A., 1957, *The estates of Ramsey Abbey: a study of economic growth and organization*, Toronto.

Richard, J.M., 1892, 'Thierry d'Hireçon, agriculteur artésien', *Bibliothèque de l'école des chartes* 53, pp. 383–416 and 571–604.

Roberts, B.K., 1973, 'Field systems of the west midlands', pp. 188–231 in Baker and Butlin, eds., *Studies of field systems.*

Roden, D., 1973, 'Field systems of the Chiltern Hills and their environs', pp. 325–74 in Baker and Butlin, eds., *Studies of field systems.*

Saunders, H. W., 1930, *An introduction to the obedientiary and manor rolls of Norwich Cathedral Priory*, Norwich.

Searle, E., 1974, *Lordship and community: Battle Abbey and its banlieu 1066–1538*, Toronto.

Sheppard, J.A., 1966, 'Pre-enclosure field and settlement patterns in an English township: Wheldrake, near York', *Geografiska Annaler* 48B, pp. 59–77.

Sheppard, J.A., 1976, 'Medieval village planning in northern England: some evidence from Yorkshire', *Journal of Historical Geography* 2, pp. 3–20.

Slicher van Bath, B.H., 1960, 'The rise of intensive husbandry in the Low Countries', pp. 130–53 in J.S. Bromley and E.H. Kossmann, eds., *Britain and the Netherlands: papers delivered to the Oxford-Netherlands historical conference 1959*, London.

Smith, C.T., 1967, *An historical geography of western Europe before 1800*, London.

Smith, R.M., 1978, 'Population and its geography in England, 1500–1730', pp. 199–238 in R.A. Dodgshon and R.A. Butlin, eds., *An historical geography of England and Wales*, London.

Smith, R.A.L., 1943, *Canterbury Cathedral Priory: a study in monastic administration,* Cambridge.

Spufford, M., 1964, *A Cambridgeshire community: Chippenham from settlement to enclosure*, University of Leicester, Department of English Local History, Occasional Paper 20, Leicester.

Thirsk, J., 1964, 'The common fields', *Past and Present* 29, pp. 3–25.

Thirsk, J., 1966, 'The origin of the common fields', *Past and Present* 33, pp. 142–7.

Titow, J.Z., 1965, 'Medieval England and the open field system', *Past and Present* 32, pp. 86–102.

Titow, J.Z., 1969, *English rural society 1200–1350*, London.

Titow, J.Z., 1972, *Winchester yields: a study in medieval agricultural productivity*, Cambridge.

Vries, J. de, 1974, *The Dutch rural economy in the golden age, 1500–1700*, New Haven, CT.

White, L., 1962, *Medieval technology and social change*, Oxford.

Wrigley, E.A., 1966, 'Family limitation in pre-industrial England', *Economic History Review*, 2nd series, 19, pp. 82–109.

Wrigley, E.A., 1968, 'Mortality in pre-industrial England: the example of Colyton, Devon, over three centuries', *Daedalus* 97, pp. 246–80.

Wrigley, E.A., 1969, *Population and history*, London.

V

Commonfield Agriculture: The Andes and Medieval England Compared

Bruce M. S. Campbell and Ricardo A. Godoy

Commonfield agriculture is one of the most distinctive and intriguing manifestations of common-property resource management. Distinctive, because of its peculiar blend of private and communal endeavors and its complex patterns of decision making and interaction. Intriguing, because farmland is inherently divisible, there being no technical or physical reason why individual holdings should not be managed on an entirely private basis. That this has not been the case in many parts of the world over remarkably long periods of time is consequently a matter of considerable interest.

Four key attributes define the core dimensions of commonfield agriculture (Thirsk 1964). First, the holdings of individual cultivators comprise many separate parcels scattered among unenclosed commonfields. Second, after the harvest, and usually during fallow years, these commonfields revert from private farmland to communal pasture ground, as all villagers exercise their customary right to graze their animals on the herbage temporarily available on the arable land. In commonfield agrarian regimes, villagers also enjoy the collective right to gather peat, timber, and firewood from common pastures and fallow fields. Finally, regulation and supervision of the entire system is provided by an "assembly of cultivators."

Any of these features may be found in isolation in other farming systems. Many pastoralists, for instance, graze their stock communally. Village councils rule Himalayan, Swiss, Andean, Japanese, and Vietnamese peasant communities (Rhoades and Thompson 1975; Popkin

1979). The simultaneous occurrence of all four traits, however, is rarer; we see it only in selected parts of Europe, colonial New England, the central Andean highlands, Mesoamerica, India, the Middle East, and West Africa.[1] The two- and three-field system of England is probably the best known of these systems and certainly one of the most systematized and regularized. In its case, historians now believe that the four elements noted above coalesced only after a long gestation period, possibly in the tenth and eleventh centuries. Thereafter, the system endured in some parts of the country until well into the nineteenth century.[2]

No single definition is likely to capture all the subtleties of an agrarian system found over so wide a geographical area and so long a period of time. Nor is any one theory likely to explain the causal factors responsible for the emergence of such a complex system in so many geographical areas and sociopolitical environments. What this exploratory essay offers, therefore, is a systematic comparison of the technical and physical attributes, the decision-making arrangements, and the patterns of interaction among users of commonfield systems found in two widely separated parts of the world: the central Andean highlands and medieval England. This comparison is less farfetched than might at first appear for, despite vast differences in the material underpinnings of these two commonfield regimes, they manifest striking similarities in their functional attributes, demographic patterns, and evolutionary trajectories.

Current knowledge and understanding of Andean commonfield systems may be deficient, but, as Marc Bloch once remarked, there are times when synthesis, comparisons, and the formulation of interesting problems contribute more to an understanding of cultural phenomena than further detailed case studies. Accordingly, this essay is offered as a first step in the development of a genuinely cross-cultural understanding of commonfield systems.[3] It also serves the more immediate function of helping to frame questions and rank priorities for further research. Above all, in the wider context of this book, it furnishes two illuminating cases of highly developed and successful common-property management systems.

Historical Background of the Two Systems

English and Andean commonfields are far removed from each other in time and in space. In England, commonfield farming is a thing of the past. Today, only a solitary, consciously preserved, commonfield township survives—at Laxton in Nottinghamshire (Beckett 1989). In contrast, in

Peru and Bolivia, commonfield farming continues to be practiced over an extensive geographical area.

Precisely when and how commonfield farming came into being in England remains a matter of considerable debate. Nevertheless, there is general agreement that the system reached its heyday during the early Middle Ages, from approximately the tenth to the fourteenth centuries. Throughout that period, commonfields were expanding and developing; by its close, approximately two-thirds of England's population lived in commonfield townships (Baker and Butlin 1973; Dodgshon 1980; Campbell 1981a; Rowley 1981). Thereafter, the prevailing trend, with certain exceptions, involved increasing consolidation and enclosure, so that by the close of the seventeenth century England had become a country in which farming in severalty (that is, with land held by an owner in his own right and not jointly or in common with others) predominated (Wordie 1983).[4]

It was at this same time that a fully fledged commonfield system seems to have been crystallizing in the Andes, as native systems of husbandry were transformed under Spanish colonial influence. Almost everything remains to be learned about the history of these commonfields, but fragmentary evidence suggests that it was during the seventeenth and eighteenth centuries that the system became most widespread (Chevallier 1953, 60; Gade 1970; Custred and Orlove 1974; Gade and Escobar 1982; Málaga Medina 1974). Since then, these commonfields have also begun to succumb to alternative methods of land management.[5]

Technical and Physical Attributes of English and Andean Commonfields

Entirely different though their respective chronologies of development may be, both commonfield systems share the same fundamental physical attribute: arable fields made up of myriad unenclosed and intermixed parcels. This is the one abiding feature of all commonfield systems and it is from this that associated decision-making arrangements and patterns of interaction spring. Thus, rights of stubble grazing and the communal regulation of cropping are most satisfactorily interpreted as responses to the problems of farming in subdivided fields (Dahlman, 1980). That being said, subdivided fields could exist independently of such rights and regulations: the latter were not an invariable concomitant of the former.

In Andean and medieval English commonfields, the degree of subdivision was often extreme. The community of Irpa Chico in Bolivia, for instance, possesses six great fields in which W. Carter and M. Mamani

(1982, 26–27) noted some 11,000 separate parcels. These mostly ranged from 1,200 to 3,000 square meters, with some diminutive plots and others as large as 24,000 square meters. In England, the size range of plots was narrower, although some diminutive plots did exist. At Martham, in Norfolk, for example, the land held by the peasantry was divided into at least 2,500 separate plots at the end of the thirteenth century, with an average plot size of 2,000 square meters and a significant number of plots measuring 1,000 square meters or less (Campbell 1980).

There has been much discussion of the reasons for this most distinctive form of field layout. It has recently been suggested that dispersed holdings may represent a strategy of risk minimization. The Andean evidence lends some support to this interpretation, insofar as plot scattering increases with altitude, which is positively linked to higher natural risk factors (McCloskey 1976; Dodgshon 1980, 22–25, 45–46; McPherson 1983; Figueroa 1982, 127, 129, 132; Bentley 1987). Yet, although plot scattering may reduce the risk of wholesale crop failure, there is no unequivocal, empirical evidence to show that it was actually undertaken with this express purpose.[6] Indeed, it may have arisen from entirely different motives. Thus, in England, several studies have demonstrated that piecemeal colonization by groups of cultivators, together with the repeated partitioning of holdings between heirs and the sale and exchange of portions of land between different cultivators, were all capable, over a period of time and under conditions of population growth, of creating subdivided fields from formerly consolidated holdings (Bishop 1935; Baker 1964; Sheppard 1966; Campbell 1980). When the rules governing the transference of land permitted, population growth was likely to lead at one and the same time to an extension of the cultivated area and the fragmentation of established holdings. As population expanded, so holdings proliferated, individual parcels became smaller, and the degree of scattering increased. In this context, it is significant that partible inheritance, whose contribution to the formation of subdivided fields in medieval England is now well established, is still practiced in many Andean commonfield communities today. Other things being equal, such an inheritance system is likely to ensure the persistence of a highly subdivided field layout.[7]

Notwithstanding the high degree of parcellation in both English and Andean commonfields, it would be misleading to represent their physical appearance as at all similar. The shape of the parcels and the way in which they were organized into fields differed due to the contrasting technological and ecological circumstances under which the two systems evolved.

English commonfields were developed for the most part on level or gently undulating terrain and in conjunction with a plow technology and mixed grain and stock economy. Indeed, the plow was arguably the single most formative influence upon the morphology of English commonfields. As J. Langdon has shown (1986), three main types of plow were in use by the thirteenth century: the wheeled, foot, and swing varieties.

Wheeled plows were more likely to be drawn by horses than were the other varieties (on the lightest soils, a team of only two horses would sometimes suffice), and in distribution were confined to the southeastern counties and parts of East Anglia. This pattern is partly a function of soil conditions, but it also reflects social, economic, and institutional factors, insofar as the adoption of horse traction entails a greater emphasis upon the production of fodder crops, notably oats. This, in turn, is associated with higher labor inputs and the kind of intensive cultivation system that, at this date, was found only in conjunction with the more loosely regulated commonfield systems.[8]

Elsewhere in the country, swing and foot plows predominated, the ox was the principal plow beast, and plow teams were often large—usually eight animals, but sometimes ten, or even as many as twelve. Again, this is partly because of physical conditions, as large, slow, ox teams were a necessity on the heavy clay soils of much of lowland England; but it also correlates with lower population densities and cultivation systems that placed greater emphasis upon fallowing, with a corresponding dependence upon natural rather than produced fodder. These conditions obtained in much of those parts of central and southern England where commonfield farming was most strongly developed, so there was a general association between foot and swing plows, large ox teams, and regular commonfield systems. Finally, it was for the simple but obvious reason that these large teams were cumbersome to manage and awkward to turn that individual parcels within the commonfields acquired their characteristically long, sinuous, and strip-like shape (Eyre 1955).

The prevailing plow technology can also be credited with creating the equally characteristic micro-relief pattern known as "ridge and furrow." This resulted from the repeated turning of the sod inwards, toward the center of the strip, which the fixed moldboards (it was the moldboard that turned the sod) of medieval plows made unavoidable. The boundaries between strips thus became marked by furrows, which had the additional advantage on heavy soils of assisting drainage (Beresford 1948; Kerridge 1951). A buildup of soil also resulted at the end of each strip from the action of turning the plow: the resultant "headlands" often became so massive that they may still be identified from aerial photographs, even where the associated strip pattern has long since been plowed out (Hall 1981).

Land hunger during the thirteenth century pushed the cultivated area of most commonfield townships to its physical limit, so that property boundaries became clearly demarcated (double furrows, grass balks, and marker stones and posts were all used for that purpose) and property rights—private as well as communal—became jealously guarded.[9] Odd patches of ground were sometimes left untilled within the commonfields for reasons of shape, accessibility, or soil conditions; these were generally utilized as a valuable supplement to the otherwise meager pasturage resources. For the most part, however, commonfields were very regular and only appeared fragmented and haphazard in areas of broken relief, poor soil, or bad drainage (Elliott 1973). Even then, they still bore little physical resemblance to the commonfields of the Andes.

Andean commonfields occur in high mountains, where the terrain is extremely fractured, and where cattle, and especially plow oxen, have difficulty in adapting to the altitudes. The commonfields are distributed over a very extensive geographical area: they have been found as far north as Huanuco in Peru and as far south as Macha, Department of Potosi, in Bolivia (Orlove and Godoy 1986). Within this zone, they lie at 3,000–4,000 meters above sea level on both the eastern and western flanks of the Andes, including the *altiplano*.

Throughout this area, yoked oxen are employed for plowing only in the lands surrounding Lake Titicaca and on the flatter patches of the Bolivian plateau. Elsewhere, plow animals are precluded by the rugged topography, easily degradable soils, and risk of hypoxic stress (Guillet 1981). For these reasons, and partly because the principal crops are roots and tubers (notably potatoes), the predominant tool of cultivation is not the plow but the digging stick (*chakitaclla*), supplemented by picks, shovels, and scythes for planting and harvesting (Gade and Rios 1976). Individual parcels of land are therefore free to assume every conceivable size and shape, a phenomenon that is encouraged by the steep and broken slopes and stony soils.

In consequence, the typical appearance of Andean commonfields is a mosaic of irregular parcels, many of them Lilliputian in scale. The boundaries between these parcels are often vague; they include natural features, untilled land, marginal pasture grounds, and up-ended sod blocks. The same applies to the commonfields themselves and the boundaries between them, which tend to be zones rather than precise lines and are usually demarcated by small piles of stones (*mojones*) or natural landmarks (Godoy 1985). This endows commonfield agriculture with an element of flexibility, for cultivation can be expanded or contracted as required according to demographic changes and altered land requirements (Mamani 1973, 93). It also produces a different agricultural landscape from the neatly aligned arable strips of lowland England.

Decision-Making Arrangements in England and the Andes

Pronounced though outward differences may have been, both English and Andean subdivided fields presented their dependent cultivators with the same basic problem: how were cropping and grazing to be organized in fields that were so parcellated? In particular, how was advantage to be taken of the valuable opportunity that fallow land afforded for feeding livestock and fertilizing soils? In medieval England, the need to utilize the fallow grazings was especially acute, for in many townships (especially in the counties of the East Midlands) the area of arable land had been so expanded that permanent grassland was scarce (Fox 1984). Yet livestock, both for traction and manure, remained an indispensable adjunct of arable production. In the plowless Andes, the need was different. What was important here was the conservation of soil fertility in a mountain environment where soils are deficient in nitrogen, phosphate, and potassium, and easily degradable (Eckholm 1976; Crawford, Wishart, and Campbell 1970; Orlove 1977, 119; Ravines 1978a, 3–74; Thomas 1979; Brush 1980). Indeed, adequate dunging of the soil (usually by flocks of sheep and llamas), is essential to the successful cultivation of one of the region's main staples, the potato (LaBarre 1947; Browman n.d.; Winterhalder, Larsen, and Thomas 1974; Camino, Recharte, and Bidegaray 1981).[10]

It is in the context of these ecological requirements that the adoption of communal decision-making arrangements must be interpreted. The precise nature of these arrangements depended upon environmental, technological, demographic, and sociopolitical circumstances, which is why England, for example, contained so many different types of commonfield systems (it remains to be established whether the same applied in the Andes, although it is a priori likely).[11] These systems differed from one another in both form and function, possessed distinctive geographical distributions, and followed separate chronologies of development. Apart from subdivided fields that were devoid of communal decision-making arrangements, two basic generic types of field systems can be identified: irregular commonfield systems, and regular commonfield systems. Within the former category, further distinctions can be drawn among systems in which there was no regulation of cropping, systems in which there was some regulation of cropping, and those where, very occasionally, there was complete regulation of cropping. This last feature was, however, more typical of regular commonfield systems, whose distinctiveness lay in the superimposition of communally enforced rotations upon a regular layout of holdings, with the result that each peasant holding, large or small, effectively became a microcosm of the entire arable area.[12] S. Fenoaltea has recently argued that this offers the additional, and in his view more profound, advantage of optimizing the allocation of village labor to the village land, as though the township were a single village-wide farm, "*without* loss of effort to shirking or supervision, as each household optimizes the allocation of its own labor to its own land" (1988, 191).

Although many different commonfield types existed, it is important to recognize that there have always been some subdivided fields within which individual holdings have been managed without reference to any wider framework of decision making. J. Thirsk believed that attempts to herd and farm in subdivided fields were so prone to conflict that "the community was drawn together by sheer necessity to cooperate in the control of farming practices" (1964, 9). Commonfield agriculture is thus regarded as offering lower transaction costs than the continuance of farming in severalty. In fact, this was by no means necessarily so, as England and the Andes both demonstrate.

Examples may be found in both countries of intensely subdivided fields with little or no communal regulation of cropping and herding. This was particularly the case in environmentally favored areas of relatively high population density and intensive agriculture. A high population means that labor is available for the fencing and policing of individual plots and private tethering, herding, and folding of livestock. At low population densities, as B. C. Field has demonstrated for seventeenth-century New England, such exclusion costs, and especially the costs of fencing, were a major factor promoting communal herding. On the other hand, as Field also observes (1985, 104–7), the tendency of population growth is to push property-rights institutions in the direction of individual tenures. Intensification of cultivation also means that the area left fallow and available for pasturage is usually either small or nonexistent.

The husbandry systems where this occurred in medieval England—notably in parts of Sussex, Kent, and Norfolk—were characterized by the cultivation of fodder crops and associated stall-feeding of livestock, coupled with labor-intensive methods of fertilizing the land. When fallowing occurred, its sole purpose was to cleanse the land of weed growth by means of multiple plowings, a practice that would have been in direct conflict with any attempt to utilize fallows as a source of forage. Wherever these husbandry methods were employed, rights of common grazing on the arable fields were therefore either restricted to the period immediately after the harvest (the one time in the year when the fields were free from standing crops) or absent altogether (Baker 1973; Campbell 1981c, 1983).

In the Andes, the counterparts of these intensive grain-producing districts are the areas of irrigated maize production, at lower altitudes than the main area of commonfields, where a warmer and more stable climate and a more benign topography permit a greater intensity of cultivation and correspondingly higher densities of population. Here, too, the organization of cultivation is largely on an individual basis, as communal supervision of grazing is precluded by the intensity of cropping (Donkin 1979, 120; Guillet 1981; Platt 1982). Cultivators make their own private arrangements for feeding the plow oxen employed in these areas.

The opposite extreme is represented by the classic commonfield system of the English Midlands. Here, demographic, economic, and environmental circumstances were less conducive to the kind of intensification of production outlined above, insofar as this area supported only moderate population densities, was at some distance from major urban markets, and lacked cheap and ready access to external supplies of nutrients for the maintenance of soil fertility. As a result, there were both greater incentives and fewer obstacles to the adoption of collective controls upon agriculture. In fact, communal management of an integrated system of cropping and grazing was taken further in this system than in any other. What made this possible was an artificially regular layout of holdings, whereby an equal amount of land was held in each of the commonfields of the township. This was essential since a regular, and communally enforced, rotation of crops was superimposed upon the entire arable area, effectively transforming the village into a single large farm (Fenoaltea 1988).

The furlong—a bundle of adjacent strips—was the basic unit of cropping, with the result that individual commonfields frequently carried a range of different crops. Nevertheless, when it came to fallowing, the field retained a central place in the whole system of rotation: "Whatever changes in cropping were rung on the furlongs of the sown field or fields, the fallow field remained inviolate" (Fox 1981, 74). Under the two- and three-field system, each field was fallowed either every second or every third year. The basic rotation was either a combination of winter- and spring-sown cereals and legumes followed by fallow, or a season of winter-sown crops, a season of spring-sown crops, and then fallow. The purpose of the fallow was to rest the soil so that it might recuperate its fertility, to allow the land to be fertilized with the dung of grazing livestock and, above all, to supply the livestock with forage, which was in such short supply in many of the townships that followed this system. Since the need to find grazing for the livestock was, ecologically, the raison d'être of the entire system, there was no question of subjecting the fallow to repeated plowings: on the contrary, it was left to sward over with weeds and grasses and only put back under the plow shortly before it was returned to cultivation.

Andean commonfields share many affinities with these English arrangements and display the same association with a moderate density of population and intensity of land use. They, too, employ communal controls to rationalize the distribution of sown and unsown plots, thereby facilitating common grazing of the fallows. On the other hand, the crops involved are very different from those grown in medieval England, as are the functions and organization of fallows. These differences ensure that Andean commonfields possess considerable individuality in their decision-making arrangements and attendant patterns of interaction.

Within the Andes, commonfield lands are generally sown for up to four consecutive years and there is no distinction between winter-sown and spring-sown crops. The normal rotational sequence is: first, potatoes fertilized with llama or sheep dung; second, native chenopods (*guinua* and *cañahua* and the tubers *ocas* and *ullucus*); and then cereals and leguminous crops in the third and fourth years. Having deep roots, the cereals and legumes seek nutrients below the shallow surface layer of the soil, whose fertility is rapidly depleted during the first two years of cropping (Freeman 1980). Thereafter, soil nutrients are allowed to build up because the fields are allowed to rest for as long as thirty years, but the mean fallow duration is three or four years. The length of fallow is never fixed; it varies according to soil conditions and cropping requirements, and this proves the key to the whole system. The higher the altitude, the longer must be the fallow period, because of reduced soil fertility and slower rates of growth (Caballero 1981; Orlove and Godoy 1986).

In very few cases are fields sown for more than five consecutive years or fallows reduced to one year. The exceptions include areas undergoing intensification, or those communities situated on the shores of Lake Titicaca that, thanks to a more benign climate and richer soils, plant on what approaches a continuous basis (Carter 1964; Mamani 1973, 89, 110; Urioste 1977, 43; Lewellen 1978, 16, 49; LeBaron 1979; Godoy 1985). It remains to be established how far such intensifications of production have led to modifications in conventional Andean commonfield arrangements and to what extent they have led to irreversible ecological degradation. Nevertheless, that intensification has occurred at all does demonstrate that commonfield systems are nowhere a direct adaptive response to environmental factors; various nonecological considerations have always been important.

Once a field is designated for cropping, the precise pattern and sequence of crops sown is a matter of individual choice. There is thus no Andean equivalent of the furlongs found in medieval England. The range of crops grown within any field is usually quite wide, as cultivators tend to diversify their pattern of planting as a hedge against environmental hazards and the risk of wholesale harvest failure (Brush 1981, 71). Nevertheless, as far as the decision-making arrangements of these commonfields are concerned, it is not so much what crops are sown that matters, but rather, which fields are to be left fallow and for how long. Such important decisions are taken at a village level. To accommodate the relatively long fallow period required in this high mountain environment, the arable land of each community will usually be divided into at least seven or eight commonfields, and sometimes as many as fifteen. Whenever any of these fields lie fallow, villagers exercise a customary entitlement to pasture their livestock, collect firewood, and cut turf. At the same time, arrangements are made for the systematic dunging of the land by sheep and llamas penned in movable folds. This ensures that all the land is adequately manured before it is eventually returned to cultivation: the dung, urine, and treading of the animals are all highly beneficial to these upland soils. Analogous arrangements occurred in certain English commonfield systems on the light soil of East Anglia, where there was likewise a tendency for fallow periods to be of several years' duration (Postgate 1973; Bailey 1990).

Two aspects of these Andean arrangements require further comment. The first concerns the household's entitlement to common pasturage. In some cases, households have rights to graze only portions of the stubble of the commonfields. When this occurs, the location of this grazing ground is often independent of the distribution of parcels making up the holding. W. E. Carter has described this distinctive arrangement (Carter 1964, 68; see also Platt 1982; Godoy 1983). According to him, the section of the commonfields reserved to each household as pasture for its flocks is known as an *unta* (literally meaning "that which one can see"), which is a prolongation of the houseplot into the commonfields. This *unta* privilege directly overlays the normal rights of cultivation that apply to individual plots in the fields and sometimes applies to such uncultivable land as mountaintops or swamps. Such demarcation of each family's own grazing zone within the village's territory is an Andean peculiarity and reflects a desire for private control of their own animals by individual community members. It finds no counterpart in the commonfield villages of lowland England, with their greater emphasis on arable farming, scarcer pasturage, and much smaller flocks and herds.

A second, much more significant, characteristic of Andean arrangements is that a communally determined system of cropping and fallowing coexists with an irregular layout of holdings. Such a state of affairs carries with it the obvious penalty that each year some households will be obliged to leave a disproportionate amount of their land uncultivated. There is thus an inherent inequity within the system, a deficiency that is avoided in the English midland system by the equal distribution of a holding's strips among all the fields of a township.[13] Such a regular layout of holdings would be of no practical advantage in the Andean situation, where the timing and duration of fallows are, perforce, subject to such flexibility. It is certainly true that (as in those few English instances where there was a similar mismatch between holding layout and rotations) Andean people are sometimes able to use land held outside the commonfield system to offset the inequities arising within that system. Thus all commonfield holdings include a houseplot that is held in severalty and capable of intensive cultivation. Although such plots are occasionally quite substantial, most have been much reduced in size through the application of a custom of male partible inheritance (Carter 1964, 65; Heath, Buechler, and Erasmus 1969, 177; Rodríguez-Pastor 1969, 84–86). For the tenants of these diminutive houseplots, the most effective supplements to commonfield land are therefore valley plots. Not only are the latter not subject to communal decisions, they are also environmentally more favored and can consequently be cropped much more intensively than the commonfields (Guillet 1981; Platt 1982).

When houseplots, valley lands, and pasture grounds are all taken into account, it transpires that commonfields generally constitute between 20 and 70 percent of total landholdings, the proportion rising with altitude (Figueroa 1982, 133). Nevertheless, many individuals remain dependent upon the commonfields for the basic staples of daily life. For them, the only solution when they are temporarily disadvantaged by the system is to come to some kind of reciprocal arrangement with those who are temporarily advantaged. It is upon this kind of social exchange between members of the same agricultural community that the commonfield system ultimately depends for its success.

The nearest equivalent in England was the commonfield system of parts of East Anglia, noted above. Here a similar coexistence occurred between common rotations and an irregular layout of holdings. The rotations in question assumed the form of flexible cropping shifts which were capable of variation from plot to plot and year to year. Their object was likewise to concentrate fallow strips for sheep folding. Since this periodically placed certain individuals at a disadvantage, successful operation of the system, as in the Andes, depended upon the establishment of a satisfactory method of compensation. As control of the system was vested in the manorial lord (who, as principal flockmaster, was also usually the major beneficiary of it), a tenant thus placed might receive part of the lord's crop, temporary use of a portion of the lord's demesne, or financial compensation in the form of a cash handout or rent rebate. Even so, this system was particularly prone to conflict, as is testified by the large number of resultant court cases (Allison 1957, 1958; Simpson 1958; Postgate 1973; Bailey 1990). It was also one of the issues that provoked Kett's Rebellion of 1549 (MacCulloch 1979).

On the whole, both English and Andean commonfield systems made good practical sense in a situation where land was cropped with only moderate intensity, and where population levels were such that substantial dividends were to be derived from pooling scarce labor and organizing basic farming tasks in common. Savings in exclusion costs were obviously to be made by eliminating the need for fencing and by appointing a few guards to watch over the field and stock of all the villagers. Moreover, information and transaction costs were reduced when decisions were taken at a village level as to when and where to plant and pasture. Real gains in agricultural labor productivity may consequently have resulted (Fenoaltea 1988).

Such arrangements may also have proved advantageous to subjects faced with heavy labor-tribute liabilities, a relevant point in both a medieval English and an Andean context. In the former, lords were entitled to exact labor services from their tenants through the institution of serfdom; these services characteristically assumed the form of agricultural work on the lord's demesne. In fact, under the conditions of labor scarcity that probably prevailed when serfdom was first instituted, lords would have had a vested interest in promoting the development of a system of husbandry that enabled them to redeploy labor to their own ends. Certainly a general association existed between areas of strong lordship and fully developed commonfield systems.[14]

Likewise in the Andes, the Spanish instituted a system of forced labor to work the silver and mercury mines of Potosi and Huancavelica. This assumed the form of an annual migration of able-bodied males (the *mita*) drawn from a very extensive area. At the end of the sixteenth century, this migration totaled some 13,000 workers per year, some of whom came from so far away that they had to walk for an entire month to reach the mines. As E. Tandeter (1981) has pointed out, a migration on this scale must have had major repercussions for the accumulation and reproduction of the communities that were being exploited, the more so as these heavy labor demands coincided with a prolonged and massive reduction in population.[15] It seems, therefore, that in the Andes too there is a coincidence between the area of heaviest labor-tribute liabilities and the area where commonfield agriculture appears to have attained its most complex form and survived the longest.

Patterns of Interaction

The common denominator of all these commonfield systems is a reversion from private use of the soil for tillage to communal rights for grazing on the herbage of the fallow fields. This communal arrangement places a premium upon the collective management of resources. For instance, since commonfields remain unfenced, individual householders face incentives to steal crops from adjacent plots and encroach upon neighboring lands. This potential threat fosters collective action, as isolated households by themselves would be less effective in opposing interlopers (Gade 1970, 51; Orlove 1976, 213; Albo 1977, 23; Platt 1982, 45). That is why villages appoint guards and other officials. Furthermore, the movement across time and space of different flocks and herds, and the designation of fields to be sown and fallowed, involve complex scheduling problems affecting all villagers.

These logistical problems are therefore frequently decided upon by village assemblies; it is they who determine the date and place of planting, harvesting, and grazing. In the Andes, these village councils, as noted by McBride, constitute the "*de facto* government of a community, though its operation is so silent and its deliberation so carefully guarded that its existence is seldom even suspected" (1921, 9). Much the same is true of similar assemblies in England, whose existence is often barely hinted at in the historical record. Yet although there is a clear association between commonfield agriculture and strongly developed corporate village communities, the precise causal connection between them is enigmatic. At any event, the outward physical expression of the strong corporate character possessed by these commonfield communities in both countries is the nucleated village, in which the dwellings of the cultivators and other inhabitants are concentrated into a single settlement cluster. Such villages are now recognized by archaeologists as having made a relatively late appearance on the rural scene, their arrival coinciding, it would appear, with that of commonfield agriculture in the ninth and tenth centuries (Astill 1988b; Fox 1992). In contrast, more dispersed and more ancient forms of settlement—loosely clustered hamlets, isolated farmsteads, and a mixture of villages and scattered messuages—tend to prevail in areas without a commonfield system.[16]

The Andean evidence demonstrates that the corporate sense of these commonfield communities is usually strong enough to override even quite substantial inequalities of holding size among cultivators.[17] As is to be expected, the larger landholders do tend to exert a disproportionate influence within village assemblies and dominate the principal village offices. On the other hand, all household heads serve as field guards by yearly turns. This rotational incumbency possibly had colonial origins, but it still functions (Rasnake 1988).

These officers, known variously as *pachacas, campos, muyucamas, arariwas, camayoqs,* or *rigidores de varas,* are in charge of supervising fields, preventing animals from straying onto cultivated lands, guarding against crop theft and trespass, punishing and levying fines on miscreant shepherds, and performing rituals to protect crops when hail, drought, and other natural calamities threaten. Their honesty is ensured because they are answerable to the higher-level authority of the village and charged with responsibility for any crops stolen from the fields. In recompense, if the harvest proves successful, they receive the produce of a few furrows from each family, or are allowed to plant in uncultivated plots of the commons.

As B. Thomas observes, this system of incumbency by yearly turns symbolizes total community involvement in the decision-making process of the entire community (Thomas 1979, 161; see also Gade 1970, 12; Degregori and Golte 1973, 42; Preston 1973; Fujii and Tomoeda 1981, 54). Household heads also sponsor village festivals at one or more times in their life cycle. These festivals confer prestige upon those who sponsor them, but are also essential for validating the individual household's rights of access to village assets in the eyes of the community (Platt 1982; Godoy 1983).

The need to establish who belongs to a community and has a stake in its resources is critical: that it is perceived as such is demonstrated by the symbolic reapportionments and public reconfirmations executed under the supervision of *hacienda* officials of a household's rights to land.[18] If anything, the issue of who is entitled to land rights has become more prominent in recent years as expanding populations have brought resources under increasing pressure.

Today, as in the past, a pronounced social stratification is apparent within many of these villages. In Bolivia, the true insiders (*originarios*) tend to have more parcels within the commonfields than do later arrivals (*agregados*). Below these two groups lie the *kantu runas* ("people of the margin"), peasants who settled in the village during the nineteenth century and who obtained indirect access to common land in exchange for services rendered to wealthier households (Platt 1982; Godoy 1983).[19] These divisions tend to be perpetuated by rules that proscribe the renting or selling of commonfield land to outsiders, although they permit cultivators to rent or mortgage their parcels to other members of the community (Guillet 1979; Fujii and Tomoeda 1981, 53; McBride 1921, 14; Metraux 1959; Carter 1964, 68; Godoy 1985; Custred 1974, 258). Such entry and exit rules are enforced by the village council, which, if need be, employs expulsion as the ultimate sanction. That village councils should have acquired such powers is a function of the historic weakness of national power structures in the areas of commonfield agriculture.

In medieval England, the administrative structure and patterns of interaction of these commonfield communities are more difficult to ascertain, filtered as they are through the historical record. Most of what is known is provided by the proceedings of manorial courts (the lowest level of courts with legitimate legal jurisdiction). It was in these courts that commonfield bylaws were enacted and enforced, and their proceedings usually record innumerable boundary disputes and prosecutions for trespass and crop theft (Ault 1965). The election of village officials was also usually enrolled in the courts. Some of these officers, like the "pinder" and "hayward," are close equivalents of the Andean field guards. They watched over the livestock feeding on the commons, and when necessary impounded them and assessed the damage done by cattle and trespassers, after which a fine was imposed by the manor court on those responsible. Effective operation of these courts was obviously partly a function of the strength of seignorial authority, but it also depended upon the cooperation of the village community. Although a good deal of friction often existed between the villagers and their lord, it is clear that they derived considerable benefits from such ready access to a means of resolving local disputes.

The role of the manor court was important in the operation of the commonfield system, but the prerogative of overseeing the regular routine of commonfield husbandry and ensuring that cultivators conformed to its discipline was probably reserved to informal village assemblies.[20] Effectively, all those who owned land in the commonfields had a say in their management and enjoyed an entitlement to the appurtenant common rights that was usually in proportion to the size of their landholding. The only exceptions were various landless but long-established families within the community who sometimes retained a customary claim upon its resources through retention of the ancient houseplot. All those holding such rights were known as "commoners."

On the large Worcestershire manor of Halesowen, there were no less than twelve separate commonfield communities, each of which was represented in the central manor court by two villagers elected by its members, an arrangement implying that they must each have possessed some kind of well-organized self-governing machinery. As in the Andes, these "assemblies" were almost certainly dominated and run by the richer peasants, for it was they who usually fulfilled the majority of manorial offices. Patterns of social and economic interaction reconstructed by Z. Razi from these court records indicate a high incidence of reciprocity between peasants, its precise nature varying according to socioeconomic status. It is his view that in the late thirteenth and early fourteenth centuries the manor of Halesowen was characterized by "a high degree of cohesiveness, cooperation, and solidarity as a result of the requirements of an open-field husbandry, a highly developed corporate organization, and a sustained and active resistance to the seignorial regime" (1981, 16).

Nevertheless, the strong corporate sense manifested by these commonfield communities should not be mistaken for rural egalitarianism. Nor should commonfields be regarded as an expression of such principles.

Cooperation, a shared identity, and a sense of common purpose at a village level were perfectly compatible with the existence of sharp inequalities between peasants and marked intragroup rivalry. Moreover, in the long term, these internal divisions were potentially disruptive to the commonfield regime, particularly in view of any changes in the wider political or economic context.

Documentation of the long-entrenched social stratification that existed within these rural communities is now becoming increasingly available (Dewindt 1972; Britton 1977; Smith 1979; Razi 1980). Thus, Halesowen village society may have functioned as a community but it was also highly monetized and competitive. From his reading of the evidence, Razi was in no doubt that the well-to-do villages were exploiting the needs of their less well-off neighbors to maximize their profits. Equivalent studies of villages in other parts of England have come to much the same conclusion. There was no question of arable lands being periodically reallotted (all attested cases of reallotment relate to meadowland, a common resource, like most other sources of herbage). From at least the middle of the thirteenth century, it is plain that peasants had attached strong individual ownership rights to their land. According to customary law, even villein land, which theoretically belonged to the lord rather than the tenants who held and worked it, descended according to the prevailing rules of inheritance within the same family; only in default of heirs did it revert to the lord, who might then reallocate it among his tenants. Moreover, an active market in peasant land was already established by this date in much of lowland England. Its effect was generally to encourage the emergence of socioeconomic differences between individual peasant families (Smith 1984b; Harvey 1984). Nevertheless, through the observation of certain commonsense safeguards, this land market proved in no way inimical to the effective operation of the commonfields.

Conclusions

That two commonfield systems with such a strong functional affinity should have developed under such fundamentally different technological conditions is highly significant, for several writers on the origin of English commonfields have placed great stress on the role of technology. F. Seebohm (1883) and the Orwins (1967), for instance, have all attributed the creation of commonfields to the practice of coaration or joint plowing with a heavy mold-board plow (Dodgshon 1980, 30–34). Yet the culture that evolved such similar agricultural arrangements in the Andes was effectively plowless.

Environmentally, too, there was a vast difference in the circumstances under which these two commonfield systems developed. However, despite the obvious physical differences between the high Andes and lowland England, both environments presented cultivators with an analogous problem. In each case, the productivity of the agricultural system rested upon the maintenance of a delicate ecological balance that required the reconciliation, on the same land, of the conflicting requirements of animal and pastoral husbandry. The need to supply forage to the animals and dung to the soil was the link between them. Even so, that the same basic need should have elicited such a similar institutional response says as much about the sociopolitical conditions prevailing when commonfields emerged, as it does about environmental considerations per se.

As has been shown, English and Andean commonfields make complete sense only when viewed in the context of a specific combination of economic, demographic, social, and political circumstances. Field (1988) has emphasized the connection between cultivation practices at one extreme and sociopolitical institutions at the other, although this is difficult to demonstrate empirically. Nevertheless, we believe that it was a real and vital link in both of the cases discussed above. Such a conclusion is not merely of relevance to students of commonfield systems: it also provides a warning against adopting an approach to the whole question of common-property resources that is either too environmentally or too economically deterministic.

These two cases also demonstrate the capacity of common-property resource management systems to take on an existence of their own, independent of the circumstances that may have led to their creation. Commonfield systems were self-perpetuating. This was partly because the system could only be dismantled if the common rights that applied to it were dissolved first, a step that entailed considerable costs because it required a consensus particularly difficult to obtain where there were so many vested interests. The process of parliamentary enclosure in England provides a graphic illustration of this and demonstrates that the intervention of a superior legal authority was sometimes required before long-established common rights could be finally extinguished (Tate 1967; Yelling 1977; Turner 1984).

Strict adherence to the specific agricultural routine imposed upon a community by the commonfield system was a further source of inertia. It was not that progress was impossible—the system could not have survived for so long had this been the case—but rather that substantial changes, such as in the number of rotational courses, were cumbersome to achieve (Havinden 1961; Dahlman 1980, 146–99). Communal consent was required before any deviations could take place from established crop rotations, or before alterations could be made in the existing ratio of pasture to tillage. Radical changes in the techniques and intensity of cultivation were consequently to be avoided. As Fenoaltea observes, where technologies were in flux and the agricultural population poorly educated, "the critical advantage of enclosed farms was that they could introduce advanced techniques without the need to convince a village full of cautious peasants that the new methods were better than the old" (1988, 197). For these various reasons, commonfield systems had a bias toward the maintenance of the economic and demographic status quo, and their dependent communities adopted social and cultural values and demographic strategies that actually retarded population growth and moderated technological change (Homans 1941; Howell 1975; Goody, Thirsk, and Thompson 1976). The resultant symbiosis between commonfield regime and sociodemographic behavior sometimes endured for centuries.

Nevertheless, commonfield systems were by no means immutable, and it would be a mistake to presume that seventeenth- and eighteenth-century commonfields were carbon copies of medieval ones. Over time they furnish much evidence of adaptation to new technologies and socioeconomic circumstances. Parcels have been altered in shape and size, and fields in layout; new crops have been incorporated into rotations, and increases made in the number of rotational courses; livestock stints have been reassessed; and modifications have been made to the management of fallows. Provided that the pace of change has been gradual, it has usually been possible for commonfields to adapt themselves to it.

Problems have, however, arisen when the pace and nature of change have been more revolutionary. In England, for instance, although economic and technological developments rendered commonfield agriculture increasingly anachronistic from the fifteenth century onward, so that enclosure by agreement began to make quiet but steady progress, the final demise of the system did not come until the nineteenth century. Even then, it took powerful economic forces and strong vested interests, combined with the facility of enclosure by Act of Parliament, before the last bastions of the system fell. During this final period, commonfield agriculture was much castigated by agricultural writers so that it became widely regarded as a moribund and inefficient system. This verdict has tended to color contemporary Western attitudes to communally managed resources in general and has been used to support a strong preference for privatized property systems in particular. Yet recent research into the productivity of English agriculture before and after enclosure has begun to challenge this traditional orthodoxy. Thus, R. C. Allen has argued that English commonfield farmers achieved major productivity gains during the seventeenth and eighteenth centuries, and has further argued that the

V

major economic consequence of enclosure was not a gain in productivity, but the redistribution of agricultural income in favour of landlords (1982, 1991).

Yet in the Andes, the question of enclosure remains very much a live issue. Much privatization of former commonfield land has already taken place, by one means or another, in those areas where agriculture has been most strongly exposed to commercial penetration (Heath, Buechler, and Erasmus 1969, 192; Rodríguez-Pastor 1969, 86; Mamani 1973, 87–88; Preston 1974, 247; Mayer 1981, 82; Figueroa 1982, 133). But away from the influence of the Peruvian coastal cities and the chief towns and mining centers of the Bolivian interior, traditional commonfield agriculture continues largely unaffected. Coincidentally, it is in these same areas that the terrain is most rugged and rural poverty greatest, and this poses a major dilemma for those working in development (Eckholm 1976; Thomas 1979; Guillet 1981; Godoy 1983). Should contemporary Andean commonfields be condemned, like their erstwhile English counterparts, as an obstacle to progress and a cause of rural poverty and backwardness? Or should a lesson be drawn from recent reassessments of the English evidence and stress be placed upon the delicate ecological balance that they undoubtedly maintain in this high mountain environment, the moderate rates of technological progress and productivity growth which they are potentially able to sustain, and the sense of corporate identity and solidarity that they nurture in these isolated, materially deprived, and agriculturally dependent communities?

NOTES

1. For the European distribution, see Bloch 1967, 69. For commonfields in New England, see Walcott 1936, 218–52, Bidwell and Falconer 1925, and Field 1985. For Andean commonfields, see Orlove and Godoy 1986. Mesoamerican systems are briefly discussed in Wolf 1966, 20–21. Indian commonfields are described in Chapter 9 of this book. Middle Eastern systems are discussed in Goodell 1976, 60–68; and in Poyck 1962.

2. For an up-to-date review of the literature on English commonfield origins, see Fox 1981, 64–111; and Fenoaltea 1988, 171–240. For a comprehensive treatment of the development of English commonfields, see Baker and Butlin 1973.

3. Pleas for comparative research into commonfield systems have been made by Bloch (1967, 70), Thirsk (1966), and McCloskey (1975, 91).

4. For the late medieval antecedents of the enclosure movement, see Fox 1975, 181–202; and Campbell 1981a.

5. See Godoy 1991 for a consideration of the evolutionary trajectory of Andean commonfield systems.

6. In this context, it should be noted that the main reason that the villagers of Vila Vila, in the north of Potosi, abandoned commonfield tillage was that a frost would kill everyone's potatoes, since all the villagers planted potatoes in the same great commonfield (Mamani 1973, 88).

7. The most usual arrangement is for male coheirs to work a holding as a group: each brother receives the right to work some parcels within each commonfield, and further plot fragmentation is thereby halted (Mamani 1973, 91–92).

8. The complex economics of the changeover from natural to produced fodder, and thus from ox to horse plowing, are discussed in Boserup 1965, 36–39.

9. For agrarian conditions at this time, see Miller and Hatcher 1978.

10. One author has suggested that simple fallowing may not be enough to build up nutrients in the Andes (Yamamoto 1988, 130). For an analysis of the corresponding situation in England see Shiel 1991.

11. These different British field systems are surveyed in Baker and Butlin 1973.

12. For a fuller specification of the diagnostic features of these different commonfield systems, see Campbell 1981b, 112–29.

13. Households will also be faced with a periodic seasonality of agricultural surpluses or deficits, depending upon the amount of land held in the commonfields open to use.

14. This argument is elaborated more fully in Campbell 1981b. On the strength of seignorial power at the time that European commonfields were crystallizing, see Duby 1974. For an illustration of the coincidence between variations in lordship and variations in field systems, see Harley 1958, 8–18; and Roberts 1973, 188–231. For the factors that promoted the institutions of serfdom, see Hatcher 1981, 3–39.

15. The population collapse after the conquest is discussed in Dobyns 1963, 493–515; Smith 1970, 453–64; and Shea 1976, 157–80.

16. For the pattern of rural settlement in Britain, see Roberts 1979 and Astill 1988a.

17. See, for instance, Albo 1977 and Isbell 1978. Earlier echoes of the same theme may be found in the *indigenista* literature, as in Valcarcel 1925 and Castro-Pozo 1936.

18. These reapportionments seem to date from Inca times, when they were used to ensure that all households had the means of meeting tribute obligations to the kings in Cuzco (Murra 1980a, xv). The system was adapted by the Spanish to serve a similar purpose (see Rowe 1957, 182). This practice survives in fossilized form today. The shift from true reallotment to a system of nominal or symbolic reallotment, wherein households continue to use the same parcels year after year, probably reflects a growing shortage of land and concomitant increased specification of individual land rights (Carter 1964, 69; Buechler 1969, 179; Preston 1973, 3).

19. For the existence of unequal holdings among commonfield farmers in Peru, see also Mishkin 1946, 421–22; Soler 1958, 190; Matos Mar 1964, 130–42; and Guillet 1981, 146.

20. For a review of the literature on this subject, see Smith 1984a.

V

REFERENCES

Alberti, G., and E. Meyer, eds. 1974. *Reciprocidad e intercambio en los Andes peruanos*. Lima: Instituto de Estudios Peruanos.

Albo, X. 1977. *La paradoja aymara*. La Paz: Centro de investigación y promoción del campesinado (CIPCA).

Allen, Robert C. 1982. "The Efficiency and Distributional Consequences of Eighteenth-Century Enclosures." *Economic Journal* 92:937–53.

———. 1991. "The Two English Agricultural Revolutions, 1450–1850." In *Land, Labour and Livestock: Historical Studies in European Agricultural Productivity*, ed. Bruce M. S. Campbell and Mark Overton, 236–54. Manchester: Manchester University Press.

Allison, K. J. 1957. "The Sheep-Corn Husbandry of Norfolk in the Sixteenth and Seventeenth Centuries." *Agricultural History Review* 5:12–30.

———. 1958. "Flock Management in the Sixteenth and Seventeenth Centuries." *Economic History Review*, 2d ser. 11:98–112.

Arguedas, J. M., ed. 1964. *Estudios sobre la cultura actual del Perú*. Lima: Universidad Nacional Mayor de San Marcos.

Astill, Grenville. 1988a. "Rural Settlement: The Toft and Croft." In *The Countryside of Medieval England*, ed. G. Astill and A. Grant, 36–61. Oxford and New York: Blackwell.

———. 1988b. "Fields." In *The Countryside of Medieval England*, ed. G. Astill and A. Grant, 62–85. Oxford and New York: Blackwell.

Ault, W. O. 1965. "Open-Field Husbandry and the Village Community: A Study of Agrarian By-Laws in Medieval England. *Transactions of the American Philosophical Society*, new ser., 55, no. 7.

Bailey, Mark. 1990. "Sand into Gold: The Evolution of the Foldcourse System in West Suffolk, 1200–1600. *Agricultural History Review*, 38:40–57.

Baker, A. R. H. 1964. "Open Fields and Partible Inheritance on a Kent Manor." *Economic History Review*, 2d ser. 17:1–23.

———. 1973. "Field Systems of Southeast England." In *Studies of Field Systems in the British Isles*, ed. A. R. H. Baker and R. A. Butlin, 393–419. Cambridge: Cambridge University Press.

Baker, A. R. H., and R. A. Butlin, eds. 1973. *Studies of Field Systems in the British Isles*. Cambridge: Cambridge University Press.

Baker, A. R. H., and D. Gregory, eds. 1984. *Explorations in Historical Geography: Interpretive Essays*. Cambridge: Cambridge University Press.

Beckett, J. V. 1989. *A History of Laxton*. Oxford: Blackwell.

Bentley, J. W. 1987. "Economic and Ecological Approaches to Land Fragmentation: The Defense of a Much-Maligned Phenomenon." *Annual Review of Anthropology* 16:31–67.

Beresford, M. W. 1948. "Ridge and Furrow and the Open Fields." *Economic History Review*, 2d ser., 1:34–35.

Biddick, K., ed. 1984. *Archaeological Approaches to Medieval Europe*. Kalamazoo, Mich.: Medieval Institute, Western Michigan University.

Bidwell, P., and J. Falconer. 1925. *History of Agriculture in the Northern United States 1620–1860*. New York: Smith.

Bishop, T. A. M. 1935. "Assarting and the Growth of the Open Fields." *Economic History Review* 6:13–29.

Bloch, M. 1967. *Land and Work in Medieval Europe*. Berkeley: University of California Press.

Boserup, E. 1965. *The Conditions of Agricultural Growth: The Economics of Agrarian Change under Population Pressure*. London: Allen and Unwin.

Britton, E. 1977. *The Community of the Vill: A Study in the History of the Family and Village Life in Fourteenth Century England*. Toronto: Macmillan.

Browman, D. L. n.d. "Llama Caravan *Fleteros*, and Their Importance in Production and Distribution." Unpublished manuscript.

Brush, S. 1980. "The Environment and Native Andean Agriculture." *América Indígena* 40:163.

———. 1981. "Estrategías agrícolas tradicionales en las zonas montanosas de América Latina." Seminario internacional sobre producción agropecuaria y forestal en zonas de ladera de America tropical. Informe Téchnico 11. Turrialba, Costa Rica: Centro Agronómico Tropical de Investigación y Ensenanza.

Buechler, H. C. 1969. "Land Reform and Social Revolution in the Northern Altiplano and *Yungas* of Bolivia." In *Land Reform and Social Revolution in Bolivia*, ed. D. Heath, H. Buechler, and C. Erasmus, 179. New York: Praeger.

Caballero, J. M. 1981. *Economía agraria de la sierra peruana*. Lima: Instituto de Estudios Peruanos.

Camino, A., J. Recharte, and P. Bidegaray. 1981. "Flexibilidad calendárica en la agricultura tradicional de las vertientes orientales de los Andes." In *La tecnología en el mundo andino*, ed. H. Lechtman and A. M. Soldi, 169–94. Mexico City: Universidad Nacional Autonoma de México.

Campbell, B. M. S. 1980. "Population Change and the Genesis of Commonfields on a Norfolk Manor." *Economic History Review*, 2d ser., 33:174–92.

———. 1981a. "The Extent and Layout of Commonfields in Eastern Norfolk." *Norfolk Archaeology* 38:5–32.

———. 1981b. "Commonfield Origins—the Regional Dimension." In *The Origins of Open Field Agriculture*, ed. T. Rowley, 112–29. London: Croom Helm.

———. 1981c. "The Regional Uniqueness of English Field Systems? Some Evidence from Eastern Norfolk." *Agricultural History Review* 29:16–28.

———. 1983. "Agricultural Progress in Medieval England: Some Evidence from Eastern Norfolk." *Economic History Review*, 2d ser., 36:26–46.

Carter, W. E. 1964. *Aymará Communities and the Bolivian Agrarian Reform*. Social Science Monograph no. 24. Gainesville: University of Florida.

Carter, W., and M. Mamani. 1982. *Irpa Chico*. La Paz: Juventud.

Castro-Pozo, H. 1936. *Del ayllu al cooperativismo socialista*. Lima: Mejia Baca.

Chambers, J. D., and G. E. Mingay. 1966. *The Agricultural Revolution 1750–1880*. London: Batsford.

Chevallier, F. 1953. *La formation des grands domaines au Mexique: Terre et société XVI–XVII siècles*. Paris: Institut d'Ethnologie, Musée de l'Homme.

Crawford, R. M. M., D. Wishart, and R. M. Campbell. 1970. "A Numerical Analysis of High Altitude Scrub Vegetation in Relation to Soil Erosion in the Eastern Cordilerra." *Journal of Ecology* 58:173–81.

Custred, G. 1974. "Llameros y comercio inter-regional." In *Reciprocidad e intercombìo en los Andes peruanos*, ed. G. Alberti and E. Meyer, 252–89. Lima: Instituto de Estudios Peruanos.

Custred, G., and B. Orlove. 1974. "Sectorial Fallowing and Crop Rotation Systems in the Peruvian Highlands." Paper presented at 41st International Congress of Americanists.

Dahlman, C. J. 1980. *The Open Field System and Beyond; A Property Rights Analysis of an Economic Institution*. Cambridge: Cambridge University Press.

Degregori, C., and J. Golte. 1973. *Dependencia y desintegración estructural en la comunidad de Pacaraos*. Lima: Instituto de Estudios Peruanos.

Denevan, W. N., ed. 1976. *The Native Population of the Americas in 1492*. Madison: University of Wisconsin Press.

Dewindt, E. B. 1972. *Land and People in Holywell-cum-Needingworth: Structures of Tenure and Patterns of Social Organization in an East Midlands Village 1252–1457*. Toronto: Pontifical Institute of Mediaeval Studies.

Dobyns, H. 1963. "An Outline of Andean Epidemic History to 1720." *Bulletin of the History of Medicine* 37:493–515.

Dodgshon, R. A. 1980. *The Origin of British Fields Systems: An Interpretation*. London: Academic Press.

———. 1981. "The Interpretation of Subdivided Fields: A Study in Private or Communal Interests?" In *The Origins of Open Field Agriculture*, ed. T. Rowley, 130–44. London: Croom Helm.

Donkin, R. A. 1979. *Agricultural Terracing in the Aboriginal New World*. Tucson: University of Arizona Press.

Duby, G. 1974. *The Early Growth of the European Economy: Warriors and Peasants from the Seventh to the Twelfth Century*. London: Weidenfeld and Nicolson.

Eckholm, E. 1976. *Losing Ground*. New York: Norton.

Elliott, G. 1973. "Field Systems of Northwest England." In *Studies of Field Systems in the British Isles*, ed. A. R. H. Baker and R. A. Butlin, 42–92. Cambridge: Cambridge University Press.

Eyre, S. R. 1955. "The Curving Ploughstrip and its Historical Implications." *Agricultural History Review* 3:80–94.

Fenoaltea, S. 1988. "Transaction Costs, Whig History, and the Common Fields." *Politics and Society* 16, no. 2–3:171–240.

Field, Barry C. 1985. "The Evolution of Individual Property Rights in Massachusetts Agriculture, 17th–19th Centuries." *Northeastern Journal of Agricultural and Resource Economics* 14, no. 2:97–109.

———. 1988. "The Evolution of Property Rights." Department of Agricultural and Resource Economics, University of Massachusetts, Amherst, Mass.

Figueroa, A. 1982. "Production and Market Exchange in Peasant Economies: The Case of the Southern Highlands in Peru." In *Ecology and Exchange in the Andes*, ed. D. Lehmann, 126–56. Cambridge Studies in Social Anthropology no. 41.

Fox, H. S. A. 1975. "The Chronology of Enclosure and Economic Development in Medieval Devon." *Economic History Review*, 2d ser., 28:181–202.

———. 1981. "Approaches to the Adoption of the Midland System." In *The Origins of Open Field Agriculture*, ed. T. Rowley, 64–111. London: Croom Helm.

———. 1984. "Some Ecological Dimensions of Medieval Field Systems." In *Archaeological Approaches to Medieval Europe*, ed. K. Biddick, 119–58. Kalamazoo, Mich.: Medieval Institute, Western Michigan University.

———. 1992. "The Agrarian Context." In *The Origins of the Midland Village*, 36–72. Papers prepared for a discussion session at the Economic History Society's annual conference. Leicester, April.

Freeman, P. 1980. "Ecologically Oriented Agriculture." Unpublished manuscript.

Fujii, T., and H. Tomoeda. 1981. "Chacra, laime y auquénidos." In *Estudios etnográficos del Perú meridional*, ed. S. Masuda, 33–63. Tokyo: Tokyo University.

Gade, D. 1970. "Ecología del robo agrícola en las tierras altas de los Andes centrales." *América Indígena* 30:3–14.

Gade, D., and M. Escobar. 1982. "Village Settlement and the Colonial Legacy in Southern Peru." *Geographical Review* 72:430–49.

Gade, D., and R. Rios. 1976. "La chaquitaclla: herramienta indigena sudamericana." *América Indígena* 36:359–74.

Godoy, R. A. 1983. "From Indian to Miner and Back Again: Small-Scale Mining in the Jukumani Ayllu, Northern Potosi, Bolivia." Ph.D. diss., Columbia University.

———. 1985. "State, Ayllu, and Ethnicity in Northern Potosi." *Anthropos* 80:53–65.

———. 1990. *Mining and Agriculture in Highland Bolivia*. Tucson: University of Arizona Press.

———. 1991. "The Evolution of Common-Field Agriculture in the Andes: A Hypothesis." *Comparative Studies in Society and History* 33, no. 2:395–414.

Goodell, G. 1976. "The Elementary Structures of Political Life." Ph.D. diss., Columbia University.

Goody, J. R., J. Thirsk, and E. P. Thompson, eds. 1976. *Family and Inheritance: Rural Society in Western Europe, 1200–1800*. Cambridge: Cambridge University Press.

Guillet, D. 1979. *Agrarian Reform and Peasant Economy in Southern Peru*. Columbia: University of Missouri Press.

———. 1981. "Land Tenure, Ecological Zone, and Agricultural Regime in the Central Andes." *American Ethnologist* 8:139–56.

Hall, D. 1981. "The Origins of Open-Field Agriculture—the Archaeological Field Evidence." In *The Origins of Open Field Agriculture*, ed. T. Rowley, 23–25. London: Croom Helm.

Harley, J. B. 1958. "Population Trends and Agricultural Developments from the Warwickshire Hundred Rolls of 1279." *Economic History Review*, 2d ser., 11: 8–18.

Harvey, P. D. A., ed. 1984. *The Peasant Land Market in Medieval England*. Oxford: Oxford University Press.

Hatcher, J. 1981. "English Serfdom and Villeinage: Towards a Reassessment." *Past and Present* 90:3–39.

Havinden, M. 1961. "Agricultural Progress in Open-Field Oxfordshire." *Agricultural History Review* 9:73–88.

Heath, D., H. Buechler, and C. Erasmus. 1969. *Land Reform and Social Revolution in Bolivia*. New York: Praeger.

Homans, G. C. 1941. *English Villagers of the Thirteenth Century*. Cambridge, Mass.: Harvard University Press.

Howell, C. 1975. "Stability and Change 1300–1700: The Socio-economic Context of the Self-perpetuating Family Farm in England." *Journal of Peasant Studies* 2:468–82.

Hoyle, B. S., ed. 1974. *Spatial Aspects of Development*. London: Wiley.

Isbell, B. J. 1978. *To Defend Ourselves: Ecology and Ritual in an Andean Village*. Austin: University of Texas Press.

Kerridge, E. 1951. "Ridge and Furrow and Agrarian History." *Economic History Review*, 2d ser., 4:14–36.

Langdon, John. 1986. *Horses, Oxen, and Technological Innovation: The Use of Draught Animals in English Farming from 1066–1500*. Cambridge: Cambridge University Press.

LaBarre, W. 1947. "Potato Taxonomy among the Aymara Indians of Bolivia." *Acta Americana* 6:83–103.

LeBaron, A., et al. 1979. "An Explanation of the Bolivian Highlands Grazing Erosion Syndrome." *Journal of Range Management* 32:201–8.

Lechtman, H., and A. M. Soldi, eds. 1981. *La tecnología en el mundo andino*. Mexico City: Universidad Nacional Autonoma de México.

Lehmann, D., ed. 1982. *Ecology and Exchange in the Andes*. Cambridge Studies in Social Anthropology, no. 41. Cambridge: Cambridge University Press.

Lewellen, T. 1978. *Peasant in Transition: The Changing Economy of the Peruvian Aymara. A General Systems Approach*. Boulder, Colo.: Westview Press.

MacCulloch, Diarmaid. 1979. "Kett's Rebellion in Context." *Past and Present* 84: 36–59.

Málaga Medina, A. 1974. "Las reducciónes en el Perú durante el Virrey Francisco de Toledo." *Anuario de Estudios Americanos* 31:819–42.

Mamani, M. P. 1973. *El rancho de Vila Vila*. La Paz: Consejo Nacional de Reforma Agraria.

Masuda, S., ed. 1981. *Estudios etnográficos del Perú meridional*. Tokyo: Tokyo University.

Matos Mar, J., ed. 1958. *Las actuales comunidades de indígenas: Huarochirí en 1955*. Lima: Instituto de Etnologia y Arqueología.

———. 1964. "La propiedad en la isla de Taquile (Lago Titicaca)." In *Estudios sobre la cultura actual del Peru*, ed. J. M. Arguedas. Lima: Universidad Nacional Mayor de San Marcos.

Mayer, E. 1981. *Uso de la tierra en los Andes: ecología y agricultura en el valle del Mantaro del Perú con referencia especial a la papa*. Lima: Centro Internacional de la Papa.

McBride, G. 1921. *The Agrarian Indian Communities of Highland Bolivia*. New York: Oxford University Press.

McCloskey, D. 1975. "The Persistence of English Common Fields." In *European Peasants and Their Markets*. ed. W. Parker and E. Jones, 73–119. Princeton, N.J.: Princeton University Press.

———. 1976. "English Open Fields as Behaviour toward Risk." *Research in Economic History* 1:124–70.

McPherson, M. F. 1983. "Land Fragmentation in Agriculture: Adverse? Beneficial? and for Whom?" Development and Discussion Paper no. 145. Cambridge, Mass.: Harvard Institute for International Development.

Metraux, A. 1959. "The Social and Economic Structure of the Indian Communities of the Andean Region." *International Labour Review* 74:231.

Miller, E., and J. Hatcher. 1978. *Medieval England—Rural Society and Economic Change 1086–1348*. London: Longman.

Mishkin, B. 1946. "The Contemporary Quechua." In *Handbook of South American Indians*, vol. 2, Bulletin no. 143, ed. J. H. Steward, 411–70. Washington, D.C.: Bureau of Ethnology.

Murra, J. V. 1980a. "Waman Puma, etnógrafo del mundo andino." In J. V. Murra, *El Primer Nueva Coronica y Buen Gobierno*, xiii–xix. Mexico City: Siglo Veintiuno.

———. 1980b. *El Primer Nueva Corónica y Buen Gobierno*. Mexico: Siglo Veintiuno.

Orlove, B. 1976. "The Tragedy of the Commons Revisited: Land Use and Environmental Quality in High-Altitude Andean Grasslands." In *Hill Lands: Proceedings of an International Symposium*, 208–14. Morgantown: West Virginia University Press.

———. 1977. *Alpaca, Sheep, and Men: The Wool Export Economy and Regional Society in Southern Peru*. New York: Academic Press.

Orlove, B., and R. Godoy. 1986. "Andean Sectorial Farming System." *Journal of Ethnobiology* 6:169–204.

Orwin, C. S., and C. S. Orwin. 1967. *The Open Fields*. 3d ed. Oxford: Clarendon Press.

Parker, W., and E. Jones, eds. 1975. *European Peasants and Their Markets*. Princeton, N.J.: Princeton University Press.

Platt, T. 1982. "The Role of the Andean Ayllu in the Reproduction of the Petty Commodity Regime in Northern Potosi (Bolivia)." In *Ecology and Exchange in the Andes*, ed. D. Lehmann, 27–69. Cambridge Studies in Social Anthropology no. 41. Cambridge: Cambridge University Press.

Popkin, S. 1979. *The Rational Peasant*. Berkeley: University of California Press.

Postgate, M. R. 1973. "Field Systems of East Anglia." In *Studies of Field Systems in the British Isles*, ed. A. R. H. Baker and R. A. Butlin, 281–322. Cambridge: Cambridge University Press.

Poyck, A. P. G. 1962. *Farm Studies in Iraq*. Wageningen, Holland: Mededelingen van den Landbouwhogeschool te Wageningen, Nederland 62.

Preston, D. 1973. "Agriculture in a Highland Desert: The Central Altiplano of Bolivia." Department of Geography, Working Paper 18:6. Leeds, England: University of Leeds.

———. 1974. "Land Tenure and Agricultural Development in the Central Altiplano, Bolivia." In *Spatial Aspects of Development*, ed. B. S. Hoyle, 231–51. London: Wiley.

Rasnake, R. 1988. *Domination and Cultural Resistance*. Durham, N. C.: Duke University Press.

Ravines, R. 1978a. "Recursos nasturales de los Andes." In *Tecnología Andina*, ed. R. Ravines. Lima: Instituto de Estudios Peruanos.

———. ed. 1978b. *Tecnología Andina*. Lima: Instituto de Estudios Peruanos.

Razi, Z. 1980. *Life, Marriage, and Death in a Medieval Parish: Economy, Society, and Demography in Halesowen (1270–1400)*. Cambridge: Cambridge University Press.

———. 1981. "Family, Land, and the Village Community in Later Medieval England." *Past and Present* 93:3–36.

Rhoades, R. E., and S. I. Thompson. 1975. "Adaptive Strategies in Alpine Environments: Beyond Ecological Particularism." *American Ethnologist* 2:535–51.

Roberts, B. K. 1973. "Field Systems of the West Midlands." In *Studies of Field Systems in the British Isles*, ed. A. R. H. Baker and R. A. Butlin, 188–231. Cambridge: Cambridge University Press.

———. 1979. *Rural Settlement in Britain*. London: Hutchinson.

Rodríguez-Pastor, H. 1969. "Progresismo y cambios en Llica." In *La Comunidad Andina*, ed. J. R. Sabogal Wiesse, 73–143. Mexico City: Instituto Indigenista Interamericano.

Rowe, J. H. 1957. "The Incas under Spanish Colonial Institutions." *Hispanic American Historical Review* 37:182.

Rowley, T., ed. 1981. *The Origins of Open Field Agriculture*. London: Croom Helm.

Sabogal Wiesse, J. R., ed. 1969. *La Comunidad Andina*. Mexico City: Instituto Indigenista Interamericano.

Seebohm, F. 1883. *The English Village Community*. London: Longmans, Green.

Shea, D. 1976. "A Defense of Small Population Estimates for the Central Andes in 1520." In *The Native Population of the Americas in 1492*, ed. W. N. Denevan, 157–80. Madison: University of Wisconsin Press.

Sheppard, J. A. 1966. "Pre-enclosure Field and Settlement Patterns in an English Township, Wheldrake, Near York." *Georgrafiska Annaler*, ser. B, 48:59–77.

Shiel, Robert S. 1991. "Improving Soil Fertility in the Pre-Fertiliser Era." In *Land, Labour and Livestock: Historical Studies in European Agricultural Productivity*, ed. Bruce M. S. Campbell and Mark Overton, 51–77. Manchester: Manchester University Press.

Simpson, A. 1958. "The East Anglian Foldcourse; Some Queries." *Agricultural History Review* 6:87–96.

Smith, C. T. 1970. "Depopulation of the Central Andes in the 16th Century." *Current Anthropology* 11:453–64.

Smith, R. M. 1979. "Kin and Neighbours in a Thirteenth Century Suffolk Community." *Journal of Family History* 4:219–59.

———. 1984a. "'Modernization' and the Corporate Medieval Village Community in England: Some Skeptical Reflections." In *Explorations in Historical Geography: Interpretive Essays*, ed. A. R. H. Baker and D. Gregory, 140–94. Cambridge: Cambridge University Press.

———, ed. 1984b. *Land, Kinship, and Life-Cycle*. Cambridge: Cambridge University Press.

Soler, E. 1958. "La comunidad de San Pedro de Huancaire." In *Las actuales comunidades de indigenas: Huarochirí en 1955*, ed. J. Matos Mar, 167–257. Lima: Instituto de Etnología y Argueología.

Steward, J. H., ed. 1946. *Handbook of South American Indians*. Vol. 2, Bulletin no. 143. Washington, D.C.: Bureau of Ethnology.

Tandeter, E. 1981. "Forced and Free Labour in Late Colonial Potosi." *Past and Present* 93:98–136.

Tate, W. E. 1967. *The English Village Community and the Enclosure Movements*. London: Gollancz.

Thirsk, J. 1964. "The Common Fields." *Past and Present* 29:3–9.

———. 1966. "The Origins of the Common Fields." *Past and Present* 33:143.

Thomas, B. 1979. "Effects of Change on High Mountain Adaptive Patterns." In *High-Altitude Geoecology*, ed. P. J. Webber, 139–88. Boulder, Colo.: Westview Press.

Turner, M. 1982. "Agricultural Productivity in England in the Eighteenth Century: Evidence from Crop Yields." *Economic History Review*, 2d ser., 35:489–510.

———. 1984. *Enclosures in Britain 1750–1830*. London: Macmillan.

Urioste, M. 1977. *La economía del campesinado altiplánico en 1976*. La Paz: Universidad Católica Boliviana.

Valcarcel, L. 1925. *Del ayllu al imperio*. Lima: Editorial Garcilaso.

Walcott, R. 1936. "Husbandry in Colonial New England." *New England Quarterly* 9:218–52.

Webber, P. J., ed. 1979. *High-Altitude Geoecology*. Boulder, Colo.: Westview Press.

Winterhalder, B., R. Larsen, and B. Thomas. 1974. "Dung as an Essential Resource in a Highland Peruvian Community." *Human Ecology* 2:89–104.

Wolf, E. 1966. *Peasants*. Englewood Cliffs, N.J.: Prentice-Hall.

Wordie, J. R. 1983. "The Chronology of English Enclosure, 1500–1914.: *Economic History Review*, 2d ser., 36:483–505.

Yamamoto, N. 1988. "Papa, llama, y chaquitaclla. Una perspectiva etnobotánica de la cultura andina." In *Recursos Naturales Andinos*, ed. S. Masuda, 111–52. Tokyo: University of Tokyo.

Yelling, J. A. 1977. *Common Field and Enclosure in England, 1450–1850*. London: Macmillan.

Towards an Agricultural Geography of Medieval England

[A review article of John Langdon, *Horses, Oxen and Technological Innovation: The Use of Draught Animals in English Farming from 1066–1500*, CUP, 1986. xvi + 331 pp. 42 figures. £30.]

Horses, Oxen and Technological Innovation makes a major contribution to the study of a period whose basic economic outlines are progressively changing. In establishing an appropriate data-base for his chosen subject, Dr Langdon draws upon an impressive and imaginatively assembled array of published and unpublished sources, and uses them to obtain significant new insights into the husbandry and economy of this period.[1] With the publication of this book there can be little doubt that medieval agricultural history has at last come of age. Not that considerable scholarly endeavour in this field has been absent during the past thirty years. But the valuable crop of individual and estate studies which have been produced, though they have added greatly to existing knowledge, have afforded few radical new insights and have tended to reinforce traditional assessments of the agriculture of the period.[2] Moreover, the picture of medieval agriculture that has emerged has been patchy and unsystematic, a function of the adherence to an estate-orientated methodology. Most of what has been written continues to relate almost exclusively to the demesne rather than the peasant sector and is disproportionately concerned with those leading ecclesiastical estates for which there are

1 Over 30 years ago R H Hilton drew attention to the wide range of sources relating to medieval agriculture: 'The Content and Sources of English Agrarian History before 1500', *Ag Hist Rev*, III, 1955, pp 3–19. Significantly, Dr Langdon was trained in Hilton's own Birmingham stable, where his supervisor was Dr C C Dyer.

2 Notable examples include J A Raftis, *The Estates of Ramsey Abbey*, Toronto, 1957; E M Halcrow, 'The Administration and Agrarian Policy of the Manors of Durham Cathedral Priory', unpublished BLitt thesis, University of Oxford, 1959; J Z Titow, 'Land and Population on the Bishop of Winchester's Estates 1209–1350', unpublished PhD thesis, University of Cambridge, 1962; I Keil, 'The estates of the Abbey of Glastonbury in the Later Middle Ages', unpublished PhD thesis, University of Bristol, 1964; I Kershaw, *Bolton Priory*, Oxford, 1973; E Searle, *Lordship and Community: Battle Abbey and its Banlieu*, Toronto, 1974; B F Harvey, *Westminster Abbey and its Estates in the Middle Ages*, Oxford, 1977; C C Dyer, *Lords and Peasants in a Changing Society: The Estates of the Bishopric of Worcester, 680–1540*, Cambridge, 1980. Current orthodoxy largely derives from the writings of M M Postan, most notably 'Medieval Agrarian Society in its Prime: England', pp 548–632 of M M Postan (ed), *The Cambridge Economic History of Europe*, I, Cambridge, 2nd edition, 1966. See also J Z Titow, *English Rural Society 1200–1350*, London, 1969; and E Miller and J Hatcher, *Medieval England: Rural Society and Economic Change 1086–1348*, London, 1978. For an alternative view see H E Hallam, *Rural England 1066–1348*, London, 1981.

well-preserved archives.[3] Some parts of the country have been well served in this respect, but geographically the coverage has been uneven. The forthcoming volumes II and III of the *Agrarian History of England and Wales* – conceived as long ago as 1956 – should eventually go a long way towards rectifying this, but may well suffer from having been too long in the making.[4]

I

The methodological challenge confronting all analyses of estate or regional farming systems is to reconcile the individual case study with the creation of a consistent and systematic picture of the country as a whole. Without some national frame of reference no full evaluation is possible of what is typical or unusual about individual farms or regions. There is an obvious limitation to any national picture of agriculture which is built up piecemeal from independent regional studies. Dr J Thirsk makes explicit acknowledgement of this in her introduction to Volume V, Part I, of *The Agrarian History of England and Wales*, when she states (of the national map of farming types):

> some boundaries which separate farming types are also the boundaries between chapters; in other words, authors on either side of a county boundary have not always agreed in their identification of the dominant local farming type.[5]

This problem is compounded when, as in these volumes, the picture is built up qualitatively rather than quantitatively. It is here that Langdon's study breaks important new ground, and not just in a medieval context. He takes as his unit of analysis the country as a whole and establishes regional variations in the technology of haulage and traction by a process of disaggregation.[6] Since the labour required in such an exercise is considerable – the PhD thesis upon which the book is based took almost six years – sampling and quantification provide two of his most essential tools.[7] This applies particularly to one of the most original achievements of the book: the creation of a national data-base from information contained in manorial accounts.

The rich store of agricultural information contained in the annual accounts drawn up each Michaelmas by the reeve or bailiff of a manor has long been extensively used by historians.[8] What is not so fully appreciated is the fact that these accounts survive in their thousands. The small number of estates, like those of the bishops of Winchester and abbots of Westminster, with long and relatively complete runs of accounts, have tended to divert attention away from the much more typical situation in which an individual demesne is represented by perhaps just one or two stray *compoti*. For Norfolk alone almost 2000 accounts are extant for the period 1238–1450, representing some 219 different

3 Exceptions are F J Davenport, *The Economic Development of a Norfolk Manor, 1086–1565*, Cambridge, 1906; K Ugawa, *Lay Estates in Medieval England*, Tokyo, 1966; D Postles, 'Problems in the Administration of Small Manors: Three Oxfordshire Glebe-Demesnes, 1278–1345', *Midland History*, IV, 1977, pp 1–14; R R Davies, *Lordship and Society in the March of Wales 1282–1400*, Oxford, 1978; R H Britnell, 'Minor Landlords in England and Medieval Agrarian Capitalism', *Past & Pres*, 89, 1980, pp 3–22; M Mate, 'Profit and Productivity on the Estates of Isabella de Forz (1260–92)', *Econ Hist Rev*, 2nd ser, XXXIII, 1980, pp 326–34.

4 Publication of Volume II, edited by H E Hallam, is imminent: it deals with the period 1066–1350. Contributions to Volume III, which deals with the period 1350–1500, are currently being finalized. Its publication may be anticipated in 1988 or 1989.

5 J Thirsk (ed), *The Agrarian History of England and Wales, V, 1640–1750. I. Regional Farming Systems*, Cambridge, 1984, p xxi.

6 Regional farming systems have attracted increasing attention in recent years, although they have hitherto lacked any wider frame of reference. Examples include: D Roden, 'Demesne Farming in the Chiltern Hills', *Ag Hist Rev*, XVII, 1969, pp 9–23; P F Brandon, 'Demesne Arable Farming in Coastal Sussex during the Later Middle Ages', *Ag Hist Rev*, XIX, 1971, pp 113–34; B M S Campbell, 'Agricultural Progress in Medieval England: Some Evidence from Eastern Norfolk', *Econ Hist Rev*, 2nd ser, XXXVI, 1983, pp 26–46; M Mate, 'Medieval Agrarian Practices: The Determining Factors', *Ag Hist Rev*, XXXIII, 1985, pp 22–31.

7 J Langdon, 'Horses, Oxen, and Technological Innovation: The Use of Draught Animals in English Farming from 1066 to 1500', unpublished PhD thesis, University of Birmingham, 1983.

8 For a discussion of manorial accounts etc see P D A Harvey, 'Agricultural Treatises and Manorial Accounting in Medieval England', *Ag Hist Rev*, XX, 1972, pp 170–82, and especially his editor's introduction to *Manorial Records of Cuxham, Oxfordshire circa 1200–1359*, Oxfordshire Records Society, 50, 1976.

demesnes, 60 per cent of them with fewer than five *compoti*.[9] Not all counties are as well documented as this, and in the remote north-west and south-west of the country it takes much hard work in the archives to turn up any account rolls at all. But for England as a whole the total number of extant grange accounts is possibly in excess of 20,000. These are scattered through numerous public and private archives and it is from these Langdon has drawn his sample of 1565 accounts (including several which are available in print), providing him with information on 637 demesnes for the period 1250–1320 and 399 demesnes for the period 1350–1420. With more diligent searching and a larger sample size the geographical coverage of both these data-sets could undoubtedly have been improved.[10] Yet it is unfair to quibble, for the achievement is considerable and the samples do yield a clear and consistent picture with the regional detail sketched in bold but simple strokes. More rigorously researched local and regional studies will no doubt refine this picture and add much detail, but it is unlikely that they will alter its basic outlines. In the long term the potential of this kind of approach is considerable, especially when harnessed to modern methods of data storage and analysis, for it is capable of transforming the scale and sophistication of analysis.[11] On the one hand, even the smallest and most miscellaneous scraps of evidence become usable; on the other, the opportunity exists to obtain a real insight into the organization and development of the demesne economy at large.

Just as significant as the transformation of scale and method employed in this study is the equal attention which is bestowed upon both lords and peasants. Many researchers would have been tempted by the abundance of explicit and relatively straightforward information available for seignorial husbandry to shirk the responsibility of investigating that of the peasantry, for which available evidence is altogether more haphazard, equivocal, and limited. Instead, Langdon puts this documentary imbalance to advantage by using the strength of seignorial documentation to establish a comparative framework, within which the more scrappy and miscellaneous evidence available for the peasantry can be evaluated. As is plain, the trends thus established for the peasantry can never be as clear as those identified for the demesnes, but considerable confidence does attach to the relative relationship between them. Moreover, sources do exist which cast at least some light on peasant agriculture: inventories, lay subsidies, heriots, maintenance agreements, surveys, tithe returns, and extents. Langdon displays considerable resourcefulness in seeking out these data and thereby demonstrates the very real potential which exists for learning more about the husbandry of this neglected but crucial group.[12]

9 There is no official listing of these Norfolk accounts, although a handlist is available on application to the author. The accounts themselves are dispersed among twenty-five public and private archives.

10 A summary register of manorial records may be consulted at the National Register of Archives, Quality House, Quality Court, Chancery Lane, London. A search of this register showed that Langdon's sample could be considerably improved in such counties as Staffordshire, Shropshire, and Herefordshire. In Cornwall, Cheshire, Lancashire, Cumberland, Westmorland, and Northumberland, on the other hand, there is a genuine paucity of records.

11 A computer analysis of seignorial agriculture in England 1250–1450 is currently in hand based upon data partly supplied by Dr Langdon. The results of this analysis will appear in my forthcoming book.

12 Studies of the social and economic relations of the peasantry, by contrast, proliferate. Recent examples include Z Razi, *Life, Marriage and Death in a Medieval Parish: Economy, Society and Demography in Halesowen, 1270–1400*, Cambridge, 1980; R M Smith (ed), *Land, Kinship, and Life-cycle*, Cambridge, 1984; B A Hanawalt, *The Ties that Bound: Peasant Families in Medieval England*, Oxford, 1986; J M Bennett, *Women in the Medieval English Countryside: Gender and Household in Brigstock before the Plague*, Oxford, 1987; M K McIntosh, *Autonomy and Community: The Royal Manor of Havering, 1200–1500*, Cambridge Studies in Medieval Life and Thought, 5, 1987; A R DeWindt, 'Redefining the Peasant Community in Medieval England: The Regional Perspective', *Jnl Brit Stud*, 26, 1987, pp 163–207. Agricultural production on peasant holdings has received rather more attention on the continent, for instance: E Le Roy Ladurie and J Goy, *Tithe and Agrarian History from the Fourteenth to the Nineteenth Centuries: An Essay in Comparative History*, Cambridge, 1982; G Bois, *The Crisis of Feudalism: Economy and Society in Eastern Normandy c1300–1550*, Cambridge, 1984.

II

That methods of investigating medieval agriculture are now being transformed is timely, for there is a mounting body of evidence which suggests that the performance of the agrarian sector is itself due for reassessment. Until now there has been a strong tendency to regard the medieval rural economy as relatively undeveloped and undifferentiated. Insofar as different farming regions existed it is thought that they derived less from a process of economic differentiation arising from interaction with the market, than from the influence of ecological, cultural, and institutional features. They were formal rather than functional regions and reflected such things as natural resource endowment, field systems, culture and ethnicity, and the degree of manorialization.[13] This view of the medieval rural economy tends to be most fully articulated by those who view it from the perspective of the sixteenth century. For instance, according to Dr J Langton:

> the spatial system of medieval times, one of almost discrete local economies loosely linked through London, was transformed [after 1500] by the greater cohesion imposed by London's growth and by the inter-position of hierarchical steps and whole regional economies between the two old levels.[14]

Yet such a verdict sits increasingly uncomfortably with the available evidence, which suggests that the economy of *circa* 1300 had certain very important things in common with that of *circa* 1600.

In the first place, historical opinion increasingly favours a medieval population at peak – *circa* 1300 – equal to or greater than the early seventeenth-century demographic maximum of 5.5 millions.[15] This shift in opinion stems from a reassessment of the accuracy of the 1377 Poll Tax returns coupled with the steady accumulation of comparatively small pieces of evidence, all of which point to extremely high rural population densities in much of the country by the end of the thirteenth century.[16] Scholars of the period have long been familiar with the notion that it was experiencing conditions of relative over-population, but that the population may have been absolutely so large raises obvious questions about the organization of the agricultural sector and, in particular, the extent to which it was geared towards exchange rather than self-sufficiency. The thirteenth century is now known to have experienced a tremendous proliferation of trading institutions, and in certain localities it has been shown that all members of rural society were deeply involved in the market nexus, with all that this implies about market specialization.[17] The possibility that a more specialized and integrated pattern of food production and supply may have been evolving during the thirteenth century is lent further credibility by the revised population estimates which have recently been made of several of the larger medieval towns at this time. London, in particular, has been shown by Dr D Keene to have attained a maximum *circa* 1300 of at least 90–100,000 – twice the conventionally accepted estimate.[18] The task of provisioning such a city, plus the other leading cities of the realm (whose accepted population

13 Studies which lay stress on these factors include H L Gray, *English Field Systems*, Cambridge, Mass, 1915; D C Douglas, *The Social Structure of Medieval East Anglia*, Oxford Studies in Social and Legal History, IX, 1927; G C Homans, *English Villagers of the Thirteenth Century*, Cambridge, Mass, 1941; W G Hoskins, 'Regional Farming in England', *Ag Hist Rev*, II, 1954, pp 3–11; J C Jackson, 'Regional Variations in Agriculture in Medieval England', *Northern Universities' Geographical Journal*, 1, 1960, pp 41–53.

14 J Langton, 'Industry and Towns 1500–1730', Chapter 7 in R A Dodgshon and R Butlin (eds), *An Historical Geography of England and Wales*, 1978, p 194.

15 M M Postan, *The Medieval Economy and Society: An Economic History of Britain in the Middle Ages*, 1972, p 30; Miller and Hatcher, *op cit*, p 29; Hallam, *op cit*, pp 246–7; E A Wrigley and R S Schofield, *The Population History of England 1541–1871: A Reconstruction*, 1981.

16 For an up-to-date review of this evidence see: R M Smith, 'Human Resources in Rural England', Chapter 3 of G Astill and A Grant (eds), *The Medieval Countryside*, Oxford, forthcoming.

17 R H Britnell, 'The Proliferation of Markets in England, 1200–1349', *Econ Hist Rev*, 2nd ser, XXXIV, 1981, pp 209–221; K Biddick, 'Medieval English Peasants and Market Involvement', *Jnl Econ Hist*, XLV, 1985, pp 823–31.

18 D Keene, *Cheapside before the Great Fire*, London, 1985.

estimates must all now be open to doubt), cannot but have exercised an influence over a very extensive area.[19] Certainly, the lure of such large-scale urban demand would help to account for the highly specialized and intensive demesne farming systems which are now known to have developed in certain parts of the country, systems which were more complex and technologically advanced than was previously thought possible at this date.[20]

III

Langdon's work lends further support to the notion that English agriculture underwent important organizational changes during the course of the twelth and thirteenth centuries, over and above the general expansion in production which it has long been accepted took place over this period. In the first place, a significant transformation was effected in the technology of traction and, more particularly, of haulage. At the time of Domesday horses provided little more than 5 per cent of total animal draught force on the demesne, and no more than 10 per cent in any of the regions for which there are figures. There are signs that the level of horses was already higher among peasant draught stock, but this cannot be quantified. In contrast, by the beginning of the fourteenth century horses accounted for at least 20 per cent of the animal draught force on demesnes and almost 50 per cent on peasant farms, and these figures exceeded 50 per cent and 75 per cent respectively in certain regions. The changeover from oxen to horses was greatest for hauling, to the extent that horses easily dominated the carriage of goods by vehicle by the end of the thirteenth century.[21] The horse's adoption into ploughing was less spectacular: mixed teams gained in popularity and some all-horse farms emerged, with progress being most pronounced on the smallest holdings. As Langdon acknowledges, the pace of innovation was comparatively slow, and must to some extent have been conditioned by the supply of animals, but the scale of the eventual transformation was no less considerable for that.

Substitution of the horse for the ox made particular progress in certain regions and played an important part in establishing regional variations in agriculture. This extended to more than just the adoption of horses *per se*, for it tended to encourage a shift from natural to produced fodder, and had an important influence upon the number and type of other animals kept. Among all regions it was East Anglia which stood in the van of progress. Here, a shift towards a greater use of horses was already taking place by the first half of the twelfth century and it was only towards the end of that century that horses began to be employed in increasing numbers in the Home Counties and the east midlands. The spread of horses was very much a diffusion process and in Norfolk – one of the counties where they appeared earliest and in which their adoption eventually proceeded furthest – account rolls are numerous enough, and survive from a sufficiently early date, to allow the process to be reconstructed in some detail. Table 1 and Figure 1 illustrate the changing ratio of oxen to horses in this county over the 200-year period 1250–1449 (these trends are derived from a survey of all extant Norfolk grange accounts).[22] As Langdon

19 Norwich may have had a population of at least 18,000: T H Hollingsworth, *Historical Demography*, 1969, pp 363–4. The population of Winchester circa 1300 (approximately the 17th ranking town in the urban hierarchy) has been estimated at 10–12,000: D Keene, *Winchester Studies, 2, Survey of Medieval Winchester*, I, part 1, 1985, pp 366–70.

20 P F Brandon, 'Cereal Yields on the Sussex Estates of Battle Abbey during the Later Middle Ages', *Econ Hist Rev*, 2nd ser, XXV, 1972, pp 403–20; Campbell, *op cit*; Mate, *op cit*, 1985.

21 J Langdon, 'Horse Hauling: A Revolution in Vehicle Transport in Twelfth- and Thirteenth-Century England?', *Past & Pres*, 103, 1984, pp 37–66.

22 These results comprise part of a comprehensive analysis of Norfolk demesne farming which will be presented in full in my forthcoming book. I am grateful to J Orr for research assistance.

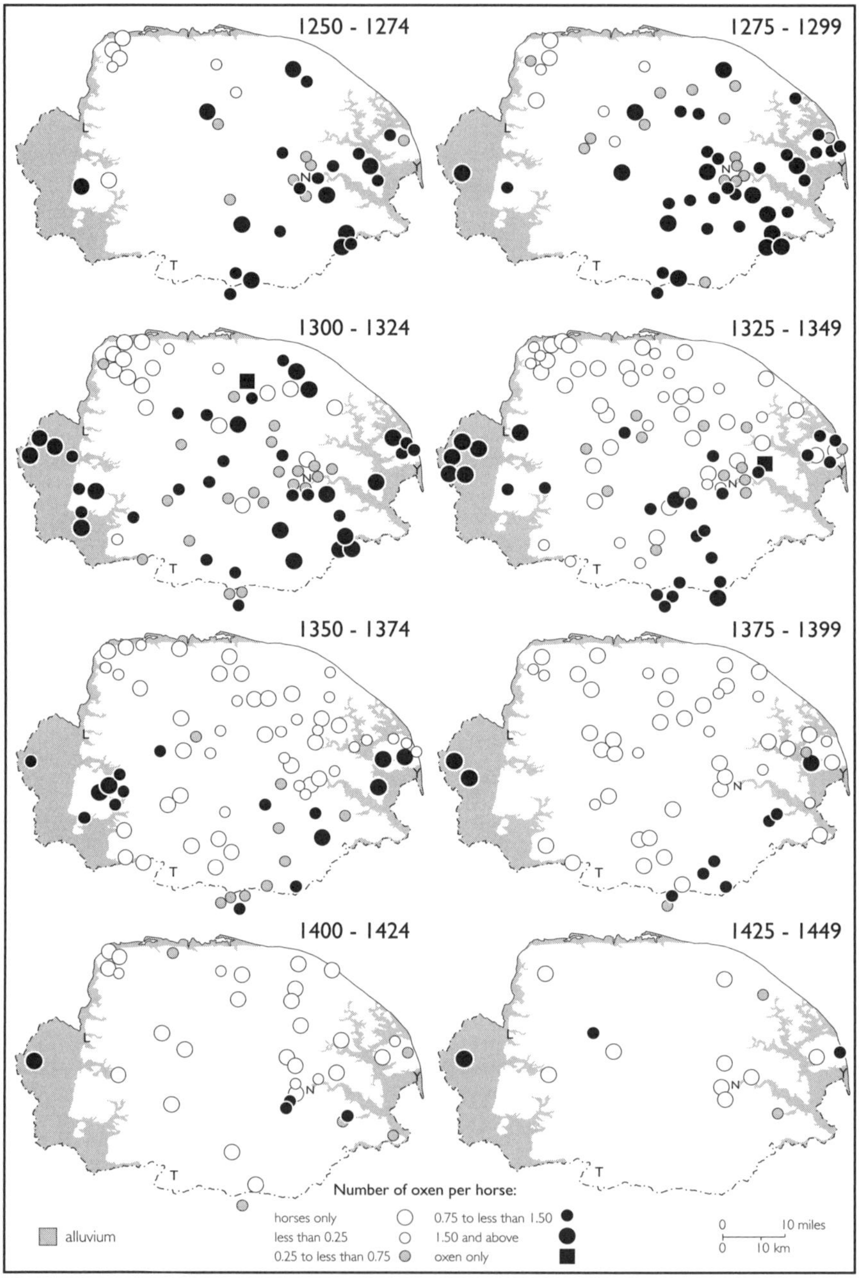

FIGURE I
Norfolk: the changing ratio of oxen to horses, 1250–1449

TABLE I
Mean Number of Oxen per Horse on Demesnes in Norfolk and England, 1250–1449[23]

Years	*Norfolk*	*England*
1250–1299	0.84	4.98
1275–1324	0.88	4.48
1300–1349	0.64	4.04
1325–1374	0.45	3.96
1350–1399	0.41	4.35
1375–1424	0.34	4.20
1400–1449	0.27	4.89

observes, horses seem first to have been utilized for farm traction on the light, dry soils of the extreme north-west of Norfolk. They are recorded in significant numbers here in the reign of Henry I on the Ramsey Abbey demesnes of Brancaster with Deepdale, Ringstead, and Holme-next-the-Sea.[24] As early as the mid-thirteenth century demesnes in this locality had converted to all-horse teams and in the county as a whole horses already significantly outnumbered oxen. In this respect Norfolk was far ahead of the rest of the country. At a national level oxen consistently outnumbered horses by at least four to one, and although horses steadily gained in relative numbers down to the mid-fourteenth century, they lost ground thereafter. In Norfolk, however, the advance of the horse experienced no such setback: demesnes converted first from all-ox teams to mixed teams, and then from mixed teams to all-horse teams. By the middle of the fifteenth century it was horses which outnumbered oxen by almost four to one. The stages by which this was achieved are shown in Figure 1; the areas which were most resistant to change tending to be those which were remotest from the initial source of innovation, especially where forage was abundant and soils were relatively heavy. Thus oxen remained in a majority on the demesnes of the Norfolk fenland throughout this two-hundred-year period, and they long retained an important role on the heavy soils of the south-east of the county, where mixed teams persisted throughout. Elsewhere, however, the horse was consistently in the ascendant and as its use spread so plough-teams got smaller. Towards the end of the fourteenth century, when demesne profits began to be squeezed by rising wage rates and depressed corn prices, small two-horse and one-man teams began to become a common sight on demesnes in all the lighter soil districts. It is probably no coincidence that in the mid 1390s Chaucer chose to sit Oswald, the reeve of Bawdeswell:

upon a ful good stot, that was all pomely grey and highte Scot.[25]

The figure of the reeve reminds us, as Langdon demonstrates quite emphatically, that the horse was much more a peasant than a demesne animal. Indeed, it is the medieval peasant, for long regarded as too backward and under-capitalized a creature to have made much contribution to the development of agriculture and the raising of its productivity, who emerges from this study as technologically the most active member of medieval society. Some recent research has suggested that this might have been the case but until now the necessary hard evidence has been lacking.[26] Patently, the economics of peasant small-holdings were very different from those of seignorial demesnes, hence the differential which is apparent from the very beginning in the respective rates at which the work horse was adopted. Whether or not the same applied to other aspects of contemporary

23 The results for England have been recalculated from Langdon's original, unpublished data. I am very grateful to him for supplying these to me. These figures indicate a proportion of working horses *circa* 1300 lower than that calculated by Langdon and, contrary to his impression, some reduction in work-horse levels after 1350.

24 *Horses, Oxen and Technological Innovation*, p 43.

25 J Winny (ed), *The General Prologue to the Canterbury Tales*, Cambridge, 1966, p 70, lines 617–18.

26 Campbell, *op cit*, pp 39–41. The view that output per acre on peasant holdings was less than that obtained by a well-managed demesne in the same locality, is stated most explicitly in Postan, *op cit*, 1966, p 602.

technology now becomes an open question and it is time that it was addressed. Certainly, a more innovative, dynamic, and productive peasant sector would go a long way towards explaining how medieval demographic and urban growth managed to proceed as far as they did; for it was peasant cultivators who tilled the greater part of the land in medieval England.

For Langdon, all of these developments – technological change, the emergence of regional variation, and the greater innovativeness of the peasantry – are ultimately explicable in terms of changes in the market economy. Lords and peasants were both increasingly drawn into this economy during the twelfth and thirteenth centuries and this seems to have been the incentive which promoted the substitution of the quicker but costlier horse for the cheaper but slower ox. The rise of horse-hauling proceeded concurrently with the rise of the market economy, since horse haulage increased the speed and range of market transactions. The resultant increased circulation of goods should have been a direct stimulant to the economy and undoubtedly helped to sustain the urban growth which is such a conspicuous feature of the period. In the case of horse traction, this was associated with increasing regional differentiation and the evolution of farming systems of differing degrees of intensity. This is symptomatic of a market system of growing sophistication in which land use and its intensity are increasingly a function of economic rent. The progressive diffusion of the horse, therefore, derived not just from the expansion of the economy but from the fact that the economy was also becoming more integrated and complex.

IV

The problem is to establish the precise contribution which greater use of horses made to this general economic process of expansion and elaboration. On the face of it the economic benefits of horse haulage seem clear enough. Other things being equal the more rapid transit of goods should have reduced transport costs, extended the sphere of the market, and increased the rate of circulation. The only problem is that the heavy, oat-fed cart horse favoured on many demesnes was extremely costly to maintain. On the estates of Peterborough Abbey, for instance, Dr Biddick has calculated that in the opening decade of the fourteenth century more was invested in cart horses and transport (fodder, shoeing, maintenance of carts, wages of carters etc) than was made in wool sales – the abbey's principal cash crop – from its flock of 4000–9000 sheep.[27] Of course, on many demesnes and virtually all peasant holdings ordinary working horses served the dual function of traction and haulage, and this would have helped to keep transport costs down, but the fact remains that improved haulage was frequently only obtained at a high financial price.

When it comes to the advantages of horse traction Langdon is at something of a loss, since he can find little evidence that greater use of horses led to an improvement in agricultural productivity. Improved speeds of ploughing (the horse could work up to 50 per cent faster than the ox) may have led to some reduction in costs and hence an improvement in profit margins, since the same amount of work could be done by fewer animals, but there is no evidence that higher yields were ever a significant outcome. Of itself this is hardly surprising, since the method of traction had no direct bearing upon the fertility of the soil. Had Langdon considered the issue of productivity in all its aspects, however, he would have realized that one of the main benefits of replacing oxen with horses was the greater intensity

27 K Biddick, *The Economy which was not One: Pastoral Husbandry on the Estate of Peterborough Abbey*, forthcoming.

of cultivation which it allowed. One of the most effective ways of expanding arable output was not to raise yields but to crop the land more frequently.[28] This meant reducing fallows to a bare minimum and utilizing them solely as a means of cleansing the land of weed growth (by means of repeated summer ploughings). Since fallows no longer supplied forage the cultivation of fodder crops became unavoidable. Because this imposed a significantly increased workload on the labour force it became important to convert that fodder into traction with the maximum degree of efficiency; hence the substitution of the horse for the ox.[29] Such a changeover was also encouraged by the fact that the greater intensity of cropping meant a much more demanding ploughing schedule with, often, a major seasonal imbalance between autumn and spring. Intensive arable regimes such as these had evolved in both Norfolk and Kent by the end of the thirteenth century and in both areas demesnes employed either mixed or all-horse teams.[30]

Furthermore, grain output is not the only element in agricultural productivity, and it is important not to overlook the gains which may have accrued to the livestock sector from greater use of the horse. Most medieval farms were mixed farms. Except on the very smallest holdings animals were essential to provide haulage and traction and resources accordingly had to be devoted to their upkeep as well as to the maintenance of breeding stock to supply replacements. Only after these needs had been satisfied could additional animals be kept exclusively for their meat, dairy produce, fleeces or hides. Since horses worked faster than oxen fewer resources therefore needed to be reserved exclusively for the maintenance of the plough. Fodder and especially forage could be diverted to the use of other stock, allowing higher stocking densities (with consequent benefits for the arable in terms of increased manure supplies) and the development of much more specialized forms of livestock production. Langdon, it is true, recognizes this possibility and speculates that it may well account for the prominence of horses in Essex, whose heavy soils might otherwise have led oxen to be preferred.[31] As he points out, London's voracious demand for meat may have encouraged farmers in the Home Counties and East Anglia to sell their ploughing and hauling oxen as meat cattle and to replace them with horses instead. Thereafter, female cattle would have been retained as breeding and milking stock and the unwanted bullocks used to supply the meat demands of the metropolis. This explanation works very well and can be applied to a much more extensive geographical area, as is borne out by a general analysis of the demesne livestock population in the country as a whole.

Figure 2 illustrates the broad regional variations which existed in the composition of demesne livestock during the period 1250–1349, on the basis of a national sample of 741 demesnes. Much of the relevant data was supplied by Langdon himself and this has been supplemented by additional material drawn from various published and unpublished sources (including original account-roll information relating to 126 demesnes in Norfolk).[32] Map A shows the number of oxen per horse whilst

28 M Overton, 'Agricultural Revolution? Development of the Agrarian Economy in Early Modern England', Chapter 4 in A R H Baker and D Gregory (eds), *Explorations in Historical Geography: Interpretative Essays*, Cambridge, 1984, pp 125–7; B M S Campbell, 'Arable Productivity in Medieval England: Some Evidence from Norfolk', *Jnl Econ Hist*, XLIII, 1983, pp 379–404.

29 For an elaboration of this point see E Boserup, *The Conditions of Agricultural Growth*, London, 1965, pp 35–39.

30 Campbell, *op cit*, 1983 ('Agricultural Progress'); R A L Smith, *Canterbury Cathedral Priory*, Cambridge, 1943; Mate, *op cit*, 1985.

31 *Horses, Oxen and Technological Innovation*, pp 261–2; for the involvement of one Essex community in livestock production for the metropolitan market see McIntosh, *op cit*.

32 Again, I am grateful to Dr Langdon for making his original research notes available and to G Alexander, J Orr, and J Power for research assistance. A fuller discussion of these and other results will appear in my forthcoming book.

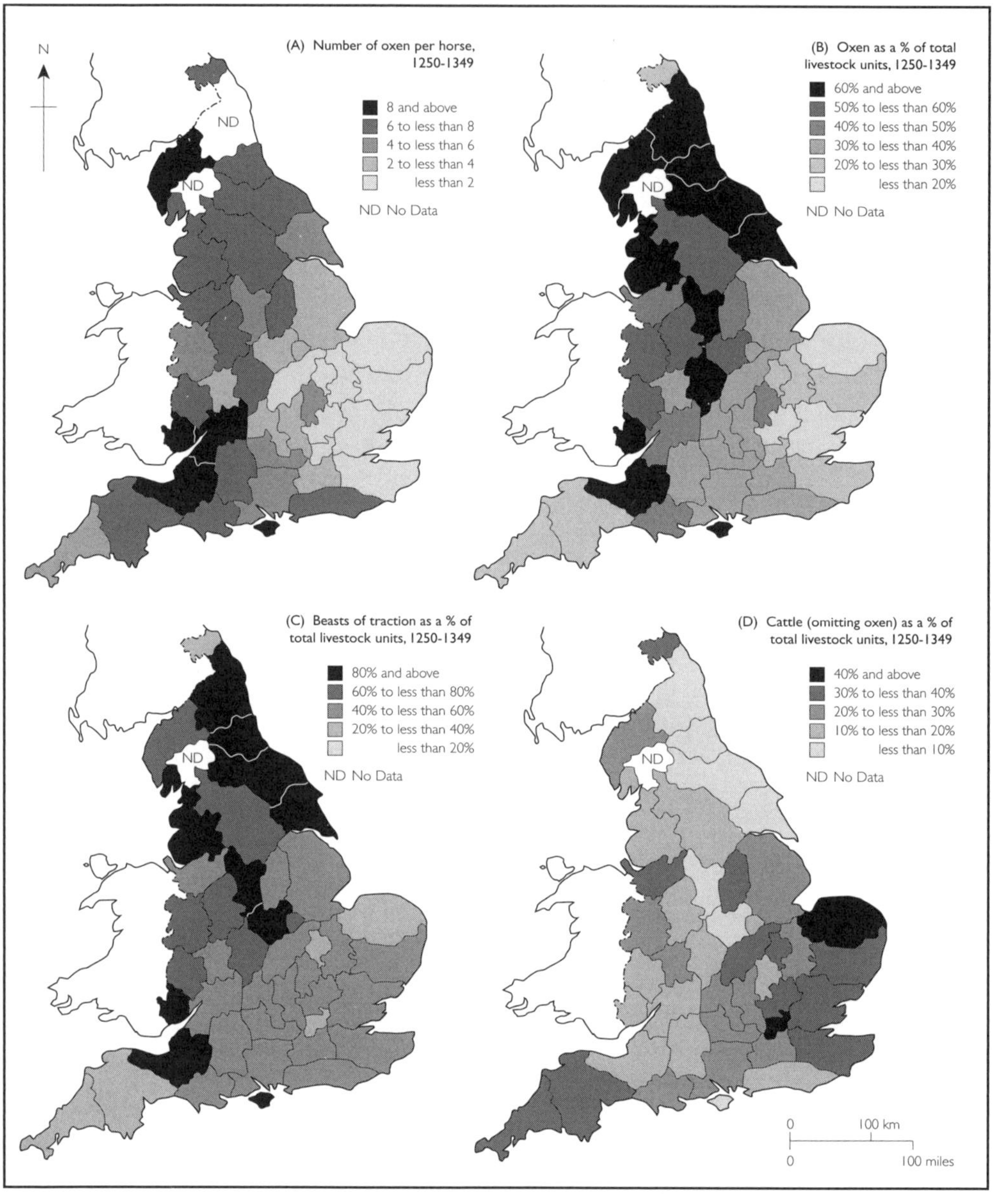
N
(A) Number of oxen per horse, 1250-1349
8 and above
6 to less than 8
4 to less than 6
2 to less than 4
less than 2
ND No Data
ND
(B) Oxen as a % of total livestock units, 1250-1349
60% and above
50% to less than 60%
40% to less than 50%
30% to less than 40%
20% to less than 30%
less than 20%
ND No Data
(C) Beasts of traction as a % of total livestock units, 1250-1349
80% and above
60% to less than 80%
40% to less than 60%
20% to less than 40%
less than 20%
ND No Data
(D) Cattle (omitting oxen) as a % of total livestock units, 1250-1349
40% and above
30% to less than 40%
20% to less than 30%
10% to less than 20%
less than 10%
ND No Data
0
100 km
0
100 miles

FIGURE 2

Maps B, C, and D show, respectively, oxen, beasts of traction, and cattle (omitting oxen), as a percentage of total livestock units (the latter calculated using a weighting of 1.0 for horses, oxen, and cattle, and 0.1 for sheep and swine).[33] The patterns thus revealed should not be regarded as definitive, but they do suggest a number of highly significant relationships. In particular, it was in the most horse-dominated counties – notably Norfolk, Essex, and Hertfordshire – that oxen were least in evidence and beasts of traction (ie horses plus oxen) accounted for the smallest proportion of total livestock. Non-working animals made up over 60 per cent of all demesne livestock in these counties, with breeding and dairying herds especially to the fore.[34] There were other parts of the country where the ratio of livestock to crops was higher, but very few where non-working animals assumed such relative importance. East Anglia and the Home Counties are popularly perceived as arable districts *par excellence*, yet at the beginning of the fourteenth century they supported highly intensive mixed farming systems in which cattle assumed a unique prominence. In this respect, the outstanding position of cattle in the livestock economies of Middlesex and Norfolk speaks eloquently of the influence of urban demand from London and Norwich.[35] The ox may have dominated in the north and west, but the south and east was the preserve of the horse and cow. Indeed, this regional pattern suggests that the more pastoral counties of the extreme north-west and south-west were serving as reservoirs of working animals for those counties further south and east which could not satisfy their own breeding requirements.[36]

Contrary to Langdon's conclusion, adoption of the horse does therefore seem to have allowed the evolution of more intensive and productive mixed husbandry systems, with consequent benefits for the economy. Such systems naturally incurred considerable costs; hence they only developed when justified by the prevailing level of economic rent.[37] This helps to explain why they were so clearly orientated towards the main centres of urban demand, as well as why diffusion of the horse remained so geographically circumscribed. Nor did the productivity benefits of horse traction necessarily end there, for at least as important as the physical productivity of the land was the relative productivity of labour.

In all pre-industrial economies the limits to economic growth were set by the productivity of labour in agriculture. It was output per person rather than output per acre that determined the release of resources – food supplies, raw materials, labour, and capital – to other sectors of the economy. Direct measures of labour productivity are hard to obtain (although demesne accounts contain much of the necessary information); hence indirect measures are normally preferred. Of these, the level of urbanization is the one most

33 These weightings are based on those used by J T Coppock, *An Agricultural Atlas of England and Wales*, 1964, p 213; J A Yelling, 'Probate Inventories and the Geography of Livestock Farming: A Study of East Worcestershire, 1540–1750', *Trans Inst Brit Geog*, 51, 1970, p 115; an equivalent method is used in R C Allen, *The 'Capital Intensive Farmer' and the English Agricultural Revolution: A Reassessment*, Department of Economics, University of British Columbia, Discussion Paper No 87–11, 1987, pp 27–33. For alternative weightings see J Z Titow, *Winchester Yields*, Cambridge, 1972.

34 An analysis of herd demography in these counties reveals a strong bias towards female, adult animals (Campbell, forthcoming).

35 In Norfolk demesnes with well-developed cattle herds were especially prominent in a belt of country approximately 5–15 miles north of Norwich. Examples include Alderford, Attlebridge, Hainford, Haveringland, Hevingham, Kerdiston, Marsham, and Wroxham.

36 C Skeel, 'The Cattle Trade between Wales and England from the Fifteenth to the Nineteenth Century', *Trans Roy Hist Soc*, 4th ser, IX, 1926, pp 137–8; R Cunliffe Shaw, *The Royal Forest of Lancaster*, Preston, 1956; Davies, *op cit*, pp 115–16. The presence of surplus oxen, reared for trade, may lead Langdon to over-estimate mean plough-team size in some of the counties of the extreme north and west.

37 On the effect of economic rent upon land-use and farming intensity see M Chisholm. *Rural Settlement and Land-use: An Essay on Location*, London, 1962, especially Chapter 2, 'Johann Heinrich von Thunen', pp 20–32.

commonly used. It is on this basis that E A Wrigley has recently demonstrated that English agriculture experienced a marked increase in labour productivity during the late seventeenth and eighteenth centuries, with the result that the urban population rose from 8.0 per cent of the total in 1600 to 27.5 per cent in 1801. No such estimates have yet been made for the Middle Ages, and the uncertainty surrounding all population estimates for this period may mean that they are never possible. Yet the twelfth and thirteenth centuries do afford clear evidence of urban growth and there can be little doubt that society was more urbanized by the end of this period than it had been at the beginning. This is certainly a development to which more widespread use of the horse may have made an indirect contribution. Regardless of whether yields rose, labour productivity in agriculture should have risen, simply because adoption of the horse would have enabled existing tasks to be performed faster. This effect was even more pronounced when the number of ploughmen was reduced from two to one, a development which was already taking place in a few localities at this time and which can be presumed to have been especially characteristic of peasant holdings.

V

The technology of haulage and traction is, for even more reasons that those advanced by Langdon, a subject of central importance to our understanding of the medieval economy. This book tells us much about how that economy developed and affords new insights into the process of medieval technological change. Along the way it also has a lot of useful things to say about such secondary matters as the incidence of coaration, the size of plough teams, types of plough, and the various forms of farm vehicle and their distributions. It represents a major new methodological departure and draws attention to several major areas where further research is needed. Obviously, the whole question of livestock productivity should be high on future research agendas.[39] So, too, should urban food supply in general and interregional trade in animals in particular, since there is a very real possibility that certain aspects of livestock production were organized at a national level, with the major centres of urban demand the ultimate focus. Moreover, much more needs to be known about regional farming systems and the place of haulage and traction within them. Several technical mysteries also remain, notably whether reductions in plough-team size and the eventual introduction of one-man ploughs were contingent upon further refinements in plough design and the breeding of stronger animals. Finally, Langdon's data are tantalizingly thin for the later Middle Ages, a period which may have witnessed significant changes in the use of horses and oxen. During the fifteenth century some farms, he believes, went over completely to horses whereas others bolstered the use of oxen. Since this runs counter to expectations, and to the trend implied by the ox:horse ratio given in Table 1, it would clearly repay closer investigation. The potential for further research is therefore considerable and it is a tribute to the scholarship of this book that this should emerge so clearly.

38 E A Wrigley, 'Urban Growth and Agricultural Change: England and the Continent in the Early Modern Period', *Jnl Interdisciplinary Hist*, XV, 1985, pp 683–728. Also, R C Allen, *The Growth of Labour Productivity in Early Modern English Agriculture*, Department of Economics, University of British Columbia, Discussion Paper No 86–40, 1986.

39 The only systematic treatment of medieval livestock husbandry remains R Trow Smith, *A History of British Livestock Husbandry to 1700*, London, 1957.

The diffusion of vetches in medieval England[1]

The long-standing debate about the technological proficiency of medieval agriculture has been lent new impetus by recent research.[2] Detailed study of one region, east Norfolk, has demonstrated that, where soils, location, and social structure were all favourable, remarkably advanced and productive arable farming systems could and did evolve during the course of the thirteenth century.[3] On a larger geographical scale Langdon has shown that from the mid-twelfth century an increased usage of the horse began to have a significant impact upon traction and haulage.[4] Substitution of the horse for the ox in ploughing made particular progress on peasant holdings and allowed an intensification of arable production, the development of more specialized types of livestock husbandry, and an improvement in labour productivity. Its effect upon road transport was still greater. Not only was the horse more widely adopted in carting than in ploughing, but the economic dividends of this substitution were more immediate and direct: it accelerated the circulation of goods and extended the range of market transfers. In short, it was integral to the growing commercialization of the economy and for that reason made most progress in those parts of the country—notably East Anglia and the south-east—which were most deeply involved in the burgeoning market nexus.[5]

These developments in agriculture are symptomatic of an emergent rural-urban complex and this ties in with what is now known about the scale of late thirteenth-century English urbanism. Thus, Keene has shown convincingly that London *c.* 1300 was possibly the second largest European city north of the Alps, with a population of perhaps 90-100,000.[6] Winchester and Norwich have also been shown to have been much more populous than was formerly thought.[7] These substantial urban populations cannot but have

[1] I am grateful for research assistance received from Gill Alexander, Jenitha Orr, and John Power. Part of the research upon which this paper is based was undertaken whilst in the tenure of a Personal Research Fellowship of the Economic and Social Research Council.

[2] The case for technological progress in medieval agriculture is most strongly stated by White in his *Medieval technology*; see also Hallam, *Rural England*. Nevertheless, it is the contrary view which is implicit in much writing on the subject, most notably Fussell, 'Social change but static technology'; Titow, *English rural society*, and Postan, *The medieval economy*.

[3] Campbell, 'Agricultural progress'.

[4] Langdon, *Horses, oxen and technological innovation*; *idem*, 'Horse hauling'.

[5] Britnell, 'The proliferation of markets'; Biddick, 'Medieval English peasants'.

[6] Keene, *Cheapside before the Great Fire*.

[7] At the opening of the fourteenth century Winchester ranked about twelfth in the urban hierarchy and had a population of approximately 10-12,000; Keene, *Winchester studies, 2: survey of medieval Winchester*, vol. I, pp. 366-70. For a revised estimate of Norwich's population from the tithing roll of 1311-33 see Rutledge, 'Immigration and population growth'.

had a major influence on the pattern of food production over a very extensive area, eliciting sophisticated networks of supply and stimulating greater specialization of production, with all that this implies for the process of technical change.[8] Such a conclusion would appear to accord well with the latest econometric assessment of the European economy at this time. According to Persson and Skott the thirteenth- and early fourteenth-century economy possessed a genuine potential for sustained growth.[9] In their view, demographic expansion did not outstrip economic developments but was intimately associated with crucial changes in property relations, increased dependence upon the market, and a mounting pace of technical progress.

Nevertheless, before the traditional picture of a technologically inert and substantially self-sufficient agricultural sector is discarded for one emphasizing increasing technical and economic sophistication, it is important to take account of Fox's carefully considered verdict on the adaptability of the midland commonfield system during this period.[10] In his view, and contrary to established opinion, there was no substantial conversion of two-field to three-field systems at any time between *c.* 1200 and the Black Death. To all intents and purposes the intensity of cropping in these commonfield communities remained fixed at levels commensurate with late twelfth-century population densities. The midland system was essentially a low-intensity solution to the problems of reconciling arable and pastoral husbandry where permanent grassland was scarce and land resources were decidedly limited. All medieval husbandry systems were perforce mixed: not only were livestock essential for haulage and traction, but they were the principal source of fertiliser. Progressive expansion of the arable at the expense of the pasture, so that the area devoted to cereals was maximized at the expense of all else, might therefore, as Postan long ago pointed out, have jeopardized the long-term productivity of the soil.[11]

The dilemma confronting medieval agriculturalists was how to feed a growing population without incurring seriously diminishing returns to both land and labour. In the most ecologically circumscribed commonfield communities in the counties of the south-east midlands it is possible that this dilemma was never satisfactorily resolved, for it is in precisely these counties that the *Nonarum Inquisitiones* reveal the clearest signs of economic distress and demographic retreat in the early 1340s.[12] Not until the sixteenth and seventeenth centuries, when new types of fodder crops and lay grasses were introduced, were these commonfield communities able to proceed to a higher plane of intensity. Nevertheless, fodder crops were not unknown in the Middle Ages and the conversion of cornland into fallow pasture every second or third year was by no means the only way of catering for the needs of livestock in a predominantly arable system of husbandry.[13] Medieval

[8] For an historical analogy see Fisher, 'The development of the London food market'. On the patterns of land-use induced by urban markets in an essentially pre-industrial economy see Chisholm, *Rural settlement and land-use*, especially ch. 2, 'Johann Heinrich von Thünen'.

[9] Persson and Skott, *Growth and stagnation in the European medieval economy*.

[10] Fox, 'The alleged transformation'.

[11] Postan, 'Medieval agrarian society in its prime', in *idem*, ed., *The Cambridge economic history of Europe*, vol. I.

[12] Raftis, *Assart data and land values*; Fox, 'Some ecological dimensions of medieval field systems'; Baker, 'Evidence in the *Nonarum Inquisitiones*'.

farmers may have lacked sainfoin, clover, and turnips, but they did not lack ingenuity, and the growth of the market presented them with an important range of new opportunities. It is in this context that it is instructive to consider one of medieval England's lesser crops, whose history has hitherto passed largely unsung and uncelebrated.

I

Vetches, like the peas and beans with which they were frequently mixed, are a member of the legume family. Unlike peas and beans, however, they were grown exclusively for fodder and were normally fed unthreshed to horses or threshed to working cattle. As Tusser points out, they were particularly valuable in arable districts where meadowland was scarce.[14] In common with other legumes they also helped restore the nitrogen content of the soil which was so essential for plant growth.[15] Their place within rotations was therefore a potentially important one. When sown between successive corn crops they both conserved soil fertility and helped to choke out weed growth. Alternatively, they assisted in establishing a rich and substantial sward when sown immediately before the fallow, as they did in convertible husbandry systems when sown prior to the lay.[16] Nevertheless, throughout the medieval period they were always of secondary importance to peas and, on a national scale, it is unlikely that they ever accounted for more than 4 per cent of the total cropped acreage.[17] In spite of this, they do constitute an exceptionally clear example of technological diffusion,since they were grown in increasing quantities and over a steadily widening geographical area during the course of the thirteenth and fourteenth centuries. It is herein that their principal interest lies. They provide further evidence of the pace and pattern of technical change in agriculture during these two crucial centuries.

In a useful recent note on the early history of vetches Currie has drawn attention to the antiquity of their cultivation in certain parts of south-eastern England.[18] They appear as a specified field crop on seignorial demesnes in the earliest of all surviving manorial accounts, those of the estates of the bishops of Winchester; which indicates that their initial adoption as a field crop is lost in the pre-documentary past. Nevertheless, Currie's contribution leaves unresolved the crucial question of their diffusion to other parts of the

[13] In Norfolk, for instance, there are several well-attested examples of convertible husbandry systems in which land was allowed to revert to pasture for three or four successive years before being returned to cultivation for an equivalent period of time. The proportions sown and fallowed in any one year were much the same as under a two-field system, but the quality of the grass sward was significantly better, the soil had more time to recover its fertility, and ploughing land which was due to be returned to cultivation (to cleanse it of weed growth) entailed only a relatively modest cost in terms of sacrificed grazings. For further details see my forthcoming book, *The geography of seignorial agriculture in medieval England* (Cambridge).

[14] Tusser, *Five hundred points of good husbandry*, p. 57.

[15] On the potential importance of nitrogen-fixing crops, see Chorley, 'Agricultural revolution'.

[16] Farmer was puzzled by the sowing of legumes immediately prior to the fallow on several of the Westminster Abbey demesnes: 'Grain yields on the Westminster Abbey manors', pp. 346-7.

[17] On legumes in general see, for example, the conflicting views expressed in Titow, *English rural society*, pp. 41-2, and Hallam, *Rural England*, pp. 13-4.

[18] Currie, 'Early vetches'.

country which, as will be shown, achieved its maximum spread during the course of the thirteenth and fourteenth centuries. The scale and distribution of their cultivation on the Winchester estates at the beginning of the thirteenth century is certainly commensurate with a fairly early stage in the process of innovation and is more likely to represent this than a deliberate and selective estate-wide policy of cultivation. All five of the Winchester manors that were growing vetches in 1208 lay relatively close to the south coast, with Alresford the most northerly. Of these manors, it was at Fareham—on the coastal plain itself—that vetches appear to have been most firmly established. Thereafter, they spread progressively northwards to many of the other manors on the estate. Their introduction at Harwell in Berkshire came only a decade later and is clearly documented by Currie, but is probably related to a simultaneous penetration of vetches westwards from Surrey along the line of the Thames valley.[19] Whatever the diffusion paths being followed, by 1274 vetches were being cultivated on 31 of the manors on the estate, including at least 17 of the 23 Hampshire manors.[20] Diffusion was patently taking place, but starting earlier and operating over a longer time-scale than Stacey has allowed.[21]

The pattern of innovation identified on the Winchester estate probably replicates that to be found elsewhere in the south east of England. Certainly, if the scale of later adoption is any guide to the initial pattern of innovation, vetches do appear to have spread first along the south coast—from the Isle of Wight in the west to the Isle of Thanet in the east—before penetrating inland. In the late thirteenth and early fourteenth centuries, when manorial accounts are most abundant, a clear distance-decay pattern existed away from the south coast. Only in the Isle of Wight, the southern coastal strip of Hampshire and adjacent portions of Sussex, and much of eastern and northern Kent, were vetches grown on more than a minority of demesnes: only here did they occupy a relatively substantial proportion of the sown acreage, in some localities even displacing peas as the principal legume.[22] The general scale and pattern of vetch cultivation during the period 1250-1349 can be established from a national sample of 464 demesnes largely assembled by Langdon in conjunction with his researches into the utilization of horses and oxen in medieval England.[23] The spatial coverage of this sample is less than ideal given the uneven availability of surviving accounts, but it is sufficient to allow a provisional assessment of the extent of this crop's adoption at the time when direct management of demesnes was at its height.

Demesnes upon which vetches were grown at some time between 1250

[19] Ibid.

[20] Hampshire Record Office, Winchester Bishopric MSS., Bishopric pipe rolls (Eccl. 2), no. 159302.

[21] Stacey, 'Agricultural investment', p. 930, n. 34.

[22] On the history of legume cultivation in coastal Sussex, and the prominent place of vetch, see in particular Brandon, 'Demesne arable farming in coastal Sussex', pp. 123-4. For husbandry elsewhere along the south coast see: Hockey, *The account-book of Beaulieu Abbey*; Mate, 'Profit and productivity'; *idem*, 'Medieval agrarian practices'; Searle, *Lordship and community*; Smith, *Canterbury Cathedral Priory*.

[23] I am very grateful to Dr Langdon for sharing his data with me. The bulk of the accounts are listed in appendix C of Langdon, 'Horses, oxen, and technological innovation', pp. 416-56. They have been supplemented with material of my own drawn from published and unpublished sources.

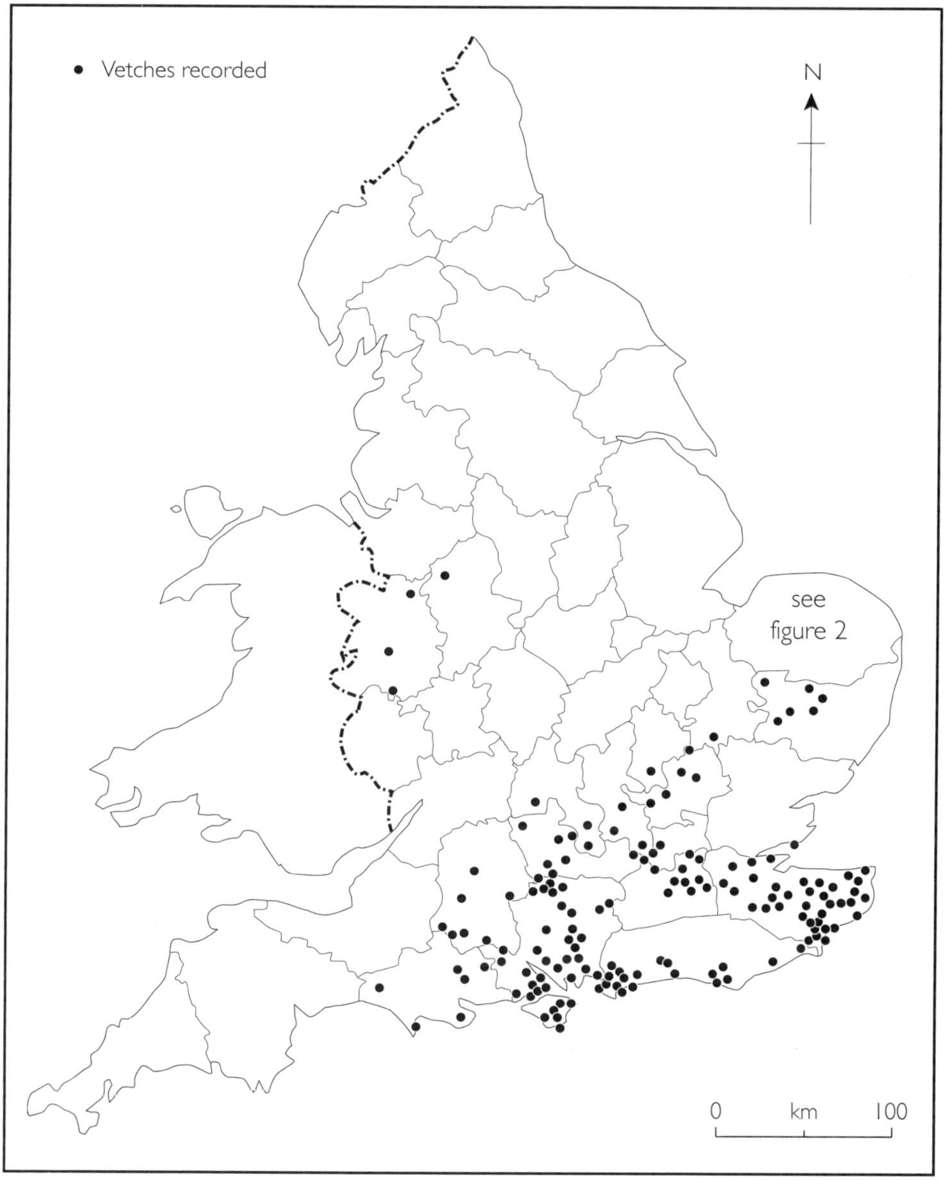

Figure 1. *Demesne cultivation of vetches recorded, 1250-1349*

and 1349 are shown in figure 1. Further detailed research will obviously refine this pattern considerably, but the single most striking feature of the distribution—the virtual confinement of vetch cultivation to the south and east of a line stretching from the Devon/Dorset border in the south west to the Wash in the north east—will doubtless always remain. The only exception is the north-west midlands (an area which was subsequently to witness a pronounced expansion in legume cultivation), where vetches are recorded in Worcestershire at Sutton Newnham in 1284-5, in Shropshire at Lydley Heys in 1312-3, at Adderley in 1318-9, and at Ludlow in 1329-30, and in Staffordshire at Keele in 1312-3.[24] Within the south east, however, the extent of vetch cultivation varied considerably. As already observed, it was most prominent in the Isle of Wight, Sussex, and Kent, where vetches occupied on average over 9 per cent of the total sown area; but in Surrey this proportion fell to 4 per cent, and in all other counties of the south and east it was invariably below 2.5 per cent, and often less than 1 per cent. In these latter counties, vetch cultivation was small in scale and limited in occurrence.

More detailed inspection of figure 1 indicates a close association between vetch cultivation and the chalk and limestone uplands of south-eastern England. In particular a band of demesnes can be identified scattered along the chalk belt, stretching north-eastwards from Berkshire, along the line of the Chiltern Hills through south Buckinghamshire, Hertfordshire, and south Cambridgeshire, into north-east Suffolk and thence, eventually, into Norfolk. The diffusion of vetch cultivation may have followed this path, with Norfolk the last and most north-easterly county to be reached. If so, vetch cultivation may have been a relatively recent innovation in most of the counties north of the Thames. This would certainly fit in with the chronology of vetch cultivation at Harwell, where vetches were first grown in 1218 but only became a permanent feature on the demesne from 1231.[25] No doubt if more mid-thirteenth-century account rolls were available many more specific instances of innovation would be found and the geographical spread of this crop could be traced. Unfortunately, in most of these counties the process of innovation took place too early to be documented. The one major exception is Norfolk, a county with excellent documentation, where vetches only arrived towards the end of the thirteenth century and mostly spread during the course of the fourteenth century.

II

A systematic survey of all extant Norfolk grange accounts allows the introduction and spread of vetches in that county to be traced in some detail.[26] No fewer than 219 Norfolk demesnes are represented by at least one extant account and over the period 1238 to 1450 almost 2,000 separate

[24] P.R.O. SC 6/1074/23; P.R.O. E 358/19; P.R.O. SC 6/967/19. Vetches were effectively a southern crop, a situation which still prevailed in the early seventeenth century: Thirsk, *Agrarian history*, IV, p. 172. Their geographical distribution was restricted by their longer ripening period as compared with peas.

[25] Currie, 'Early vetches'.

[26] Other aspects of Norfolk husbandry revealed by the same data will be discussed in my forthcoming book.

accounts are available in total.[27] The results obtained from these sources are summarized in table 1 and figure 2. As will be seen, notwithstanding the survival of a significant number of accounts from the middle decades of the thirteenth century, including several from as early as 1238-40, it is not until 1281 that there is unequivocal mention of vetches as a field crop. In that year what looks like a trial plot of 0.25 acres was sown with vetches on Lord Guydone's demesne at Newton on the chalk uplands of west-central Norfolk. (At the same time 12 acres were sown with peas, a crop which was sown in substantial quantities on many Norfolk demesnes.)[28] Six years later, in 1287-8, vetches were also being grown on Lord Thomas de Ingaldesthorpe's demesnes at Raynham and Wimbotsham, the first barely half a dozen miles to the north of Newton, the second just over a dozen miles to the west, on the edge of the peat fen.[29] From both these demesnes there are earlier accounts (for 1284-5 and 1286-7 at Raynham and 1276-7 and 1284-5 at Wimbotsham) in which vetches do not appear so it is quite possible that 1287-8 was the first occasion on which they were sown. In both instances the acreages in question were relatively substantial—10 acres at Raynham and 9.5 acres at Wimbotsham—and peas were an established feature of the demesne cropping schedule. Nevertheless, on neither demesne does the new crop appear to have found favour, for vetches are absent from all subsequent accounts.[30] In the same year Henry le Cat purchased 4 bushels of vetch seed to sow on his demesne at Hevingham in the east of the county, although later accounts suggest again that the experiment was not repeated.[31] Shortly afterwards, in 1290-1, an experimental 1 bushel of vetches was sown on the Abbot of St Edmundsbury's demesne at Hinderclay, just across the county boundary into north-west Suffolk, and the experiment was repeated the following year, when 2 bushels of seed were sown.[32] Even so, it was not until the second decade of the fourteenth century that sowings of vetches began to occur with any degree of regularity and even then they only persisted until 1344. The Hinderclay accounts run from 1251 to 1406 and form a remarkably complete series; so the chronology which they reveal can be regarded with considerable confidence.[33]

The pattern of experimentation and hesitant adoption apparent on these demesnes is entirely characteristic of the initial phase of innovation-diffusion.[34] Moreover, the fact that the early adopters were mostly located

[27] No attempt is made to list here all these accounts, which are drawn from the following archives: Public Record Office, Norfolk Record Office, North Yorkshire Record Office, Nottinghamshire Record Office, West Suffolk Record Office, Bodleian Library Oxford, British Library, Cambridge University Library, Canterbury Cathedral Library, Chicago University Library, Harvard Law Library, John Rylands Library Manchester, Lambeth Palace Library, Nottingham University Library, Eton College, Christ's College Cambridge, King's College Cambridge, Magdalen College Oxford, St George's Chapel Windsor, Elveden Hall Suffolk, Holkham Hall Norfolk, Raynham Hall Norfolk, Pomeroy and Sons Wymondham. I am grateful to the authorities concerned for access to these materials.

[28] Cambridge University Library (hereafter C.U.L.), Cholmondeley (Houghton) MSS., Reeves' and Bailiffs' Accounts, 30.

[29] Raynham Hall, Townshend MSS.; Norfolk Record Office (hereafter N.R.O.), Hare 4272/213x1.

[30] Raynham Hall, Townshend MSS.; NRO Hare 4272-4283/213x1.

[31] N.R.O. NRS 14750A 29 D 4.

[32] Chicago University Library, Bacon Rolls 425 & 426.

[33] Ibid., Rolls 405-510.

[34] Hagerstrand, *Innovation diffusion*; Gould, *Spatial diffusion*.

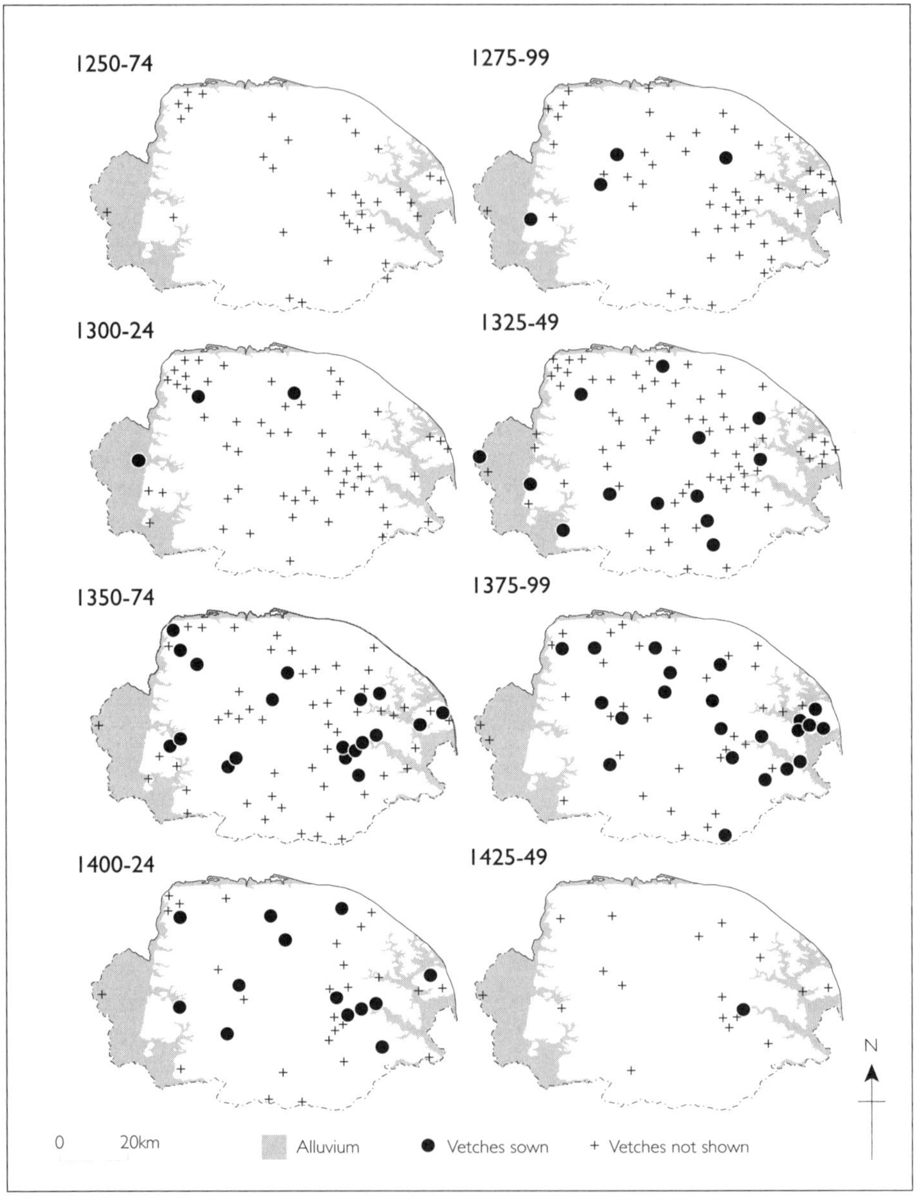

Figure 2 *The diffusion of vetches on Norfolk demesnes, 1250–1449*

Table 1. *The diffusion of vetches in Norfolk, 1238-1450*

Date	*(a) Number of demesnes without vetches*	*(b) Number of demesnes with vetches*	*(c) Total*	*(d) (b) as a percentage of (c)*
pre 1250	17	0	17	0
1250-1274	34	0	34	0
1275-1299	56	3	59	5
1300-1324	58	3	61	5
1325-1349	72	14	86	16
1350-1374	52	19	71	27
1375-1399	35	22	57	39
1400-1424	26	13	39	33
1425-1449	18	1	19	5

Source: Norfolk grange accounts.

towards the west of the county is consonant with the idea that vetches spread into Norfolk from the south and west. By 1300-1 they had appeared at Briston in north-central Norfolk and in the following year they are recorded at Wiggenhall in the Norfolk fenland, a few miles to the north west of Wimbotsham.[35] Thereafter, the pace of innovation gradually quickened as cultivation of the crop steadily diffused to most parts of the county. In the 1320s they made their first appearance at Langham in the north of the county, Bircham in the north west, Great Cressingham in the south west, and Wymondham in south-central Norfolk.[36] During the following decade they were planted at Tivetshall in south Norfolk and West Walton in the heart of the Fens.[37] Finally, from the 1340s, sporadic reference begins to be made to them on demesnes in the east and north east of the county, a locality from which—apart from their fleeting and precocious appearance at Hevingham—they had hitherto been noticeably absent. On the small and newly created demesne of the Prior of Norwich at 'Heythe', just to the east of Norwich, their introduction took place sometime between 1335 and 1345; in 1344-5 they were also being grown on the manor of Cleyhalle at Alderford, a few miles to the north west of the city; and by 1347-8 an experimental 1.25 acres were sown with them at Sloley, 10 miles to the north of the city.[38] Nevertheless, it was to be another two decades before vetches became fully assimilated into the husbandry of this distinctive area. The lateness of the arrival of vetches is especially remarkable for in other respects the agricultural technology employed in north-east Norfolk was precocious: peas had long played a prominent role in rotations and the intensive and irregular cropping regimes had much in common with those found in north-east Kent, where vetches had long been firmly established.[39]

[35] P.R.O. SC 6/931/24; N.R.O. DCN 61/62.

[36] N.R.O. MS 1554 1 C1; P.R.O. SC 6/930/5; Harvard Law Library MS. 85; N.R.O. NRS 14038 28 F6.

[37] N.R.O. WAL 1245/288x1; N.R.O. Hare 4014/210x2.

[38] N.R.O. DCN 61/35-36; N.R.O. Phi/465 577x9; C.U.L., Cholmondeley (Houghton) MSS., Reeves' and Bailiffs' accounts, 33.

[39] For a full discussion of the husbandry of this locality see Campbell, 'Agricultural progress in medieval England'.

Whereas the late thirteenth and early fourteenth centuries represent the phase of initial adoption in Norfolk, it was during the second half of the fourteenth century that vetches became generally known and cultivated throughout the county. From 1355 they were introduced on most of the demesnes of the Prior of Norwich, where they were usually sown mixed with peas.[40] Shortly afterwards they appear in the cropping schedules of demesnes in the highly productive Flegg district, just inland from Yarmouth, where legumes had always been very important.[41] By the last quarter of the century vetches were being sown on at least an occasional basis on over a third of all demesnes in the county. The only area which seems to have been immune to their penetration was the extreme south west of Norfolk, whose light, sandy soils discouraged the field cultivation of legumes. Most of the demesnes that grew vetches at this time either mixed them with peas or sowed them in relatively small quantities. Certainly, the scale of cultivation rarely matched that found in Kent. On the De Clares' demesne at Bircham, where vetches were the only legume sown, the area concerned never amounted to more than 1.75 acres.[42] On the Prior of Norwich's very much larger demesne at Sedgeford just a few miles away the acreage was always below 9 acres and frequently less than 4 acres.[43] At Plumstead, to the east of Norwich and another possession of the Prior, the acreages were somewhat larger—reaching 16 acres in 1404-5—but cultivation was much more sporadic.[44] Indeed, one of the most striking features of the Norfolk evidence is the way in which vetches were rarely consistently sown year in, year out, on the same demesne. On the manor of Panworth in Ashill, for instance, 21 accounts are available from 1320 to 1390 yet vetches are only recorded in three of them, those for 1354-5, 1355-6, and 1357-8.[45] Similarly, at Kempstone, a home farm of Castle Acre Priory, accounts run with interruptions from 1315 to 1449, and yet vetches are only recorded between 1377 and 1410.[46] Much the same situation prevailed on the Cathedral Priory manors of Great Cressingham, Hindolveston, Hindringham, and Martham.[47] On many other manors vetches put in a single fleeting appearance and then were never recorded again. This was the case at Saxthorpe, Thorpe Abbotts, West Walton, Wimbotsham, and Wymondham.[48]

[40] They had been introduced by 1355 at Sedgeford, 1358 at Eaton, 1361 at Hindolveston, 1366 at Hemsby and Newton, 1369 at Plumstead, 1373 at North Elmham, 1376 at Hindringham, and 1388 at Martham: they never seem to have been grown at Taverham. N.R.O. L'Estrange IB 1/4; Bodleian Library Oxford MS Rolls, Norfolk 29; N.R.O. DCN 60/18/38; Raynham Hall, Townshend MSS.; N.R.O. DCN 60/28/6; N.R.O. DCN 60/29/31; N.R.O. DCN 60/10/26; N.R.O. DCN 60/20/31; N.R.O. NNAS 5899 20 D1; N.R.O. DCN 60/35/1-52.

[41] Vetches are first recorded at Flegg in 1363, Hemsby in 1366, Ashby in 1379, Martham in 1388, Burgh in 1390, and Mautby at an unspecified date in the late fourteenth century. N.R.O. Diocesan Est/9; Raynham Hall, Townshend MSS.; N.R.O. Diocesan Est/9; N.R.O. NNAS 5899 20 D1; P.R.O. SC 6/931/28; N.R.O. Phi/491 578x1.

[42] P.R.O. SC 6/930/5-31.

[43] N.R.O. L'Estrange IB 1/4 & 3/4.

[44] N.R.O. DCN 60/29/31-46.

[45] N.R.O. MS 21086 34 E6, NRS 21161 45 A4, NRS 21162 45 A5, NRS 21163 45 A6.

[46] N.R.O. WIS 2-18/163x1, WIS 19-38/163x2.

[47] N.R.O. Supp. 10/12/1982 (R187A) and Harvard Law Library, MS. 85; N.R.O. DCN 60/18/1-62; N.R.O. DCN 60/20/1-39; N.R.O. DCN 60/23/1-25, NNAS 5892-5903 20 D1, NNAS 5904-5915 20 D2, and NNAS 5916-5917 20 D3.

[48] N.R.O. NRS 19650 42 D7; N.R.O. WAL 494/274x6; N.R.O. Hare 4014/210x2; N.R.O. Hare 4274/213x1; N.R.O. NRS 14038 28 F6.

From what has been said it will be plain that table 1 and figure 2, if anything, overstate the importance of vetches on Norfolk demesnes. In any one year the proportion of demesnes actually cultivating vetches must have been considerably less than these figures indicate. Many demesnes experimented with them but few incorporated them into their rotations on a permanent basis and only rarely did they supplant peas—their older and very firmly established cousin—as the principal legume. Vetches were a late arrival on the agricultural scene and this is reflected in the secondary and impermanent place which they occupied in demesne cropping schedules.[49] In the long term, in fact, many demesnes abandoned the cultivation of vetches. This seems to have been particularly the case during the first half of the fifteenth century, when direct demesne management was itself experiencing a decline. References to vetches cease at Costessey and Hindolveston in 1409, at Sedgeford in 1410, and at Great Cressingham in 1413.[50] Nevertheless, cultivation of this crop must have persisted among the peasantry, for it was an established component of Norfolk husbandry in the late sixteenth and early seventeenth centuries when probate inventories provide the next detailed insight into farming practice in the county.[51]

III

Given that the cultivation of vetches was making such advances in fourteenth-century Norfolk, especially after 1350, the question arises whether a similar process was taking place elsewhere. The detailed research has yet to be done which might reveal this, although some general impression can be formed of the direction of developments from another national sample of accounts, drawn from the period 1350-1449 and representing some 242 different demesnes.[52] The size of this second sample is barely half that of the first, with which it should be compared; hence the picture which emerges is less sharply focused. Nevertheless, it is sufficient to show that the experience of Norfolk was somewhat exceptional. Within the confines of the south east, vetches certainly seem to have gained ground. Kent remained at the head of the list of vetch-cultivating counties and continued to devote over 9 per cent of its sown area to that crop, whereas Hampshire, Wiltshire, Berkshire, Oxfordshire and Hertfordshire all registered a relative increase in the acreages sown. In absolute terms the increase was not large but proportionately the area concerned more than doubled. As far as the geographical spread of vetch cultivation was concerned, however, the process of diffusion seems to have been nearing an end. The only new areas in which the crop appears are in the Lincolnshire Fenland at Gedney in 1364-5, in north Buckinghamshire at Turweston in 1387-8, and in Gloucestershire at Chaceley, Bibury, and Horsley in 1368-9, 1371-3, and 1411-2.[53] This

[49] At the height of their popularity in Norfolk, in the last quarter of the fourteenth century, vetches accounted for less than 2 per cent of the total cropped area.

[50] N.R.O. Case 24, Shelf C; N.R.O. DCN 60/18/58; N.R.O. L'Estrange 1B 3/4; N.R.O. Suppl. 10/12/1982 (R187A).

[51] Thirsk, *Agrarian history*, IV, pp. 44, 47.

[52] See above, note 8. Again, I am indebted to Dr Langdon for sharing this material.

[53] P.R.O. DL 29/242/3888; Westminster Abbey Muniments 7828; Worcestershire Record Office, Ref. 009:1 BA 2636 160/92050 and 159/92049 4/7; P.R.O. SC 6/855/8.

penetration of vetch cultivation into Gloucestershire is possibly related to the diffusion of the crop within the west midlands and ties in with a general expansion of legume cultivation in the valleys of the Severn and the Warwickshire Avon. Much of this expansion was accounted for by increased sowings of peas but there are also numerous references to the cultivation of 'pulse' (which may or may not have included vetches); for example, at Avening, Hardwick, and Horton in Gloucestershire, and at Chaddesley, Hewell, Pensham, Pinwin, and Wadborough in Worcestershire.[54] In this context it is worth noting that vetches are again referred to in Shropshire, this time at Cleobury Barnes in 1372-3.[55] These developments in the west midlands are the nearest equivalent to those that have been described in Norfolk.

IV

Because vetches were already established as a specified field crop on seignorial demesnes in certain parts of the country well before the middle of the thirteenth century, their origin and precise provenance lie beyond the realm of documentation. Kent remains the prime candidate as the initial locus of vetch cultivation, simply because it always grew the crop in greater quantities than any other English county. It was also strategically placed in terms of a continental provenance for the crop.[56] Nevertheless, a basic distinction must be drawn between the introduction of an innovation and its more general adoption. Whatever the origin of vetches, the fact remains that during the course of the thirteenth and fourteenth centuries their cultivation became markedly more widespread. The process was long drawn out and less spectacular than the contemporaneous diffusion of peas, to which, of course, it was related.[57] Until now it has been but dimly perceived and inaccurately dated, and the story yet remains to be fleshed out in all its details. But the reality of this diffusion process cannot any longer be denied.[58] Much of it patently took place before the full advent of demesne grange accounts, but it was sufficiently protracted for us to be able to reconstruct its final stages in some detail. It has been fully documented for Norfolk and could be for other locations.

Although the pattern of diffusion has been reconstructed here from the records of demesne husbandry, it is of course a moot point whether demesne managers were the most active pioneers of this new crop. In Norfolk the small scale of initial seed purchases and sowings, and the fact that vetches might be sown on one demesne but not on a neighbour belonging to the same estate, is certainly compatible with the notion that, as with the adoption

[54] P.R.O. SC 6/856/23; Westminster Abbey Muniments 8444; P.R.O. SC 6/856/9; P.R.O. SC 6/1068/11-16; Westminster Abbey Muniments 21092, 22223, and 22285; P.R.O. SC 6/1075/17.

[55] P.R.O. SC 6/965/12.

[56] Currie, 'Early vetches'.

[57] On the evidence of a sample of 1,270 accounts representing some 601 demesnes, legumes as a class (with peas the principal component), rose from 6.3 per cent of the total cropped area in the second half of the thirteenth century to 19.1 per cent in the first half of the fifteenth century.

[58] Unlike Currie, Stacey recognizes that diffusion was taking place but fails to appreciate the full time span that was involved, 'Agricultural investment'.

of horses (much of the harvest of this crop was ultimately destined for horses), it may have been peasant cultivators who were taking the initiative.[59] At any rate, where the adoption of vetches can be traced across a whole estate, as with those belonging to the bishops of Winchester and priors of Norwich, it emerges as a fairly haphazard process. This lack of any synchronization in the timing of adoption between demesnes implies that the decision to cultivate was taken at the manorial level, in response to local knowledge and practice, rather than centrally, at the instigation of the landlord or other high-ranking official.[60] On this showing, the medieval estate does not appear to have been a very effective institution for the transmission of this kind of 'grass-roots' technology. Knowledge about vetches and their uses evidently spread by other paths.

The evolving spatial pattern of vetch cultivation corresponds with that produced by a process of contagious diffusion, spreading northwards and westwards from the initial focus of cultivation in the extreme south east. At any one time, therefore, the scale and extent of vetch cultivation were as much a function of the degree of familiarity with this crop as they were of such other relevant influences as environmental conditions, the institutional characteristics of field systems, or the nature of the prevailing farm economy. As in later centuries, vetches remained very much a crop of the south and east, where they appear to have assumed particular importance in localities relatively poorly endowed with meadow.[61] Although the existence of midland-type commonfield systems was plainly no barrier to their spread, their cultivation nevertheless made greatest progress in areas where institutional arrangements were more flexible, as they were in much of the south east, the lower Thames valley, and East Anglia.[62] Coincidentally, it was the latter areas which were most exposed to commercial influences and in which agriculture accordingly became most specialized and intensive. A growing reliance upon the horse in these areas certainly encouraged the cultivation of vetches, and together these contributed to the evolution of a relatively intensive arable-based pastoral regime. Fodder crops provided the key to this farming system. Only through their more widespread cultivation were London, Norwich, and other leading urban centres of the age able to satisfy their growing demand for meat, dairy products, hides, tallow, and other livestock products.[63]

[59] Langdon, 'Horses, oxen and technological innovation', pp. 172-253. On the enterprise of peasant farmers see also McIntosh, *Autonomy and community*, pp. 136-78.

[60] For the tendency of demesnes to conform with prevailing local husbandry practice see Campbell, 'Arable productivity' and Mate, 'Medieval agrarian practises'.

[61] A nationwide survey of demesne land-use, using extents drawn from 1,179 *Inquisitiones post mortem* for the decade 1300-9, reveals that meadow was especially scarce in many of the leading vetch-cultivating counties: less than 5 acres of meadow per 100 recorded acres of arable in the Isle of Wight, Kent, Hertfordshire, Bedfordshire, Cambridgeshire, Suffolk, Essex, Shropshire, and Staffordshire; and less than 7.5 acres in Hampshire, Sussex, Surrey, Buckinghamshire, Norfolk, Worcestershire, and Warwickshire.

[62] Gray, *English field systems*, pp. 272-402; Baker and Butlin, eds., *Studies of field systems in the British Isles*, pp. 281-429.

[63] On the importance of trade in livestock and livestock products even in a county as emphatically arable in its land-use as Bedfordshire see Biddick, 'Missing links'. For a pioneering study of the medieval trade in hides see Kowaleski, 'Town and country in late medieval England'.

Judging from the speed with which vetches diffused, the pace of agricultural change in the Middle Ages was far from rapid. Yet change there certainly was. This lends at least some support to Persson and Skott's claim that the tempo of technical advance quickened during the thirteenth century and also fits in with their notion of the pace at which this took place:

> The rate of technical progress was very low during the early part of the period (at an annual rate an estimate of around 0.3 per cent is optimistic) and even a quadrupling of the original rate of technical progress does not therefore involve a very large absolute increase in the rate of technical progress.[64]

Nevertheless, one new crop does not make an agricultural revolution and for all the claims that have sometimes been made for them, legumes as a class were no panacea for the problems confronting agriculture at this time. Higher yields and improved stocking levels were not an immediate and invariable result of their cultivation.[65] Rather, the diffusion of vetches should be interpreted as just one ingredient of a more general process of change, whereby the technological resources of medieval farmers were widened and husbandry systems became more specialized and intensive.

That vetches should be a fodder crop is especially significant, for their wider adoption is symptomatic of the improvements which medieval farmers were striving to make in the quality of mixed farming systems. Whether as an instrument of soil management, a means of upgrading fallow grazings, or simply as a protein-rich source of fodder, vetches were of primary benefit to livestock. So much attention has been lavished on grain yields that productivity gains deriving from the more intensive management of livestock have tended to be overlooked. Nevertheless, several of the technological developments which are now known to have been taking place during the thirteenth and fourteenth centuries offered significantly greater benefits for livestock than they did for crops. An increased reliance upon leguminous fodder crops, the development of systems of convertible husbandry, and the substitution of horses for oxen, thus releasing scarce supplies of forage to other classes of animal, are all cases in point.[66] Livestock and crops were combined in different ways and to different degrees in different parts of the country. In part this was a product of ecology and of prevailing institutional arrangements, but increasingly during the thirteenth century it was a function also of market influences and their effect upon prevailing levels of economic rent.[67]By the close of that century urban markets undoubtedly exercised a major influence upon the pattern of agriculture in much of eastern and

[64] Persson and Skott, 'Growth and stagnation', p. 35.

[65] See, for example, the discussions of yields and stocking densities in Campbell, 'Arable productivity', pp. 379-404, and Farmer, 'Grain yields', pp. 331-47. A recent regression analysis of crop yields in seventeenth-century Oxfordshire found that, as a determinant of yields, 'the share of land planted with beans, peas, pulses, and vetches was majestically insignificant throughout': Allen, *The 'capital intensive farmer'*, pp. 20-1.

[66] Although these methods may have helped to conserve soil fertility, speeded up certain tasks, and improved the efficiency of arable husbandry, there is no evidence that they actually raised yields *per se*. Any gains in productivity tended to assume the form of increased stocking densities, improved body weights, and heavier milk yields and fleeces etc. For a pioneering study of livestock husbandry on a medieval estate see Biddick, *The economy which was not one*.

[67] Campbell, 'Towards an agricultural geography'.

south-eastern England. Studies of the agrarian economy of the period have yet to take full account of this; when they eventually do so the diffusion of the humble vetch may not seem so insignificant. As it is, it provides yet further evidence that agricultural technology in medieval England was far from static.

Footnote references

Allen, R. C., *The 'capital intensive farmer' and the English agricultural revolution: a reassessment* (Discussion Paper no. 87-11, Department of Economics, University of British Columbia, 1987).

Baker, A. R. H., 'Evidence in the *Nonarum Inquisitiones* of contracting arable lands in England during the early fourteenth century', *Econ. Hist. Rev.*, 2nd ser., XIX (1966), pp. 518-32.

Baker, A. R. H. and Butlin, R. H., eds., *Studies of field systems in the British Isles* (Cambridge, 1973).

Biddick, K., 'Medieval English peasants and market involvement', *J. Econ. Hist.*, XLV (1985), pp. 823-31.

Biddick, K., 'Missing links: taxable wealth, markets, and stratification among medieval English peasants', *J. Interdisc. Hist.*, XVIII (1987), pp. 277-98.

Biddick, K., *The economy which was not one: pastoral economics on a medieval estate* (Los Angeles, forthcoming).

Brandon, P. F., 'Demesne arable farming in coastal Sussex during the later Middle Ages', *Agric. Hist. Rev.*, XIX (1971), pp. 113-34.

Britnell, R. H., 'The proliferation of markets in England, 1200-1349', *Econ. Hist. Rev.*, 2nd ser., XXXIV (1981), pp. 209-21.

Campbell, B. M. S., 'Agricultural progress in medieval England: some evidence from eastern Norfolk', *Econ. Hist. Rev.*, 2nd ser., XXXVI (1983), pp. 26-46.

Campbell, B. M. S., 'Arable productivity in medieval England: some evidence from Norfolk', *J. Econ. Hist.*, XLIII (1983), pp. 379-404.

Campbell, B. M. S., 'Towards an agricultural geography of medieval England', *Agric. Hist. Rev.*, 36 (1988), pp. 87-98.

Chisholm, M., *Rural settlement and land-use: an essay on location* (1962).

Chorley, P., 'The agricultural revolution in northern Europe, 1750-1880: nitrogen, legumes, and crop productivity', *Econ. Hist. Rev.*, 2nd ser., XXXIV (1979), pp. 71-93.

Currie, C. R. J., 'Early vetches: a note', *Econ. Hist. Rev.*, 2nd ser., XLI (1988), pp. 114-6.

Farmer, D. L., 'Grain yields on the Westminster Abbey manors, 1271-1410', *Can. J. Hist.*, XVIII (1983), pp. 331-47.

Fisher, F. J., 'The development of the London food market, 1540-1640', *Econ. Hist. Rev.*, 1st ser., V (1934-5), pp. 46-64.

Fox, H. S. A., 'Some ecological dimensions of medieval field systems', in K. Biddick, ed., *Archaeological approaches to medieval Europe* (Kalamazoo, 1984), pp. 119-58.

Fox, H. S. A., 'The alleged transformation from two-field to three-field systems in medieval England', *Econ. Hist. Rev.*, 2nd ser., XXXIX (1986), pp. 526-48.

Fussell, G. E., 'Social change but static technology: rural England in the fourteenth century', *History Stud.*, I (1968), pp. 23-32.

Gould, P., *Spatial diffusion* (Washington, D.C., 1969).

Gray, H. L., *English field systems* (Cambridge, Mass., 1915).

Hagerstrand, T., *Innovation diffusion as a spatial process* (Chicago, 1968).

Hallam, H. E., *Rural England, 1066-1348* (1981).

Hockey, S. F., ed., *The account-book of Beaulieu Abbey* (Camden Fourth Series, XVI, 1975).

Keene, D., *Cheapside before the Great Fire* (1985).

Keene, D., *Winchester studies, 2: survey of medieval Winchester*, vol. I (Oxford, 1985).

Kowaleski, M., 'Town and country in late medieval England: the hide and leather trade', in P. Corfield and D. Keene, eds., *Work in towns 900-1900* (Leicester, forthcoming).

Langdon, J., 'Horses, oxen, and technological innovation: the use of draught animals in English farming from 1066 to 1500' (unpublished Ph.D. thesis, University of Birmingham, 1983).

Langdon, J., 'Horse hauling: a revolution in vehicle transport in twelfth- and thirteenth-century England?', *P. & P.*, 103 (1984), pp. 37-66.

Langdon, J., *Horses, oxen and technological innovation: the use of draught animals in English farming from 1066-1500* (Cambridge, 1986).

McIntosh, M. K., *Autonomy and community: the royal manor of Havering, 1200-1500* (Cambridge, 1987).

Mate, M., 'Profit and productivity on the estates of Isabella de Forz (1260-92)', *Econ. Hist. Rev.*, 2nd ser., XXXIII (1980), pp. 326-34.

Mate, M., 'Medieval agrarian practices: the determining factors?', *Agric. Hist. Rev.*, 33 (1985), pp. 22-31.
Persson, G. and Skott, P., *Growth and stagnation in the European medieval economy* (London School of Economics and Political Science, S.T.I.C.E.R.D. economics discussion paper, 1987).
Postan, M. M., 'Medieval agrarian society in its prime: England', in *idem.*, ed., *The Cambridge economic history of Europe*, vol. I (Cambridge, 1966), pp. 548-632.
Postan, M. M., *The medieval economy and society* (1972).
Raftis, J. A., *Assart data and land values: two studies in the east midlands, 1200-1350* (Toronto, 1974).
Rutledge, E., 'Immigration and population growth in early fourteenth-century Norwich', *Urban History Yearbook*, (1988), forthcoming.
Searle, E., *Lordship and community: Battle Abbey and its banlieu, 1066-1538* (Toronto, 1974).
Smith, R. A. L., *Canterbury Cathedral Priory* (Cambridge, 1943).
Stacey, R. C., 'Agricultural investment and the management of the royal demesne manors, 1236-1240', *J. Econ. Hist.*, XLVI (1986), pp. 919-34.
Thirsk, J., ed., *The agrarian history of England and Wales*, vol. IV, *1500–1640* (Cambridge, 1967).
Titow, J. Z., *English rural society, 1200-1350* (1969).
Tusser, T., *Five hundred points of good husbandry* (O.U.P. edition, 1984).
White, Jr., L., *Medieval technology and social change* (Oxford, 1962).

VIII

Mapping the agricultural geography of medieval England

Bruce M. S. Campbell and John P. Power[1]

The annual farm accounts produced by medieval reeves and bailiffs represent one of the most remarkable compendia of agricultural information ever devised and survive in their thousands for the period 1250–1350. The potential therefore exists to reconstruct the agricultural geography of this intriguing and formative period in very considerable detail. This paper outlines the results of a pilot analysis of a national sample of these accounts. Attention is focused upon the aggregate characteristics of arable and pastoral husbandry and the spatial variations by which these were characterized are investigated using Cluster Analysis, a technique which has been hailed as possessing considerable utility for that purpose.

The task of investigating spatial variations in medieval farming systems has scarcely begun. Indeed, it has even been doubted whether farming regions as such actually existed in this period. R. E. Glasscock, for instance, in a preliminary consideration of the topic in his 1973 contribution to Clifford Darby's *A New Historical Geography of England* concluded that "technology and exchange had not progressed far enough by the early fourteenth century to allow much [agricultural] specialisation".[2] Yet we know from Glasscock's own study of the 1334 Lay Subsidy that economic conditions differed widely across the country and, as he himself observes, these variations are only partially explicable in terms of the physical controls of climate, relief, and soils.[3] Variations in the nature and intensity of husbandry there must have been and, given the absolute dominance of agriculture within the economy at this time, it is reasonable to suppose that it is these that are reflected in the varying level of assessed lay wealth. This would appear to be borne out by the result of recent research into the period which has emphasized the density and size of the late thirteenth-century population, the great scale of the leading urban communities of the day, the prominence of marketing within rural society at this time, and the technological resources available to medieval cultivators.[4] The economy of late-thirteenth century and early-fourteenth century England was evidently more developed and complex than has hitherto been supposed and a closer investigation of its agricultural geography is therefore timely.

The data available to students of medieval English agriculture is altogether remarkable, insofar as the annual farm accounts of medieval reeves and bailiffs represent one of the most encyclopaedic compendia of agricultural information ever devised. They record income and expenditure, the crops grown and livestock kept, in great detail.[5] Sometimes they survive in long series, such as the famous run of Winchester Pipe Rolls which document direct demesne management on that estate over a period of 245 years, but more usually they survive for just a handful of years, although they are none the less useful for that. The sheer number of extant accounts is something which has only recently begun to be appreciated, partly because their bulk is masked by their dispersal among a large number of public and private archives.[6] For Norfolk alone there are in excess of 2,000 extant rolls containing direct information on agriculture, and there may be as many as 20,000 for the country as a whole, with the lowland counties of the south and east markedly better served than the upland counties of the north and west. The histories of manorial accounting as a procedure and direct demesne management as a practice combine to make the late-thirteenth and early-fourteenth centuries by far the best documented period.[7] Of course the accounts are restricted in coverage to the demesne farms of seignorial lords and, within that sector, are biased towards the larger estates and particularly those in ecclesiastical ownership. On the evidence of the Hundred Rolls the demesne sector probably accounted for somewhat less than a third of the arable resources of the country.[8] As such it is hardly representative of agriculture at large although it is sufficiently distinctive and important to merit investigation in its own right.[9] Indeed, the potential exists to reconstruct the geography of seignorial agriculture in very considerable detail. This is the earliest point in European history at which such an exercise is possible.

This paper presents summary results from a national sample of 700 demesnes and approximately 1,500 accounts drawn from the period 1250–1349. This sample was largely assembled by J. Langdon in conjunction with his comprehensive analysis of the use of horses and oxen in medieval England.[10] For 465 of these demesnes there is complete crop data and for 660 complete livestock data. As a sample it suffers from a number of imperfections. Spatially it is rather uneven, so that for many counties the number of sampled demesnes is in single figures. For counties such as Cumberland, Lancashire, Cheshire, Northumberland, and Cornwall this is unavoidable given the paucity of original documents but for others, such as Shropshire, Herefordshire, and Staffordshire the accounts do exist which could have yielded fuller coverage. The sample is also biased towards demesnes belonging to the larger estates, and especially those of ecclesiastical lords, neither of which can be presumed to be representative of demesne farming as a whole. Finally, the sample is better for animals, with which Langdon was mainly concerned, than for crops. A larger and more systematically compiled sample would rectify some, if not all, of these problems, nevertheless, as it stands the sample is adequate to provide a provisional classification of farming types, to explore their geographical disposition, and, above all, to provide a context for further detailed studies of individual counties, regions, and estates.

Ideally any analysis of farming types and farming regions ought to examine together both arable and pastoral husbandry, yet crops and animals are recorded in very different units and the accounts provide no convenient common denominator whereby they can be equated with each other.[11] In this respect the problems posed by different ages and types of livestock are compounded by the frequent use of customary acres in the measurement of sown area so that ratios of livestock to sown acres cannot always be trusted. For this reason it is desirable to analyse crops and livestock in relative rather than absolute terms (i.e. the percentage of total sown area and percentage of total livestock units) and this further compounds the difficulty of investigating the two together. Various methods of executing an integrated analysis of farm enterprise have been tried but none has as yet proved entirely satisfactory. It is for this reason that these two mutually related aspects of medieval agriculture are here analysed separately. This is done using Cluster Analysis, a technique which has been hailed as possessing considerable utility for the classification of farming systems.[12]

Cluster Analysis

Cluster Analysis is less a defined statistical procedure than a particular approach to examining data which seeks to investigate differences between cases as measured across all the variables of which they are composed. There is no single way of doing this and, in fact, since the 1960s the number of clustering techniques has proliferated. The researcher is thus left with the perplexing problem of deciding which technique to select, a problem which is rendered yet more tricky by the fact that few techniques are appropriate to all types of analysis and data.[13] The choice of technique may thus determine the nature of the result obtained. Extra confidence may nevertheless be derived when consistent results are obtained from a variety of methods (although few published applications actually report the results from more than one technique). Similarly, the choice of cluster groupings can yield different solutions, particularly when examining the geographical distribution of such groupings. Merely to separate sample cases into a pre-determined number of groups, maximizing inter-group differences while minimizing those within them, is inadequate. Without proper consideration of the number of appropriate groups Cluster Analysis may either merge distinct groups, where the number of specified groups is less than that within the data, or create spurious groupings, where the number of specified groups is too great or where no real grouping exists. Some consideration should therefore always be given to what number of groupings most satisfactorily reflects the inherent divisions within the data, even though objective statistical procedures are rarely available for resolving this. Cluster Analysis cannot therefore provide definitive results: indeed, the results obtained should always be evaluated in the context of the data to which they have been applied. Rather it is a powerful exploratory technique, capable of exposing the variations within a data-set and thus of generating hypotheses worthy of further investigation.

Given the problems relating crops with livestock, the results presented here consider livestock and crop distributions separately. Each comprises six variables: horses, oxen, adult cattle, immature cattle, sheep and swine, in the case of livestock; and wheat, rye, barley, oats, grain mixtures, and legumes, in the case of crops. Due to the problems associated with customary measures, and in order to neutralize the influence of manor size, each variable is expressed as a percentage of either total livestock units or the total sown area.[14] Although manor size is undoubtedly an important agricultural factor, to include it in the analysis raised questions as to its precise importance which could not be satisfactorily resolved. Geographically, the results are examined by county and by manor. To facilitate interpretation of these results each cluster is described according to the mean characteristics of its six constituent variables, even though the actual variables entered into the Cluster Analysis were the Principal Component scores derived from these variables.[15] Principal Components Analysis was carried out to orthogonalize (i.e. reduce correlations between variables to zero) the variables entered into the formula measuring inter-case differences and, in the case of the manorial analysis, to reduce the size of the data matrix being considered. Four clustering techniques were used at a county level but only two at a manorial level due to the length of computing time required in analysing such a large number of cases. All four methods—*Ward's, Relocation, Minimum Spanning Tree*, and *Friedman and Rubins*—are available within the CLUSTAN package of clustering routines.[16] With the exception of the last, each was applied to a lower triangular matrix of inter-case differences based on squared euclidean distances and, since the techniques differ in the way they form groups and in the measurement of group differences, the relative advantages and disadvantages of each are worth noting.[17]

Ward's Method falls into the general category of hierarchical clustering techniques. It starts with as many groups as there are cases and at each step merges those which are most similar. As each step is completed the merged cases are averaged to form a new case, until the process stops when the required number of groups has been reached. This has the defect that intra-case differences become subsumed within a series of group averages so that at none but the initial step is an optimum split sought between the individual cases.[18] In some fields of analysis this tendency can be an advantage, as in biology where, for example, samples are assigned to *genera* and then to species and so on reflecting a hierarchical order. It is, however, inappropriate to the study of agriculture, where no such inherent hierarchical order exists and where the researcher is not interested in weighting groups equally with cases but, rather, in classifying cases (i.e. farms) into a given number of groups minimizing intra-group variation.[19] The three remaining techniques—*Relocation, Minimum Spanning Tree*, and *Friedman and Rubins*—all search for an optimum split between cases into a given number of groups. How optimal the solution is normally dependent upon the adequacy of the starting classification entered into the method.

Minimum Spanning Tree considers a great number of starting classifications and is the method least likely to result in a local, or sub-optimal, solution. Unfortunately, it is very expensive in computing time so that in practical terms only sample sizes of around 40 could be considered while searching for two to seven clusters. This method also has the merit of providing a useful indicator of the number of groups likely to be present within the data, namely the *Calinski Harabasz Variance Ratio Criterion* (VRC).[20] The *Relocation* and *Friedman and Rubins* methods are applied either to classifications entered by the researcher or to random classifications and are hence more likely to locate local solutions. Only if similar results are obtained from a variety of starting classifications can there be reasonable confidence that a global solution has been found. The *Friedman and Rubins* differs from the *Relocation* method in the measure of intra-group dispersion which it employs, which seeks to minimize the Wilk's lambda derived from the within-group scatter matrix $\underline{W}$. It also differs in the shape of the clusters which it seeks, which, whilst of a similar shape, are not necessarily spherical. Both the *Minimum Spanning Tree* and *Relocation* methods utilize the Error Sum of Squares to minimize intra-group dispersion and both tend to seek spherically shaped clusters and can impose such a shape even where none exists. It follows that none of these three techniques has a clear advantage over the others, although, where genuine grouping occurs the results of these techniques should correspond closely. Hence it is advisable to compare results obtained from all methods.

County and manorial analysis

Figures 1 A and 1B and Tables 2A and 2C present the results of an analysis at county level of variations in crop and livestock combinations. The county means, which provide the basis of this analysis, are the means of the individual manorial means, which are in turn the means of the annual percentages. The exercise is perforce a crude one: counties do not represent real agricultural units, variations within counties may be as great as the variations between them, and the sample of manors within each county is not always adequate to the purpose. Nevertheless, it enables some cases to be used which have to be excluded from the manorial analysis due to uncertainty over their precise location, helps to compensate for the poor representation of many of the northern and western counties in the manorial sample, and provides a broad overview of regional variation in demesne husbandry. With its relatively small number of cases—just 43 counties—it also enables all four clustering techniques to be tried (see Table 1) whereas for logistical reasons only two of these techniques can be employed in the manorial analysis.

Analysis of the mean percentage cropped in the 39 counties for which there are data resulted in a seven-cluster solution being preferred. Significantly, the VRC increased up to a seven-cluster solution, although whether it actually peaked at that point is unknown since investigation of an eight-cluster solution exceeded computing capabilities. As Table 1 demonstrates, there is a close correspondence between the results obtained from all four method which indicates that a genuine clustering pattern is present. This correspondence is never less than 75 per cent (between the *Friedman and Rubins* and *Minimum Spanning Tree* methods) and reaches approximately 95 per cent between the *Relocation* and *Minimum Spanning Tree* method. Leicestershire and Wiltshire are the only counties not to be placed in the same group by three of the four methods. In this context, the meaning of each of the seven groups does not differ greatly from one method to another. Geographically the most pleasing solution is that afforded by the *Minimum Spanning Tree* method: it is that which is presented in Fig. 1A and Table 2A. The results obtained from analysis of the mean percentage of total livestock units in 42 counties are almost as satisfactory even though livestock combinations are prone to much greater variation from demesne to demesne. In this case only six clusters were considered as there is a change in the trend of the VRC at this point. As Table 1 shows, results from three of the four methods accord fairly closely, with *Ward's* and *Relocation* being closest with over 95 per cent of counties matching. By comparison, *Friedman and Rubins* is consistently the odd one out (with a minimum correspondence of 35 per cent with *Ward's*), probably because unlike the other three methods it does not seek spherically-shaped clusters. Again, the *Minimum Spanning Tree* method provides the preferred distribution, which is shown in Fig. 1B and summarized in Table 2C.

In contrast, only two clustering methods, *Relation* and *Friedman and Rubins*, were applied at a manorial level since application of *Minimum Spanning Tree* and *Ward's* methods proved imcompatible with the space capacity of the available software. The resultant classifications corresponded closely, with almost 80 per cent of manors being assigned to the same group for both crops and livestock. For crops both methods were applied to the first four Principal Component scores, accounting for 90·84 per cent of variation in the six variables. For livestock, however, where the sample size is larger, only the

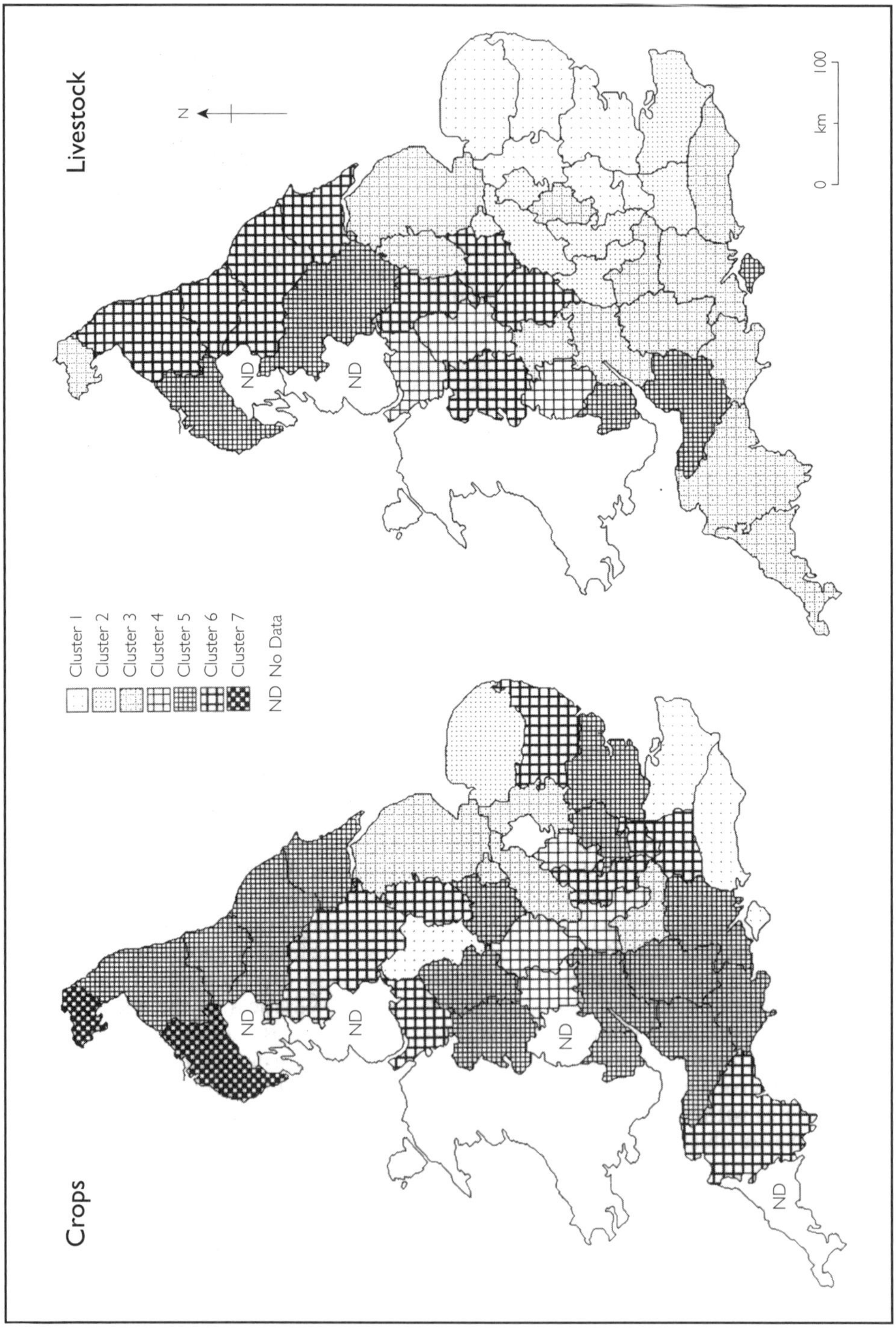

Figure 1. Cluster Analysis by county of percentage crops and percentage livestock, 1250–1349 (Method: Minimum Spanning Tree)

TABLE 1
Cluster Analysis results of % crops and % livestock by county for each of 4 clustering techniques used showing cluster group to which each county was assigned

	% Crops (methods 1–4)				% Livestock (methods 1–4)					% Crops (methods 1–4)				% Livestock (methods 1–4)			
Counties	1	2	3	4	1	2	3	4	Counties	1	2	3	4	1	2	3	4
Bedfordshire	4	4	4	4	4	4	3	3	Lincolnshire	3	3	3	3	3	3	3	2
Berkshire	3	3	3	3	3	3	3	2	Middlesex	6	6	6	6	1	1	1	1
Berwick	7	7	7	7	3	3	3	1	Monmouthshire	5	5	5	5	5	5	5	6
Buckinghamshire	6	6	6	3	3	3	2	2	Norfolk	2	2	2	2	1	1	1	1
Cambridgeshire	3	3	3	3	1	1	1	2	Northamptonshire	3	3	3	3	1	2	2	2
Cheshire	6	6	6	6	2	2	4	4	Northumberland	5	5	5	5	6	6	6	5
Cornwall	0	0	0	0	3	3	3	3	Nottinghamshire	6	6	6	3	3	3	3	1
Cumberland	7	7	7	7	5	5	5	5	Oxfordshire	3	3	4	3	3	3	2	2
Derbyshire	1	1	1	6	6	6	6	6	Rutland	3	3	3	3	1	2	2	2
Devon	6	6	6	6	3	3	3	1	Shropshire	5	5	5	5	6	6	6	6
Dorset	5	5	5	1	3	3	3	1	Somerset	5	5	5	5	5	5	5	5
Durham	5	5	5	5	6	6	6	6	Staffordshire	5	5	5	5	4	4	4	6
Essex	5	5	5	5	1	1	1	2	Suffolk	6	6	6	3	1	1	1	1
Gloucestershire	4	5	5	5	4	4	3	3	Surrey	6	6	6	1	3	3	2	2
Hampshire	5	5	5	1	3	3	3	3	Sussex	1	1	1	1	4	4	3	3
Herefordshire	0	0	0	0	4	4	4	4	Warwickshire	4	4	4	4	6	6	6	5
Hertfordshire	5	5	5	5	1	1	1	2	Westmorland	0	0	0	0	0	0	0	0
Huntingdonshire	0	0	0	0	2	2	2	1	Wiltshire	5	3	5	1	4	4	3	3
Isle of Wight	1	1	1	1	5	5	5	5	Worcestershire	4	4	4	4	4	4	3	3
Kent	1	1	1	1	2	2	2	2	Yorkshire East	5	5	5	5	6	6	5	5
Lancashire	0	0	0	0	0	0	0	0	Yorkshire North	5	5	5	5	6	6	5	6
Leicestershire	4	5	5	6	6	6	6	5	Yorkshire West	6	6	6	6	5	5	6	5

Method key: 1 = *Ward's*; 2 = *Relocation*; 3 = *Minimum Spanning Tree*; 4 = *Friedman and Rubins*

Relocation method could be applied to the first four Principal Component scores (accounting for 92·24 per cent of variation): *Friedman and Rubins* could not be applied to such a large matrix and analysis thus had to be based on the first three Principal Component scores, accounting for 78·76 per cent of variation. It is for this reason, and the fact that *Relocation* gives the most satisfactory geographical distribution, that it was finally preferred. The decision as to what comprised the most appropriate number of clusters proved more problematical as the sheer number of cases precluded utilization of such statistical aids as diagrammatical representation of inter-case distances. In the end a seven-cluster solution was sought for both crops and livestock, on the assumption that this should identify broad groupings within the data without splitting it into an excessive number of sub-groups. Seven clusters are also near the maximum representable on a single map. The statistical characteristics of the groupings according to the *Relocation* method are specified in Tables 2B and 2D and mapped in Figs 2 and 3.

The geography of seignorial arable and pastoral husbandry

The county and manorial analyses of demesne arable husbandry summarized in Table 2 and mapped in Figs 1 and 2 identify a variety of crop combinations

TABLE 2A

Seven cluster solution of crops as a percentage of total sown acreage (Minimum Spanning Tree method applied to 39 counties)

	Wheat		*Rye*		*Barley*		*Oats*		*Mixtures*		*Legumes*		*n*
	x	*s*	*x*	*s*	*x*	*s*	*x*	*s*	*x*	*s*	*x*	*s*	
Cluster 1	31·22	6·16	1·72	1·75	19·04	5·71	24·93	6·51	1·10	1·17	21·99	3·79	4
Cluster 2	15·92	0·00	11·07	0·00	40·66	0·00	15·74	0·00	1·52	0·00	13·50	0·00	1
Cluster 3	30·18	2·12	5·57	1·39	16·63	2·15	21·23	4·87	17·16	3·94	9·23	3·38	5
Cluster 4	41·57	5·19	1·81	1·60	4·64	3·14	17·07	8·98	22·94	6·25	11·99	8·03	4
Cluster 5	43·13	6·22	1·81	1·87	8·10	6·53	37·42	6·80	3·87	3·76	5·67	3·09	15
Cluster 6	25·62	4·69	9·30	2·14	8·12	6·50	47·90	11·61	3·60	4·10	5·47	2·96	8
Cluster 7	20·56	4·20	0·24	0·24	7·33	3·29	71·41	7·73	0·00	0·00	0·47	0·47	2
Overall	34·64	9·99	3·98	3·75	10·76	8·58	35·31	15·67	6·93	8·20	8·34	6·63	39

TABLE 2B

Seven cluster solution of crops as a percentage of total sown acreage (Relocation method applied to 465 manors)

	Wheat		*Rye*		*Barley*		*Oats*		*Mixtures*		*Legumes*		*n*
	x	*s*	*x*	*s*	*x*	*s*	*x*	*s*	*x*	*s*	*x*	*s*	
Cluster 1	24·63	10·81	2·58	4·31	35·27	14·04	12·48	8·14	0·98	2·43	24·06	9·26	61
Cluster 2	11·07	9·72	26·78	10·25	31·26	17·68	21·65	11·39	3·48	6·79	5·77	5·17	42
Cluster 3	27·15	14·04	1·67	4·30	7·12	9·25	18·86	15·50	36·98	13·74	8·22	8·29	53
Cluster 4	34·83	8·08	3·04	4·94	18·50	9·84	28·47	8·76	3·76	5·84	11·41	7·45	113
Cluster 5	44·76	7·84	1·24	2·62	4·38	4·86	44·83	8·12	1·86	3·59	2·94	3·24	126
Cluster 6	69·03	15·09	0·00	0·00	10·74	12·10	12·02	8·55	4·24	7·30	3·93	4·83	30
Cluster 7	13·02	9·39	10·03	11·23	3·54	4·46	68·47	13·14	2·57	7·47	2·36	4·16	40
Overall	33·49	17·68	4·89	9·25	14·94	15·31	31·47	19·23	6·57	12·88	8·64	9·37	465
% Variance	68·17		62·50		55·95		72·32		72·45		54·13		

TABLE 2C

Six cluster solution of livestock as a percentage of livestock units (Minimum Spanning Tree method applied to 42 counties)

	Horses		*Oxen*		*Adult cattle*		*Immature cattle*		*Sheep*		*Swine*		*n*
	x	*s*	*x*	*s*	*x*	*s*	*x*	*s*	*x*	*s*	*x*	*s*	
Cluster 1	22·05	5·18	22·13	7·88	25·53	5·74	12·03	3·04	15·57	4·79	2·69	0·70	6
Cluster 2	19·33	4·45	32·74	5·89	14·75	3·80	11·33	3·97	16·77	4·04	5·08	1·67	7
Cluster 3	9 67	3·66	41·13	7·27	14·17	5·10	9·55	3·93	23·17	8·08	2·32	1·26	13
Cluster 4	6·70	0·98	55·74	6·15	16·09	9·56	5·50	2·90	10·30	7·46	5·68	2·13	3
Cluster 5	8·03	3·42	68·76	10·11	8·98	4·37	5·39	1·25	8·22	5·40	0·62	0·62	5
Cluster 6	16·23	4·04	69·35	4·94	4·30	2·99	3·51	1·68	3·60	4·26	3·02	1·74	8
Overall	13·39	6·60	46·72	18·47	13·53	7·92	8·26	4·38	14·59	9·34	3·00	1·96	42

TABLE 2D

Seven cluster solution of livestock as a percentage of livestock units (Relocation method applied to 660 manors)

	Horses		*Oxen*		*Adult cattle*		*Immature cattle*		*Sheep*		*Swine*		*n*
	x	*s*	*x*	*s*	*x*	*s*	*x*	*s*	*x*	*s*	*x*	*s*	
Cluster 1	11·16	6·99	17·97	10·62	34·01	13·80	29·41	13·27	6·08	8·50	1·33	1·89	100
Cluster 2	18·72	7·83	28·15	13·46	22·81	9·51	19·73	7·34	5·63	6·96	4·98	3·62	116
Cluster 3	11·21	6·45	20·52	11·23	18·27	8·29	13·76	6·65	33·13	8·58	3·10	2·81	132
Cluster 4	9·46	7·87	25·45	13·80	3·02	4·56	2·43	3·80	57·99	13·13	1·65	2·38	75
Cluster 5	24·99	14·81	36·83	19·24	5·75	9·29	3·42	5·66	12·43	14·94	16·58	8·86	40
Cluster 6	52·46	14·53	37·45	17·67	2·11	4·72	1·27	2·95	5·04	12·14	1·66	3·61	43
Cluster 7	15·02	9·17	78·64	13·07	1·54	3·30	1·49	3·61	2·29	6·84	1·03	2·31	154
Overall	16·73	13·60	37·68	26·82	14·00	14·52	11·59	12·53	16·74	20·79	3·24	5·02	660
% Variance	57·14		75·47		67.08		67·04		79·93		53·47		

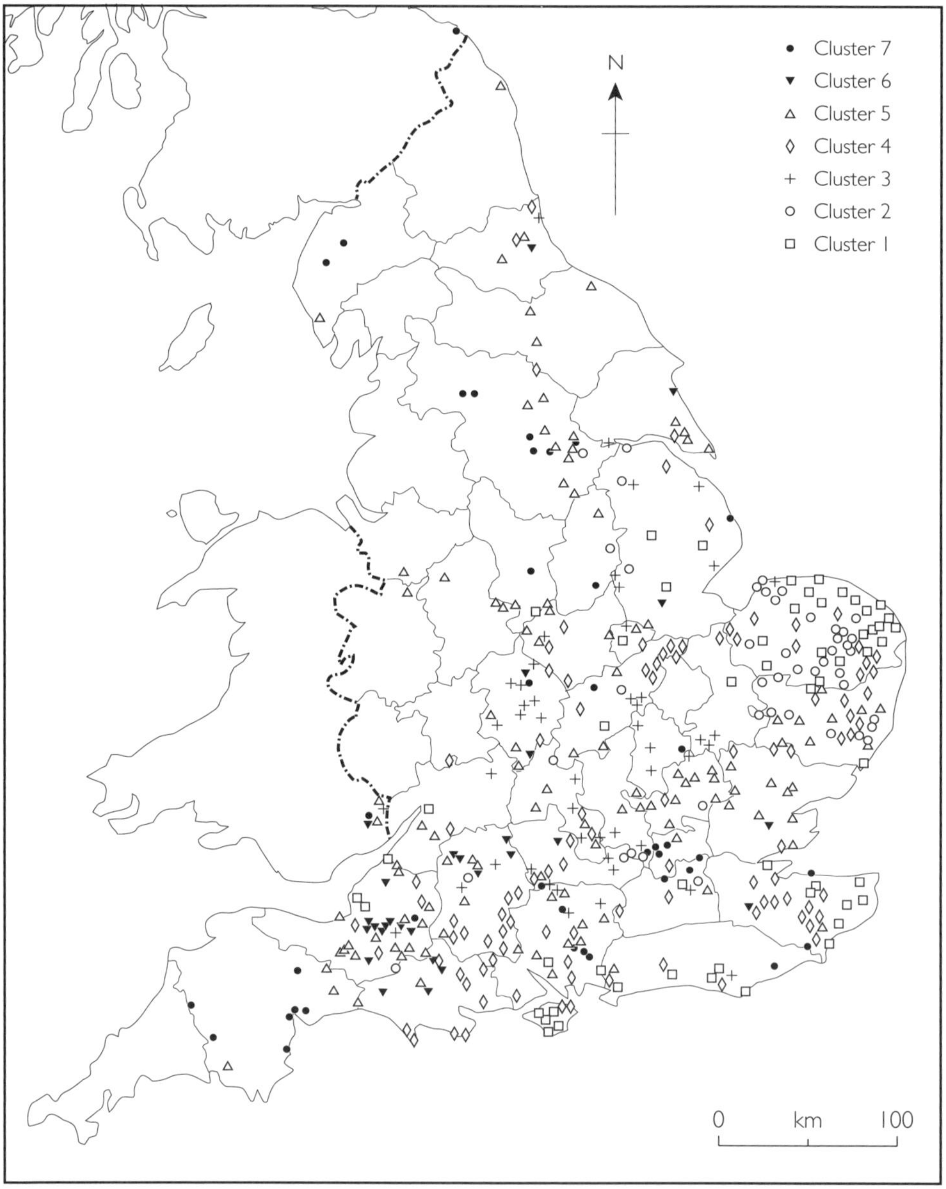

Figure 2. Cluster Analysis of crops as a percentage of total sown area by manor, 1250–1350 (Relocation Method—Seven Clusters).

ranging from the simple to the complex. At one extreme there are combinations in which a single crop predominates, either oats (manorial cluster seven and county clusters six and seven) or wheat (manorial cluster six). The former was typical of much of the north, the north-west, and the south-west of England, as well as of certain of the immediate Home Counties, where large-scale oat cultivation was undoubtedly a response to the commercial opportunities

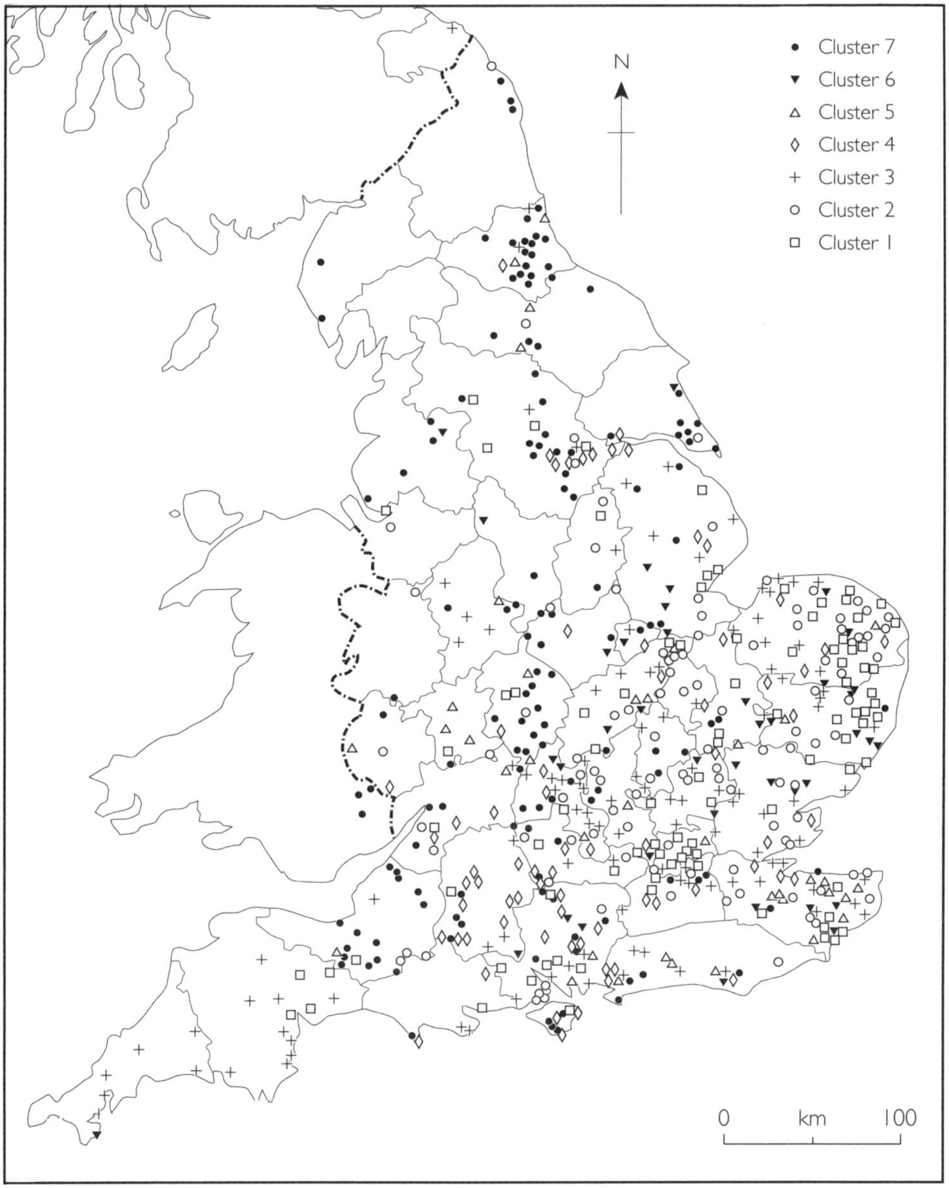

Figure 3. Cluster Analysis of percentage of total livestock units by manor, 1250–1350 (Relocation Method—Seven Clusters).

afforded by the London market (oats being a low-value crop incapable of withstanding high transport costs). The latter was very much a south-western specialism and shows up particularly in Wiltshire, Dorset, and Somerset where wheat was characteristically grown with some spring-corn in a classic two-course rotation.[21] Such relatively unintensive cultivation systems went hand-in-hand with these simple crop combinations. At a stage higher in both complexity and intensity are combinations in which wheat and oats were grown in roughly equal proportions (manorial cluster five and county cluster five). This is the most common combination of all and shows up in no less than fifteen counties and on over a quarter of all sampled demesnes, where it was probably associated with a classic three-course rotation of winter-corn, spring-corn, and fallow. Examples of this specific combination occur in many parts of the country, but they are especially prominent in the north-east, in Essex, Hertfordshire, and adjacent portions of Suffolk and Middlesex, and, to a lesser extent, in certain of the south-western counties.[22]

The remaining cluster groupings are all much more complex and feature a wider range of crops, with increasing prominence being accorded to dredge, barley, and legumes, all of which were spring sown. This transition towards a greater emphasis upon spring crops shows up in manorial cluster four, concentrations of which occur in east Suffolk, central Kent, south Wiltshire and adjacent counties in southern England, and in the east Midlands focusing upon north-eastern Northamptonshire. It also shows up in manorial cluster three, with its distinctive emphasis upon grain mixtures, examples of which are scattered throughout the midland counties. At a county level this emphasis upon grain mixtures is picked up by cluster four, whose constituent counties—Bedfordshire, Berkshire, Warwickshire, and Worcestershire—all lie within the midlands. In fact, it is in the east midlands that the greatest range of different crop combinations is to be found. Lincolnshire and Northamptonshire both furnish examples of all seven manorial cluster types and even Rutland provides examples of four. This is reflected in the wide range of crops represented in cluster three in the county analysis, to which, significantly, these three counties, plus Cambridgeshire and Berkshire, belong. Between them these counties evidently contained some of the most varied arable husbandry in England which is the more remarkable given that they all lay within the bounds of the midland commonfield system.[23] It is, nevertheless, largely outside the limits of that system that the most complex and advanced combinations of all occur (manorial clusters two and one and county clusters one and two). Outstanding in this respect are Norfolk, the south-eastern counties of Kent and Sussex, and the Isle of Wight. It is within these counties that the most complex combination of all is concentrated (manorial cluster one), in which unique prominence was accorded to barley in a four-part combination which also featured wintercorn, oats, and substantial sowings of legumes. Independent study has shown this specific combination to have been associated with a particularly intensive and productive rotational regime.[24]

These results leave little doubt that there was much local and regional variation in the pattern of demesne cropping. Naturally, analysis at a manorial and a county level fails to produce identical clusters but there is a broad measure of agreement between the detailed and the general pictures which they respectively portray. Above all, it is plain that crop combinations were much more complex and diverse in some regions than others. Thus, it is in the eastern and south-eastern counties that the greatest range of different crop combinations occurred, including several which are scarcely represented in other parts of the country. It is here that the spatial differentiation of arable husbandry was most pronounced whereas towards the north and west crop combinations were both simpler and more uniform and methods of cultivation presumably correspondingly less intensive.

In the case of livestock the county and manorial analyses both reveal an even more striking gradation of types (Figs 1B and 3). These progress from relatively simple and unintensive regimes dominated by draught animals, via regimes in which the draught and non-draught components are roughly equally balanced, to regimes in which non-working animals predominated. Generally speaking, oxen and sheep are less intensive than horses and cattle (especially dairy cattle), and in that respect too there is a progression. Demesnes in which draught animals were of overwhelming importance (manorial clusters five, six, and seven) are to be found in all parts of the country but were especially common throughout the north, in parts of the midlands, and in Monmouthshire and Somerset in the south-west (county clusters five and six). This is somewhat paradoxical given the abundant pastoral resources of many of these counties. In part this may represent a genuine emphasis upon the production of surplus oxen for meat and draught and the breeding of horses for sale, but on the bulk of the lowland demesnes which predominate within the sample it reflects the northern practice of stocking arable demesnes with only essential draught animals whilst maintaining breeding and back-up herds on specialist stock farms on the upland margins.[25] These upland vaccaries are seriously under-represented in the sample. In manorial clusters three and four this emphasis upon draught animals is more muted and cattle and especially sheep are correspondingly of greater importance. At a county level this type of mixed livestock economy is represented by clusters three and four. The twelve counties making up the former comprise all the main sheep-farming counties of England, although within this extensive area sheep-dominated demesnes (manorial cluster four) are relatively few in number and confined to localities with an abundance of suitable pastoral resources: for instance, in the vicinity of extensive marshland grazings in south Yorkshire, south-east Essex, and coastal Sussex, on the light soils of the East Anglian Breckland, and, most notably, on the chalk downlands of southern England, especially in south Wiltshire and north Hampshire.[26]

All the remaining county and manorial clusters represent more intensive livestock regimes. This greater intensity is reflected in the significantly greater contribution made by horses to draught power, in the general predominance of sheep over cattle, in a herd structure demographically skewed towards adults (indicative of a specialist interest in dairying), and, utimately, in manorial cluster one and county cluster one, in the subordination of draught to nondraught animals. In fact, manorial cluster one constitutes the most intensive and developed livestock regime of all and, significantly, notwithstanding the region's strong natural advantage as an arable producer, it is in East Anglia and the Home Counties that these demesnes occur in greatest numbers, with notable concentrations in Norfolk and east Suffolk and in the immediate environs of London. Other lesser concentrations occur around the fen edge in Lincolnshire and the Soke of Peterborough, in the vicinity of Romney Marsh in Kent, in southern Hampshire, and in east Devon. Many of the same attributes are also exhibited by manorial cluster two, which is more generally distributed throughout these eastern and southern counties. The net effect at county level is to produce a contiguous block of six East Anglian counties generally characterized by complex and advanced livestock combinations (county cluster one) surrounded by a further seven counties (county cluster two) possessing the same characteristics in less extreme form.

It will be observed that although examples of the individual livestock types can be found in most parts of the country, these local variations and specialisms are subsumed at county level into a regional pattern of remarkable clarity. As with crops, it is in the south and east, and above all in East Anglia, that the most developed, and in the north and west that the least developed, combinations occur. Between these two extremes lies a spatial gradation of pastoral types. Yet in its overall coherence this pattern of broad concentric zones is fundamentally different from that of crops, which is altogether more regional and fragmented. The latter no doubt reflects the differential impact of cheap transportation and together these two distributions imply a great deal about the contrasting economics of arable and pastoral husbandry.

Conclusions and implications

The intention of this paper has been to extend to a medieval context a method of analysing farming systems which has been pioneered in the early modern period by a number of historical-geographers, notably J. Yelling, M. Overton, and P. Glennie.[27] As has been emphasized, different methods of Cluster Analysis produce different results, so the results presented here should not be regarded as definitive. Nor is the sample of data upon which they have been based as comprehensive and systematic as might have been wished. Nevertheless, for all its shortcomings this exercise does establish a provisional classification of arable and pastoral types capable of application to the country as a whole. As such it avoids the limitations of more narrowly based classification schemes whose relevance is effectively restricted to the counties or regions for which they have been derived.[28] Future research will no doubt refine this classification and bring local variations and specialisms into sharper focus and for most parts of the country the volume of extant records is more than adequate to this purpose. In this context, a comprehensive survey of demesne husbandry in Norfolk is currently nearing completion and work has recently commenced on a study of land-use and agricultural production in ten counties within the immediate hinterland of London.[29] For these well-documented parts of the country the potential for further research is considerable and it is only in the extreme north-west and south-west that a genuine deficiency of documentation is likely to inhibit enquiry. This is unfortunate, for it would appear that these remote, upland areas possessed distinctive agrarian economies whose pastoral component may have made a unique contribution to the economy at large.[30]

When the classifications which emerge from this analysis are mapped it becomes clear that, contrary to Glasscock's supposition, seignorial husbandry had become characterized by a considerable degree of spatial differentiation by the late-thirteenth century. The reasons for this spatial diversity of husbandry systems are complex. Varied ecological opportunities are obviously part of the explanation. Yet environmental conditions, in the form of soils, climate, and terrain, conditioned rather than determined the types of farming system which could evolve. Indeed, the intensive livestock economy of much of East Anglia, a region with only very limited environmental advantages for pastoralism, cautions against adopting too deterministic an interpretation of medieval agriculture. Institutional factors in the form of field systems were also important, and helped to determine the intensity with which arable and pastoral husbandry were conducted. Thus it would appear that the most specialized and intensive forms of cropping and livestock management mostly evolved outside the bounds of the most closely regulated field systems. Nevertheless, a considerable variety of husbandry types existed even within the confines of the midland system, partly for ecological reasons but also, due to the influence of market forces. The latter operated at a variety of levels—locally, regionally, and nationally—and gave rise to local specialisms in particular crops and livestock as well as broader variations in the nature and intensity of husbandry. The most important of these markets were urban and of these none was more influential than London: with a population approaching 100,000 towards the end of the thirteenth century, its hinterland must have embraced a subtantial proportion of the country. As J. H. Von Thünen long ago demonstrated, such large central markets exercise a major influence on the pattern of economic rent and thus upon the nature and intensity of land-use.[31] It is therefore tempting to interpret the specialized and highly differentiated agriculture of much of eastern and south-eastern England in the light of such an influence. Certainly, the extent to which the pattern of medieval agricultural production reflected a process of commercialization is a subject which plainly warrants much closer investigation.

Notes

[1] We are grateful to Jenitha Orr and Gill Alexander for research assistance received

[2] R. E, Glascock, England circa 1334, in H. C. Darby (Ed.), *A new historical geography of England* (Cambridge 1973) 167

[3] R. E. Glasscock, *The Lay Subidy of 1334* (London 1975) xxvi–xxix

[4] R. M. Smith, Human resources in rural England, in G. Astill and A. Grant (Eds), *The medieval countryside* (Oxford 1988) 188–212; D. Keene, *Cheapside before the Great Fire* (London 1985); R. H. Britnells, The proliferation of markets in England 1200–1349 *Economic History Review* 2nd series **34** (1981) 209–221; Kathleen Biddick, Medieval English peasants and market involvement *Journal of Economic History* **45** (1985) 823–831; John Langdon, *Horses, oxen and technological innovation: the use of draught animals in English farming from 1066–1500* (Cambridge 1986); Bruce M. S. Campbell, The diffusion of vetches in medieval England *Economic History Review* 2nd series **41** (1988) 193–208

[5] F. B. Stitt, The medieval minister's account *Society of Local Archivists' Bulletin* **11** (1953) 2–8

[6] A summary register of manorial records may be consulted at the National Register of Archives, Quality House, Quality Court, Chancery Lane, London

[7] P. D. A. Harvey, Introduction, part II, accounts and other manorial records, in *Idem* (Ed.), *Manorial records of Cuxham, Oxfordshire circa 1200–1359* (Oxfordshire Record Society, 50, 1976) 12–71

[8] E. A. Kosminsky, *Studies in the agrarian history of England in the thirteenth century* (Oxford 1956) 87–95

[9] The literature on demesne husbandry is substantial; examples include R. A. L. Smith, *Canterbury Cathedral Priory* (Cambridge 1943); H. P. R. Finberg, *Tavistock Abbey: a study in the social and economic history of Devon* (Cambridge 1951); J. A. Raftis, *The estates of Ramsey Abbey* (Toronto 1957); P. F. Brandon, Demesne arable farming in coastal Susex during the later Middle Ages *Agricultural History Review* **19** (1971) 113–134; J. Z. Titow, *Winchester yields: a study in medieval agricultural productivity* (Cambridge 1972); B. M. S. Campbell, Agricultural progress in medieval England: some evidence from eastern Norfolk *Economic History Review* 2nd series **36** (1983) 26–46

[10] Langdon, *op. cit.*, The bulk of these accounts are listed in appendix C of John Langdon, Horses, oxen and technological innovation: the use of draught animals in English farming from 1066 to 1500 (unpubl. Ph.D. thesis, University of Birmingham, 1983) 416–456. We are very grateful to Dr Langdon for making this material available to us. The data-set also includes some additional material drawn from published and unpublished sources

[11] Students of seventeenth-century agriculture have been able to utilize the valuations ascribed to crops and livestock in the probate inventories of the period: Mark Overton, Probate inventories and the reconstruction of agricultural landscapes, in M. Reed (Ed.), *Discovering past landscapes* (London 1984) 167–94

[12] Mark Overton, *Agricultural regions in early modern England: an example from East Anglia* (University of Newcastle, Department of Geography, Seminar Paper 42, 1985)

[13] R. Gnanadesikan, J. R. Kettenring, and J. N. Landwehr, Interpreting and assessing the results of cluster analyses *Proceedings of the 41st session of the International Statistical Institute* (1977) 451–63

[14] To calculate total livestock units the following weightings have been used: horses × 1·0; oxen × 1·2; adult cattle (bulls and cows) × 1·2; immature cattle × 0·8; sheep × 0·1; swine × 0·1. These weightings are based on those used by J. A. Yelling, Probate inventories and the geography of livestock farming: a study of east Worcestershire, 1540–1750 *Transactions of the Institute of British Geographers* **51** (1970) 115, and R. C. Allen, *The 'capital intensive farmer' and the English agricultural revolution: a reassessment* (Univerity of British Columbia, Department of Economics, Discussion Paper No. 87–11, 1987) 27–33. For alternative weightings see J. Z. Titow, *Winchester yields* (Cambridge 1972)

[15] For the problems arising from the closed numbers system used in Principal Components Analysis see Leslie J. King, *Statistical analysis in geography* (New Jersey 1969) 179

[16] D. Wishart, *Clustan user manual* (3rd edn, Edinburgh 1978)

[17] Brian S. Everitt, *Cluster analysis* (2nd edn, SSRC 1980)

[18] Graphical techniques, such as Andrews plots, can be useful at uncovering such tendencies, Brian S. Everitt, *Graphical techniques for multivariate data* (London 1978)

[19] For an application of Ward's method to the analysis of farming regions see Mark Overton, Agricultural change in Norfolk and Suffolk, 1580–1740 (unpubl. Ph.D. thesis, University of Cambridge 1981)

[20] Everitt, *op. cit.* (1980) 65; Wishart, *op. cit.* 68

[21] H. L. Gray English field systems (Cambridge Mass. 1915), 17–82; H. S. A. Fox, The alleged transformation from two-field to three-field systems in medieval England *Economic History Review* 2nd series **39** (1986) 526–48

[22] For cropping patterns in central Essex see R. H. Britnell, Agriculture in a region of ancient enclosure, 1185–1500 *Nottingham Medieval Studies* **27** (1983) 37–55

[23] J. Thirsk, Field systems of the east midlands, in Alan R. H. Baker and Robin A. Butlin (Eds), *Studies of field systems in the British Isles* (Cambridge 1973) 232–80

[24] Brandon, *op. cit.*; Campbell *op. cit.* (1983); M. Mate, Medieval agrarian practices: the determining factors? *Agricultural History Review* **33** (1985) 22–31

[25] This arrangement is most clearly described in I. S. W. Blanchard, Economic change in Derbyshire in the late Middle Ages, 1272–1540 (unpubl. Ph.D. thesis, University of London, 1967) 168–74

[26] On downland sheep farming in southern England see R. Scott, Medieval Agriculture, in R. B. Pugh (Ed.), *Victoria history of the county of Wiltshire, IV* (London 1959) 19–21; J. N. Hare, Change and continuity in Wiltshire agriculture in the later Middle Ages, in W. Minchinton (Ed.), *Agricultural improvement: medieval and modern* (Exeter papers in economic history, 14, 1981) 4–9

[27] Yelling, *op. cit.* 111–26; Overton, *op. cit.* (1985); P. Glennie, Continuity and change in Hertfordshire agriculture 1550–1700: I—patterns of agricultural production *Agricultural History Review* **36** (1988) 55–76

[28] For instance, although based on the same data-source and dealing with the same period, the county-based studies of Overton and Glennie are not strictly comparable: their classification schemes are unique unto themselves and are the product of different clustering methods. The same weakness is inherent within the approach of the *Agrarian History* volumes: "authors on either side of a county boundary have not always agreed in their identification of the dominant local farming type", Joan Thirsk (Ed.), *The agrarian history of England and Wales, V, 1640–1750. 1. Regional farming systems* (Cambridge 1984) xxi

[29] Bruce M. S. Campbell, *The geography of seignorial agriculture in medieval England* (Cambridge, forthcoming); 'Feeding the city: London's food supplies 1250–1350', a research project codirected by Dr Derek Keene and Dr Bruce Campbell, funded by the Leverhulme Trust, and based at the Centre for Metropolitan History, Institute of Historical Research, University of London.

[30] They may have served as reservoirs of surplus animals to supply the more livestock-deficient areas further south and east: G. H. Tupling, *The economic history of Rossendale* (Manchester, Chatham Society, 1927) 17–41; R. Cunliffe Shaw, *The royal forest of Lancaster* (Preston 1956) 353–91; R. R. Davies, *Lordship and society in the March of Wales 1282–1400* (Oxford 1978) 115–16

[31] Johann Heinrich von Thünen, chapter 2 in M. Chisholm, *Rural settlement and land-use: an essay on location* (London 1962) 20–32

IX

THE LIVESTOCK OF CHAUCER'S REEVE: FACT OR FICTION?[1]

> *The REVE was a sclendre colerik man. . . . Wel koude he kepe a gerner and a binne; Ther was noon auditour koude on him winne. Wel wiste he by the droghte and by the reyn The yeldinge of his seed and of his greyn. His lordes sheep, his neet, his dayerie, His swyn, his hors, his stoor, and his pultrie Was hoolly in this Reves governinge, . . . Of Northfolk was this Reve of which I telle, Biside a toun men clepen Baldeswelle.*[2]

Few aspects of the *Canterbury Tales* have escaped the attention of Chaucerian scholars, but Chaucer's description of the husbandry—and particularly the pastoral husbandry—practiced by Oswald, the reeve of Bawdeswell in Norfolk, appears to be one. Oswald had in his charge his lord's horse, his dairy, neat and store cattle, his sheep, his swine, and his poultry[3]—ostensibly the range and combination of animals and

---

[1]I am grateful to Malcolm Andrew for advice on Chaucer's reeve, to John Langdon for providing data, to John Power and Jenitha Orr for research assistance, and to Gill Alexander for drawing the maps. Earlier versions of this paper were presented at seminars in the Dept. of English Local History, Univ. of Leicester, and Dept. of History, Univ. of Alberta; I am grateful to their participants for comments. Part of the research upon which this paper is based was undertaken whilst in the tenure of an ESRC research fellowship.

[2]Geoffrey Chaucer, *The General Prologue to the Canterbury Tales*, ed. James Winny (Cambridge, 1966), pp. 69–70.

[3]Chaucer, *CT, Gen. Prol.*, p. 69, lines 599–601.

poultry that might be found on almost any demesne in any part of the country in an age when, as R. H. Hilton observed in 1954, "everyone had to produce (on the whole) the same type of crop and tend the same sort of domesticated animals for meat, wool, and pulling power."[4] But was medieval agriculture so undifferentiated? Could it be instead that Chaucer's account is here more factual than fictional and that in characterizing the reeve, his dwelling, and his husbandry Chaucer had in mind a specific landscape and rural economy recognizable by repute to many in his audience? Such a view would certainly accord with the mounting body of evidence which shows that by the fourteenth century the countryside was deeply penetrated by commercial forces, with the result that farmers were increasingly specializing in what they produced and how they produced it.[5] In fact, Norfolk had already emerged as one of the most distinctive farming regions in England.[6]

Bawdeswell is situated in the Wensum Valley in the very heart of Norfolk, 14 miles northeast of Norwich. Soils in the area range from moderately heavy clays to light and free-draining sands; Faden's map of Norfolk shows that the latter, in accordance with Chaucer's description, gave rise to several extensive areas of heath.[7] The manor of which Oswald was

---

[4]"Medieval Agrarian History," in *Victoria County History: Leicestershire*, vol. 2 (London, 1954), p. 145.

[5]For a pioneering study of market specialization see Kathleen Biddick, "Medieval English Peasants and Market Involvement," *Journal of Economic History* 45/4 (1985): 823–31; eadem, "Missing Links: Taxable Wealth, Markets, and Stratification among Medieval English Peasants," *Journal of Interdisciplinary History* 18/2 (1987): 277–98.

[6]John P. Power and Bruce M. S. Campbell, "Cluster Analysis and the Classification of Medieval Demesne-Farming Systems," *Transactions of the Institute of British Geographers*, n.s., 17 (1992): 232–41; Bruce M. S. Campbell, Kenneth C. Bartley, and John P. Power, "The Demesne-Farming Systems of Post Black Death England: A Classification," *Agricultural History Review* 44/2 (1996).

[7]William Faden, *Faden's Map of Norfolk*, intro. J. C. Barringer (Dereham, 1989), sheet 15.

reeve was evidently in lay hands and "biside" rather than in Bawdeswell itself.[8] Manors in the neighboring townships of Bintree, Billingford, Foulsham, Sparham, and Themelthorpe could all fit Chaucer's description, although, if John Matthews Manly is to be believed, the prime candidate is the de Hastings manor in the immediately adjoining township of Foxley.[9] It is to this manor that a solitary manorial account of 1305/06 probably relates. At Michaelmas 1306 the stock enumerated in this account comprised five cart horses and stots, three mares, three young horses and foals, 10 oxen, 32 cows, 18 head of young cattle, 12 ducks, 13 capons, and eight hens.[10] Apart from the absence of sheep and swine, this range of livestock corresponds closely with that described by Chaucer on possibly the same manor at the end of the fourteenth century. Moreover, the payment of wages to a shepherd and swineherd and the receipt of mutton and pork by the lord's larder testify that sheep and swine must both have been present in the locality. Tallying even more closely with Chaucer's description is the stock maintained at Michaelmas 1348 on John de Gyney's manor of Guton Hall in Brandiston, five miles to the east. Here were 14 horses of varying ages, two bulls, nine oxen, five steers, five heifers, 19 cows, six calves, 67 sheep, 32 lambs, 55 swine of various ages, three swans, 19 geese, 46 capons, nine ducks, and 13 hens.[11] On this mixed-farming demesne horses and oxen evidently satisfied the draught requirements of arable husbandry, a substantial breeding and dairying herd was maintained, and sheep, swine, and assorted poultry were kept. The

---

[8]Chaucer, *CT, Gen. Prol.*, p. 70, line 622.

[9]*Some New Light on Chaucer: Lectures Delivered at the Lowell Institute* (London, 1926), pp. 84–94.

[10]PRO SC6/935/19.

[11]Magdalen College Oxford, Estate Records 166/7.

only significant departure from the list of stock enumerated by Chaucer is the presence of oxen to provide draught power. Thirty years later, however, at the time that Chaucer was writing, most demesnes in this locality (like the peasantry before them) were well advanced in the changeover to all-horse ploughing and were actively disposing of their oxen.[12]

Analysis of a national sample of 787 demesnes recorded between 1250 and 1449 helps to put the pastoral profiles of Foxley and Guton Hall into perspective: whereas 45 per cent of all Norfolk demesnes carried as wide a range of livestock, nationally less than a third of all lowland demesnes did so.[13]

For Norfolk there are over 2,000 extant manorial accounts representing more than 200 different demesnes, both lay and ecclesiastical.[14] The detailed listings of the stock remaining each Michaelmas that these provide may be analyzed statistically to reveal the range and distribution of pastoral types that existed within the county. This is most effectively undertaken by means of the technique known as cluster analysis. This is a statistically consistent method of classifying data into groups on the basis of differences between cases (i.e., individual

---

[12]Bruce M. S. Campbell, "Towards an Agricultural Geography of Medieval England," *Agricultural History Review* 36/1 (1988): 91–93.

[13]The latter proportion falls to 35 per cent during the final quarter of the fourteenth century, of which Chaucer was writing. John Langdon supplied the national sample of manorial accounts on which these figures are based.

[14]These are preserved among the following public and private archives: PRO (London), Norfolk Record Office (NRO), North Yorkshire Record Office, Nottinghamshire Record Office, West Suffolk Record Office, Bodleian Library (Oxford), British Library (BL), Cambridge Univ. Library, Canterbury Cathedral Library, Joseph Regenstein Library, Univ. of Chicago, Harvard Law Library, John Rylands Library (Manchester), Lambeth Palace Library, Nottingham Univ. Library, Eton College, Christ's College (Cambridge), King's College (Cambridge), Magdalen College (Oxford), St. George's Chapel (Windsor), Elveden Hall (Suffolk), Holkham Hall (Norfolk), Raynham Hall (Norfolk), Pomeroy and Sons (Wymondham). I am grateful to the relevant authorities for granting access to these materials. A handlist is available on request from the author.

demesnes) as measured across all the variables of which they are composed (i.e., the livestock with which they were stocked).[15] Table 1 summarizes the results of such an analysis obtained utilizing the *relocation method* as applied to the percentage of total livestock units accounted for by horses, oxen, adult cattle (cows and bulls), immature cattle, sheep, and swine.[16]

As will be noted, seven basic pastoral types are identified in the period 1250–1349 and a further seven in the period 1350–1449. The smallest of these groupings comprises a single demesne (Cluster 5, 1350–1449, distinguished by its exceptionally high proportion of swine) and the largest, a total of 35 demesnes (Cluster 1, 1250–1349). The two demesnes so far discussed—Foxley and Guton Hall in Brandiston—are placed respectively in Clusters 1 and 2 in the period 1250–1349, the difference in classification reflecting the absence of sheep and swine from Foxley and their presence at Guton Hall. As will be seen from Figure 1, other Cluster 1 demesnes occur in the immediate vicinity of Bawdeswell—at Alderford, Kerdiston, North Elmham, Gateley, and Hindolveston—while both before and after 1350 Clusters 1 and 2 form the predominant pastoral types in much of north-central and eastern Norfolk. Had Chaucer enumerated the numbers as well as the types of

---

[15]For a fuller exposition of the application of cluster analysis to data from manorial accounts see Power and Campbell, "Cluster Analysis and Classification," pp. 227–45; Bruce M. S. Campbell and John P. Power, "Mapping the Agricultural Geography of Medieval England," *Journal of Historical Geography* 15/1 (1989): 26–27.

[16]Total livestock units = (horses x 1.0) + (oxen x 1.2) + (adult cattle x 1.2) + (immature cattle x 0.8) + (sheep x 0.1) + (swine x 0.1). These weightings relate to the relative feed requirements of the various livestock and are based on those used by J. T. Coppock, *An Agricultural Atlas of England and Wales* (London, 1964), p. 213, and J. A. Yelling, "Probate Inventories and the Geography of Livestock Farming: A Study of East Worcestershire, 1540–1750," *Transactions of the Institute of British Geographers* 51 (1970): 115.

TABLE 1
NORFOLK: CLASSIFICATION OF DEMESNE PASTORAL-FARMING TYPES 1250–1349 AND 1350–1449

	Cluster	No. of Demesnes	% of Demesnes	Mean Percentage of Total Livestock Units					
				Horses	Oxen	Adult Cattle	Young Cattle	Sheep	Swine
1250–1349	1	35	27	17	10	51	17	4	2
	2	14	11	21	6	32	17	14	12
	3	31	24	15	7	30	10	37	2
	4	5	4	26	3	<1	<1	64	6
	5	32	25	19	24	26	19	10	2
	6	3	2	87	13	0	0	0	0
	7	9	7	34	58	2	1	3	2
	Overall	129	100	21	15	31	14	16	3
1350–1449	1	33	31	17	4	49	20	4	5
	2	13	12	17	13	62	5	1	2
	3	33	31	12	6	34	10	37	2
	4	12	11	10	2	4	2	81	1
	5	1	1	39	0	0	0	0	61
	6	10	9	90	1	3	4	0	3
	7	5	5	43	48	0	0	9	0
	Overall	107	100	23	7	34	11	22	3

Source: Manorial accounts
Method: Cluster analysis (relocation method), seven-cluster solution. For calculation of livestock units see n. 16 above.

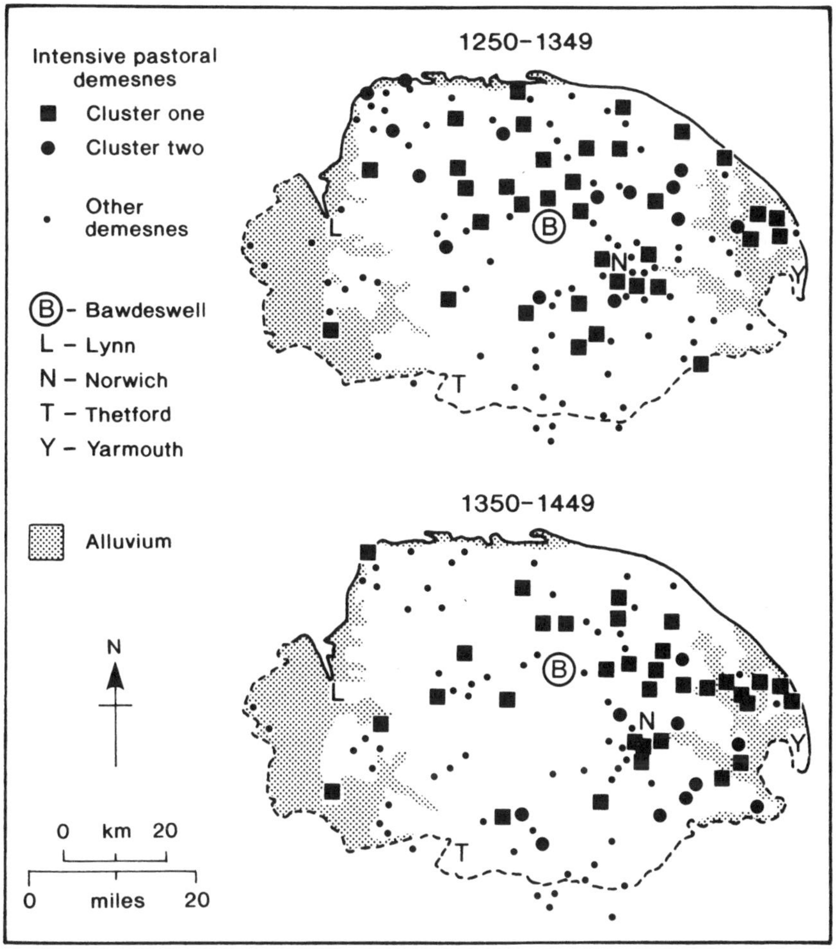

Figure 1. Norfolk: distribution of "intensive" pastoral-farming demesnes 1250–1349 and 1350–1449 (for explanation see Clusters 1 and 2, Table 1)

Oswald's livestock, and were it possible that these could have been included within the cluster analysis, it is highly likely that the Bawdeswell demesne would likewise have fallen into one or other of these two cluster groupings.

What distinguishes both these Norfolk pastoral types is their developed and intensive nature. Five features are particularly worthy of note. First, draught animals comprised barely a quarter of demesne stock, with the result that working animals were greatly outnumbered by nonworking animals. Second, within the draught sector oxen were outnumbered by horses, with a tendency for that bias to become increasingly pronounced over time as oxen were progressively replaced by horses.[17] Third, within the nonworking sector cattle predominated, accounting on average for almost half of all livestock units on Cluster 2 demesnes in the period 1250–1349 and two-thirds on Cluster 1 demesnes in the periods 1250–1349 and 1350–1449 and Cluster 2 demesnes in the period 1350–1449. Fourth, the structure of these cattle herds was demographically skewed towards adult females, so that cows and bulls outnumbered their followers by at least five to four and in many cases by more than two to one. Such ratios are indicative that dairying rather than rearing was their prime function, and this is borne out by direct evidence of butter and cheese production and the domination of butchery and sale by the disposal of decrepit adults and surplus calves. Finally, sheep, swine, and poultry occupied essentially subsidiary positions within the overall pastoral profile. This applies particularly to sheep, whose importance was, in fact, often eclipsed by swine, which were capable of more intensive forms of management.[18] The range of poultry kept was also often quite considerable: the hens, capons, geese, ducks, and swans kept on the demesne at

---

[17]For the diffusion of horses in Norfolk see John L. Langdon, *Horses, Oxen and Technological Innovation: The Use of Draught Animals in English Farming from 1066 to 1500* (Cambridge, 1986), pp. 50–53, 101–05; Campbell, "Towards an Agricultural Geography," pp. 91–93.

[18]Kathleen Biddick, "Pig Husbandry on the Peterborough Abbey Estate from the Twelfth to the Fourteenth Century," in Juliet Clutton-Brock and Caroline Grigson, eds., *Animals and Archaeology*, British Archaeological Reports, Internat'l. Ser., 227 (Oxford, 1985), pp. 161–77; eadem, *The Other Economy: Pastoral Husbandry on a Medieval Estate* (Berkeley, 1989), pp. 121–25.

Guton Hall in Brandiston were by no means unusual. Pigeons also were a feature of many demesnes. The merit of these lesser categories of livestock lay in their more rapid breeding cycles and their capacity to utilize resources either surplus or unsuited to the requirements of the other animals.[19] Ecologically, that meant that not even the smallest niches within the farm's food chain were left unoccupied.

The comparatively exceptional nature of these characteristics is brought out by a wider survey of demesne livestock within the country as a whole. Thus, Table 2 provides summary statistics at a county level of certain key diagnostic features of demesne pastoral husbandry during the period 1250–1349.[20] As will be seen from Column B, Norfolk was alone among lowland, arable counties in supporting such a small proportion of working animals. Its nearest rivals, also in the south and east, were Essex and Kent. During the course of the fourteenth century, as arable husbandry contracted and pastoral husbandry expanded, the ratio of working to nonworking animals fell almost everywhere: draught animals accounted for 46 per cent of demesne livestock in the mid-thirteenth century, 38 per cent in the mid-fourteenth century, and 29 per cent in the mid-fifteenth century. The same trend is detectable in Norfolk, but the proportion always remained consistently smaller—36 per cent, 23 per cent, and 20 per cent, respectively. Indeed, before 1350 it was only in the far north of the country, where opportunities for arable husbandry were environmentally circumscribed, that smaller proportions of working animals appear

---

[19]Martin Stephenson, "The Role of Poultry Husbandry in the Medieval Agrarian Economy, 1200–1450," *Veterinary History* 10 (1977–78): 16–24; Biddick, *Other Economy*, pp. 125–28.

[20]The data upon which this analysis is based were generously supplied by John Langdon.

TABLE 2
ENGLAND: SOME CHARACTERISTICS OF DEMESNE PASTORAL HUSBANDRY BY COUNTY, 1250–1349

COUNTY	A	B	C	D	E	F	G	H	I
Bedfordshire	10	46	24	25	6	4.5	62	15	25
Berkshire	16	43	30	25	3	4.8	112	23	50
Berwickshire	1	28	31	38	2	7.8	147	—	—
Buckinghamshire	15	46	33	17	5	2.4	93	42	34
Cambridgeshire	12	50	28	19	3	1.7	99	24	11
Cheshire	1	56	35	2	8	6.8	50	—	—
Cornwall	8	36	32	31	<1	4.8	33	40	62
Cumberland	3	23	61	16	0	9.3	59	18	21
Derbyshire	3	86	8	0	6	5.8	266	13	88
Devon	15	38	38	24	1	6.3	74	49	86
Dorset	6	45	26	27	2	6.3	156	25	53
Co. Durham	26	79	9	8	4	7.3	165	—	—
Essex	28	33	32	34	2	0.7	52	61	35
Gloucestershire	17	48	23	26	3	8.8	128	40	90
Hampshire[x]	36	47	26	25	2	4.7	84	47	44
Herefordshire	4	41	17	37	6	7.7	49	37	38
Hertfordshire	16	38	39	20	4	0.5	54	21	29
Huntingdonshire	8	50	39	14	7	1.4	158	7	38
Isle of Wight	8	52	28	19	1	10.8	113	23	18
Kent	49	34	36	23	6	1.4	72	82	29
Lancashire	6	73	26	0	0	6.4	123	—	—
Leicestershire	9	66	6	25	3	3.8	87	32	*40
Lincolnshire	31	37	40	21	2	3.5	97	48	38
Middlesex	15	38	45	15	2	1.5	53	8	50
Monmouthshire	6	73	21	6	<1	14.8	84	14	63
Norfolk	124	30	46	21	3	0.7	69	78	26
Northhamptonshire	26	44	40	12	5	1.9	138	36	36
Northumberland	4	89	10	0	1	3.5	174	19	30
Nottinghamshire	5	49	31	18	2	6.5	230	12	73
Oxfordshire	29	48	26	23	4	3.0	109	41	75
Rutland	4	48	22	25	5	2.2	127	—	+
Shropshire	2	57	38	0	5	5.1	79	46	30
Somerset	21	72	19	6	2	11.2	174	38	70
Staffordshire	9	51	21	26	2	7.7	97	14	25
Suffolk	33	37	44	17	3	1.2	111	60	37
Surrey	18	39	37	21	3	3.2	138	18	49
Sussex	22	37	24	34	5	6.9	83	42	35

TABLE 2—*Continued*

COUNTY	A	B	C	D	E	F	G	H	I
Warwickshire	20	67	23	7	4	6.3	167	21	41
Westmorland	—	—	—	—	—	—	—	—	—
Wiltshire	21	35	14	50	2	7.7	83	44	71
Worcestershire	6	47	20	30	3	3.9	53	16	43
Yorkshire E. R.	11	58	9	31	2	5.1	91	—	—
Yorkshire N. R.	6	88	6	0	6	7.5	221	25	62
Yorkshire W. R.	27	49	36	14	1	7.4	84	—	—
England	*737*	*47*	*29*	*22*	*3*	*4.6*	*91*	*1,179*	*41*

*including Rutland
+with Leicestershire
xexcluding Isle of Wight
A Number of demesnes with accounts
B Draught animals as percentage of total livestock units (from accounts)
C Cattle other than oxen as percentage of total livestock units (from accounts)
D Sheep as percentage of total livestock units (from accounts)
E Swine as percentage of total livestock units (from accounts)
F Oxen per horse (imposed maximum individual ratio of 20) (from accounts)
G Young cattle per 100 adults (from accounts)
H Number of *Inquisitiones Post Mortem*
I Pence of grassland per 100 pence of arable (from *IPMs*)

to have been supported. These were possibly the breeding areas from which non-self-sufficient lowland areas further south drew their replacement stock.[21]

Nationally, oxen remained unquestionably the single most important source of draught power throughout the period 1250–1449, consistently outnumbering horses by at least four to one. As John Langdon has shown, it was Norfolk that pioneered the application of horses to farm work, and Norfolk stands out

[21]On the strong pastoral bias to husbandry in the extreme north of England see Edward Miller, "Farming in Northern England during the Twelfth and Thirteenth Centuries," *Northern History* 11 (1975): 11–12; idem, "Farming Techniques: Northern England," in H. E. Hallam, ed., *The Agrarian History of England and Wales*, 2: *1042–1350* (Cambridge, 1988), pp. 408–11.

as the only county in which from as early as 1250 horses were consistently in the majority (Table 2, Col. F).[22] After 1350, however, it was joined by Essex and Hertfordshire, with Suffolk, Kent, and Cambridgeshire not far behind, these six counties representing a geographically contiguous block concentrated in East Anglia and the southeast.

By switching from oxen to horses these counties were relieved from the need to devote such a large share of pastoral resources to the breeding of replacement work animals, and a relative expansion in other types of pastoral enterprise became possible. Hence, the prominence of cattle other than oxen in Norfolk: these comprised no less than 46 per cent of total livestock units in the period 1250–1349, a proportion rivaled only by Middlesex and Suffolk and exceeded only by Cumberland, whose extensive pastures offered considerable scope to cattle rearing (Table 2, Col. C). The distinctiveness of this emphasis is highlighted by the fact that nationally cattle other than oxen accounted for, on average, 29 per cent of demesne livestock before 1350 and 31 per cent thereafter, by which time they accounted for almost half of all demesne livestock units in Norfolk.

Within the country at large the ratio of adult to immature cattle was only marginally in favor of the former, although the gap began to widen after 1350, as cattle herds expanded in the wake of the demographic collapse and subsequent contraction in arable cultivation.[23] Before that catastrophe Norfolk was one of only a dozen English counties in which there was a decided imbalance between adults and immatures, and of these only Cumberland, Essex, Hertfordshire, Middlesex, and Hampshire

---

[22]*Horses, Oxen and Technological Innovation*, pp. 50–53; Campbell, "Towards an Agricultural Geography," p. 93.

[23]The growing practice of farming out dairies may also mean that younger animals are increasingly masked from view.

could boast stocking densities of cattle that were as high or higher. These contrast with no less than 19 counties—the six northern counties of Berwickshire, Northumberland, Durham, Yorkshire North Riding, Lancashire, and Derbyshire prominent among them—in which immatures exceeded adults and where, therefore, the emphasis was upon rearing.[24] Indeed, if a relatively heavy emphasis upon cattle other than oxen (i.e., at least 30 per cent of total livestock units), a low proportion of oxen among cattle (i.e., less than 40 per cent of total cattle), and a low ratio of immature to adult cattle (i.e., fewer than 75 immatures per 100 adults), are taken as diagnostic features of a strong specialist interest in dairying, then only Norfolk, Middlesex, and Hertfordshire fulfill all three criteria before 1350, and only Norfolk, Suffolk, Essex, and Hertfordshire, thereafter. In the Middle Ages these were quite clearly England's premier dairying counties.

Norfolk's concentration upon cattle-based dairying (a specialism that Chaucer includes in his description of Oswald's agricultural activities) plainly eclipsed any interest that demesne lords may have taken in sheep farming, for all that it is for the latter activity that the county is most usually celebrated. This is not to deny the presence of significant numbers of sheep within the county, for their importance is confirmed by the size of Norfolk's contribution to the 1341/42 wool tax,[25]

---

[24]On commercial cattle rearing on the de Lacy estates in northern England see *Victoria County History: Lancaster*, vol. 2 (London, 1908), pp. 268–84; G. H. Tupling, *The Economic History of Rossendale*, Chetham Society, n.s., 86 (1927): 17–41; R. Cunliffe Shaw, *The Royal Forest of Lancaster* (Preston, 1956), pp. 353–91; M. A. Atkin, "Land Use and Management in the Upland Demesne of the De Lacy Estate of Blackburnshire, c1300," *Agricultural History Review* 42/1 (1994): 1–19.

[25]W. M. Ormrod, "The Crown and the English Economy, 1290–1348," pp. 149–83 in Bruce M. S. Campbell, ed., *Before the Black Death: Studies in the 'Crisis' of the Early Fourteenth Century* (Manchester, 1991), pp. 178–79.

but on the evidence of demesne stocking schedules the majority of these sheep must have been peasant owned rather than seigniorial animals. Within the demesne sector sheep tended only to assume prominence where environmental circumstances rendered cattle unsuitable: in the Broadland and fenland marshes, on the sandy soils of Breckland, and wherever soils were light and surface water scarce. On the demesne front Norfolk thus stands in the middle rank of sheep-farming counties, well behind the southern downland counties of Wiltshire and Sussex but ahead of most of central and northern England. Much the same applies to swine.[26]

Pastoral husbandry in Norfolk was therefore distinctive in almost every respect, and especially so by dint of its developed and intensive nature. The latter is the more remarkable in that as far as several grassland (i.e., grassland belonging exclusively to the demesne and excluding that held in common) was concerned, it was one of the least grassy counties in England—a point that can be readily demonstrated using the land-use information recorded in the extents attached to *Inquisitiones Post Mortem* (*IPM*s). This information relates to the estates of lay tenants-in-chief of the Crown and is available for the greater part of the country, although the consistency with which demesne resources are recorded varies according to the relative importance of the resources in question and the practices of particular groups of escheators.[27] For both sets of reasons there are grounds for believing that pastoral resources are

[26]See Table 2.

[27]Systematic use of *IPM*s to cast light on variations in land use and land values was pioneered by J. Ambrose Raftis, *Assart Data and Land Values: Two Studies in the East Midlands, 1200–1350* (Toronto, 1974). For a recent appraisal of the source and its potential see Bruce M. S. Campbell, James A. Galloway, and Margaret Murphy, "Rural Land-Use in the Metropolitan Hinterland, 1270–1339: The Evidence of *Inquisitiones Post Mortem*," *Agricultural History Review* 40/1 (1992): 1–22.

consistently under-recorded in the counties north of the River Trent. In much of the rest of the country, however, the *IPMs* provide the best available guide to the relative supply of permanent grassland. Table 2, Column I, summarizes the mean ratio of the value of arable resources to the value of pastoral resources (meadow, pasture, herbage, and other sources of forage) by county for the decade 1300–09. This decade was chosen because it probably represents the culmination of medieval demographic and economic expansion and is numerically well served by *IPMs*. The total of 1,179 extents used provides coverage of every English county, except the palatinate counties of Cheshire, Lancashire, and Durham, and for every county except Huntingdonshire and Middlesex (both small and dominated by ecclesiastical estates) the number of available extents is in double figures.

In the country as a whole the average demesne had grassland worth 40 per cent of the value of its arable. Equivalent "grassland ratios" calculated on a county basis reveal strong regional variations and highlight Norfolk's position at the most arable and least grassy end of the land-use spectrum. Norfolk lay within a block of counties concentrated in the East Midlands, East Anglia, and the extreme southeast that had grassland ratios decidedly below the national average. These ratios fell to a low of 26:100 in Norfolk, 25:100 in Bedfordshire, and a meager 11:100 in Cambridgeshire—England's least grassy county. By contrast, above-average ratios prevailed in much of the west and parts of the north of England. Thus, grassland ratios were highest of all in Derbyshire, Gloucestershire, and Devon, where they were more than three times higher than in Norfolk and the value of grassland came to within 15 per cent of the value of arable. In the neighboring counties of Nottinghamshire, Oxfordshire, Wiltshire, Somerset, and Monmouthshire, followed by Yorkshire, Cornwall, Dorset, and Berkshire, supplies of demesne grassland were also more than double those in Norfolk and worth between 50 per cent and 75 per cent of the value of the arable. Were the northern returns less defective, the same would probably be true of Northumberland, Cumberland, Staffordshire, and Shropshire, since in environmental terms they share much in common with the grassy counties of the southwest, although it is possible that more of their grassland was in common than in several ownership.

Within Norfolk, as Figure 2 shows, there was obviously some local variation in grassland supplies according to environmental circumstances and prevailing institutional arrangements. For instance, several of the demesnes on the fen-edge and in Broadland enjoyed relatively abundant supplies of rich pasturage. Demesnes on the Breck-edge in south Norfolk were also better off than most. But in much of eastern, central, and northern Norfolk several grassland was in extremely short supply and there were some quite sizeable demesnes (especially on the light soils of the northwest) that had little or no several grassland at all and were thus almost totally reliant for pasturage on the various heaths and common wastes of the locality. The situation in the immediate vicinity of Bawdeswell appears to have been only marginally less tight, and Oswald the reeve must have been grateful for the pasturage afforded by the heath beside which he lived. In 1314 an *IPM* of the demesne at Foxley of Hawise de Veer recorded 200 acres of arable worth 7*d.* an acre, six acres of meadow worth 18*d.* an acre, and eight acres of pasture worth 12*d.* an acre; 10 years later (1324) an *IPM* of the demesne, also at Foxley, of Aymer de Valence, earl of Pembroke, recorded 306 acres of arable worth 8.8*d.* an acre, 15 acres of meadow worth 24*d.* an acre, and 30 acres of pasture worth 9*d.* an acre.[28] With respective grassland ratios

[28]PRO C134/34 (7); C134/83 (99).

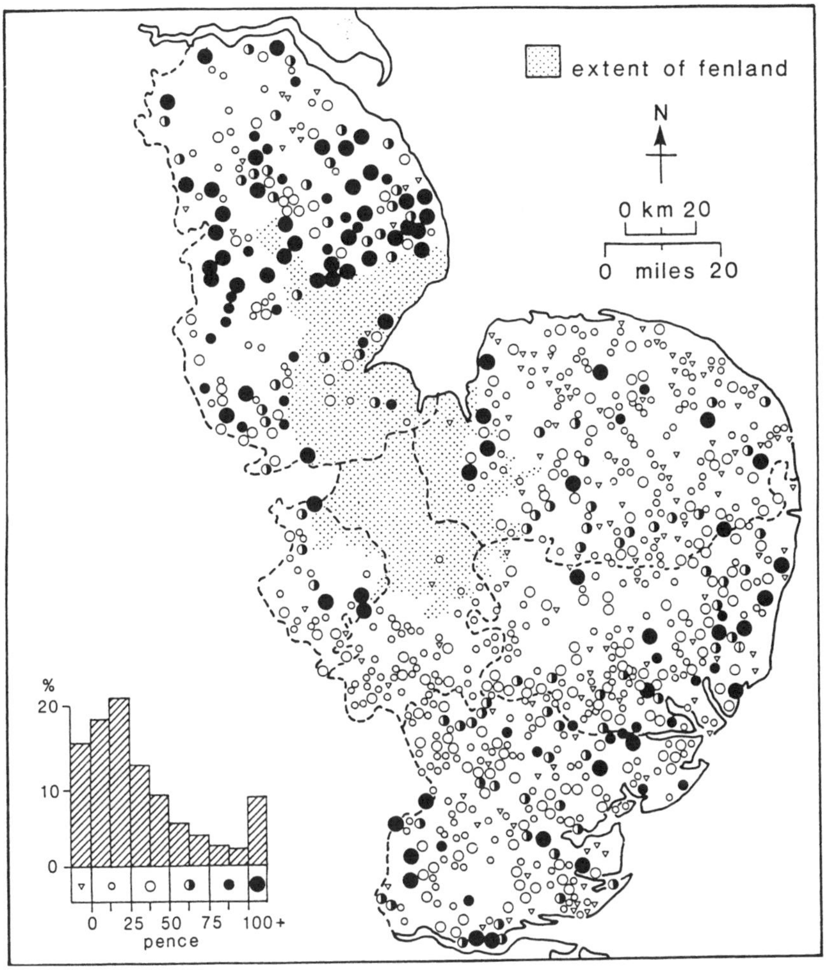

Figure 2. Eastern England: value of demesne grassland (meadow and pasture) per 100 pence of demesne arable, 1270–1349 (source: *Inquisitiones Post Mortem*)

of 15:100 and 23:100, neither of these demesnes was well provided with several grassland.

Land-use and farm enterprise thus present a striking paradox in medieval Norfolk. How was such a developed pastoral regime reconciled with such a shortage of the grassland that

M. M. Postan, for one, has argued was one of its most essential prerequisites? After all, it was his great thesis that by the end of the thirteenth and the beginning of the fourteenth centuries, "in corn-growing parts of the country taken as a whole, pasture and the animal population had been reduced to a level incompatible with the conduct of mixed farming itself."[29] For Postan, therefore, stocking densities were a direct function of the availability of permanent grassland.[30] What the Norfolk evidence shows is that this relationship was less direct than he supposed. Demesne stocking densities in medieval Norfolk were certainly by no means high, averaging 36 livestock units per 100 grain acres between 1250 and 1349 and 42 livestock units per 100 grain acres between 1350 and 1449. They were thus respectively 20 per cent and 35 per cent below the corresponding national averages of 44 and 65 livestock units per 100 grain acres, the discrepancy being greater in the later period, when livestock numbers rose significantly in many parts of the country, especially those environmentally better suited than Norfolk to convert from arable to grass. Nevertheless, Norfolk's stocking density was not disproportionately as low relative to other counties, particularly those of southern and southwestern England, as its low grassland ratio would suggest. The explanation lies in the intensity with which pastoral resources were exploited and the extent to which they were augmented by other sources of sustenance. Kathleen Biddick has recently arrived at a similar conclusion for the estates of Peterborough Abbey: "the changing composition of livestock in the herding

---

[29]*The Medieval Economy and Society: An Economic History of Britain in the Middle Ages* (London, 1972), p. 59.

[30]"We have so far assumed that it was the shortage of pasture that kept the numbers of animals down. That the assumption is right and that the shortage of pasture was great and widespread is revealed by the high and rising rents and by the prices of pastures as given in manorial surveys, custumals and similar manorial valuations of the land," Postan, *Medieval Economy and Society*, p. 59.

economy of the estate characterizes a pastoral sector of some dynamism and complexity and dispels any notion of linear relations between animal and cereal husbandry."[31] In Norfolk the careful management of meadow lands and integration of the arable and pastoral sectors via fodder cropping, the fold-course system, and convertible husbandry served to boost stocking densities.[32] The productivity of the pastoral sector further benefited from an emphasis upon those types of livestock that were most productive of draught power and food.

What recommended the horse to Norfolk farmers was that it worked faster and for longer hours than the ox and thereby allowed the size of plough teams to be reduced and the range of road transport to be extended.[33] The price of this change was higher depreciation and running costs, the former because, unlike oxen, old, worn-out horses could not be fattened and sold for meat, and the latter because horses consumed greater quantities of oats, vetches, and other legumes. The substitution of horses for oxen thus entailed a shift from natural to produced fodder. On the prior of Norwich's intensively managed demesnes of Martham and Hemsby, 51 per cent and 74 per cent, respectively, of the oats harvested between 1261 and 1335 were consumed as fodder, and 6 per cent and 10 per cent of the peas.[34] In contrast, at Sedgeford in northwest Norfolk only 47 per cent of the oats were consumed as fodder, augmented by 25

---

[31]*Other Economy*, p. 65.

[32]Bruce M. S. Campbell, "Agricultural Progress in Medieval England: Some Evidence from Eastern Norfolk," *Economic History Review*, 2nd ser., 36/1 (1983): 26–46; Mark Bailey, "Sand into Gold: The Evolution of the Foldcourse System in West Suffolk, 1200–1600," *Agricultural History Review* 38/1 (1990): 40–57.

[33]Langdon, *Horses, Oxen and Technological Innovation*, pp. 158–71; idem, "Horse Hauling: A Revolution in Vehicle Transport in Twelfth- and Thirteenth-Century England?" *Past and Present* 103 (1984): 37–66.

[34]NRO DCN60/15 and 23; DCN62/1–2; L'Estrange IB4/4.

per cent of the rye and 29 per cent of the peas.[35] Swine and poultry also were consumers of legumes and grain, and as converters of calories were more efficient meat producers than either sheep or cattle.[36]

Such a change from natural to produced fodder, as Ester Boserup has observed, represents a major increase in drudgery on the part of the labor force.[37] It also constitutes a change to a more intensive type of agricultural food chain. According to I. G. Simmons, within sedentary agriculture there are three main pastoral food chains. In the first, livestock are fed solely upon managed grassland and natural vegetation; in the second, they are fed upon a combination of managed grassland, natural vegetation, and tillage crops; and in the third, they are fed on tillage crops alone.[38] The substitution of horses for oxen and the management of swine in styes rather than as a forage animal thus represent a shift from the first to the second food chain and a corresponding step-up in the intensity of pastoral husbandry. This change constitutes a significant advance and presupposes a closer integration of arable and pastoral husbandry to the mutual benefit of both, as fodder crops contributed to the sustenance of the livestock and the latter contributed traction and manure to the arable. The next major change in the evolution of pastoral husbandry was not to take place until the late seventeenth and the eighteenth centuries, when mixed-farming systems in which livestock became almost wholly reliant upon an improved range of fodder crops were evolved. Norfolk again stood in the van of these developments.[39]

---

[35]NRO DCN60/33; DCN62/1–2; L'Estrange IB1/4 and 4/4.

[36]Ester Boserup, *Population and Technology* (Oxford, 1981), p. 18.

[37]*The Conditions of Agricultural Growth: The Economics of Agrarian Change under Population Pressure* (London, 1965), pp. 36–39.

[38]*The Ecology of Natural Resources* (London, 1974).

[39]Mark Overton and Bruce M. S. Campbell, "Norfolk Livestock Farming 1250–1740: A Comparative Study of Manorial Accounts and Probate Inventories,"

In the thirteenth and fourteenth centuries dairy cattle were the principal beneficiaries of these important changes. They became the recipients of the grass and hay released by the substitution of horses for oxen, while the reduced demand for replacement oxen allowed a greater concentration on dairying at the expense of rearing. This marked preference for dairy cattle derived from the fact that they are more productive of human food per unit area—in the form of milk, butter, cheese, and the meat of surplus calves and decrepit females—than any other class of livestock. On the evidence of modern agricultural statistics, David B. Grigg has shown that milk production is 180 per cent more productive per unit area of calories and protein than beef production, and four times more productive of calories and five times more productive of protein than mutton production.[40] It is also superior in productivity to egg and poultry production. For medieval farmers with scarce land resources, intensive dairying therefore offered the best food return per unit area, especially where labor was both relatively abundant and cheap.[41] By contrast, the inferior productivity of sheep meant that they tended to be relegated to those types of environment—particularly heath and marsh—where cattle did not thrive.

Journal of Historical Geography 18/4 (1992): 382–92; Bruce M. S. Campbell and Mark Overton, "A New Perspective on Medieval and Early Modern Agriculture: Six Centuries of Norfolk Farming *c.*1250–*c.*1850," *Past and Present* 141 (1993): 88–95.

[40]*The Dynamics of Agricultural Change: The Historical Experience* (London, 1982), p. 71.

[41]It was possibly these kinds of relationship that H. E. Hallam had in mind when he observed, "the progressive parts of England had large herds of dairy cattle whose milk the people drank and made into butter and cheese. On the Winchester estates the lord generally made cheese out of ewe's milk, a sign of a primitive economy, for a ewe gives only a quart of milk a day and is difficult to milk, whereas a goat gives a gallon and a cow two gallons. Eastern England and the east Midlands also used stots and affers (small horses) instead of slow oxen for ploughing and could therefore plough twice as much land in a given time" (*Rural England 1066–1348* [London, 1981], p. 14).

Implicit in these developments were higher inputs of capital and labor both per animal and per unit area. Labor was required to manage grassland, make hay, cultivate fodder crops, supervise flocks and herds, collect and spread manure, milk and calve cows, make cheeses, shoe horses, and attend to a host of minor tasks entailed in the management of animals. Capital was required in the form of the animals themselves, the seed for the fodder crops that they consumed and the tools and implements with which these were cultivated, and, most conspicuously, for specialist housing and equipment. The working horses were generally stabled, and some at least of the dairy herds were housed, thus facilitating stall-feeding and the accumulation of farmyard manure. Swine, too, were often housed in styes, and it was not uncommon for sheep cotes to be constructed to provide shelter during the most severe weather. Butter and cheese production required investment in pails, churns, vats, presses, and scales and regular purchases of salt and cheese-cloths.[42] On some of the larger demesnes, such as those of the prior of Norwich at Hemsby, Great Plumstead, and Newton, purpose-built dairy houses were provided.[43]

Such a system of pastoral husbandry was expensive and the outputs—in terms of traction generated, milk and meat produced, and breeding rates maintained—needed to be such as to justify the inputs. Yet the productivity of such a system is extremely hard to measure.[44] One index of the gross return

---

[42]Bruce M. S. Campbell, "Commercial Dairy Production on Medieval English Demesnes: The Case of Norfolk," in Annie Grant., ed., *Animals and Their Products in Trade and Exchange, Archaeozoologica (Quatrième Numéro Spécial)*, 16 (Paris, 1992), pp. 107–18.

[43]David Yaxley, *The Prior's Manor-Houses: Inventories of Eleven of the Manor-Houses of the Prior of Norwich, Made in the Year 1352 A.D.* (Dereham, 1988), pp. 5–7, 14–18.

[44]Mark Overton and Bruce M. S. Campbell, "Productivity Change in European Agricultural Development," pp. 12, 33–34, in eidem, eds., *Land, Labour and*

obtainable from dairy farming is provided by the rate at which cows were let at farm. The farming of demesne dairies was a profitable and reliable source of income and became increasingly common practice during the course of the fourteenth century. The terms on which the cows were leased varied. Usually the lessee was entitled to the milk of the cows, the calves that they produced, and, presumably, their manure. Occasionally the lessee was entitled to their milk (or lactage) only, and their issue was retained by the lord. Arrangements also sometimes were made whereby the calves were divided between the lessee and the lord. The rate of farm varied according to the terms agreed, but whatever the latter, as will be seen from Table 3, it is plain that cows potentially were an extremely lucrative asset. At a rate of 4*s*.–6*s*. per head when farmed for their milk and calves, one cow could be as profitable as several acres of prime arable. Since herds were usually farmed out en bloc, the payments made by the farmers were often quite considerable. At Foxley, for instance, the farm of 30 cows raised £6 15*s*.0*d*. in 1306.[45]

Cash payments of this order of magnitude presuppose that a significant part of the produce must have been sold for cash, and manorial accounts confirm that under direct management cattle, swine, and poultry were regular sources of income.[46] On the prior of Norwich's seven demesnes of Gnatingdon, Thornham, North Elmham, Taverham, Monks Granges, Plumstead,

---

Livestock: Historical Studies in European Agricultural Productivity (Manchester, 1991), pp. 1–50.

[45]PRO SC6/935/19.

[46]On demesnes within the hinterland of London, sales of animals and animal products contributed 14 per cent and 17 per cent, respectively, of gross agricultural sales income c.1300 (B. M. S. Campbell, "Measuring the Commercialisation of Seigneurial Agriculture c.1300," in Richard H. Britnell and Bruce M. S. Campbell, eds., *A Commercialising Economy: England 1086 to c.1300* [Manchester, 1995], pp. 148–49).

TABLE 3
NORFOLK: FARM OF COWS

Manor	Year	Rent per Cow
Burgh in Flegg	*1296–97*	*4*s.*0*d.
Hautbois	*1363*	*4*s.*0*d.
Horning	*1372*	*4*s.*0*d.
Wroxham	1342–43	3*s*.4*d*.
Hainford	1363–64	4*s*.0*d*.
Haveringland	1356–57	4*s*.0*d*.
Haveringland	1376–77	4*s*.0*d*.
Ludham	1355	4*s*.9*d*.
Foxley	1305–06	5*s*.0*d*.
Gimingham	1358–59	5*s*.0*d*.
Tunstead	1359–60	5*s*.6*d*.
Melton	1332–33	5*s*.6*d*.
Melton	1366–67	6*s*.0*d*.
Thurning	1319–20	6*s*.0*d*.
Horning	1372	6*s*.0*d*.
Gateley	1326–27	6*s*.8*d*.
Arminghall	1347–48	6*s*.8*d*.

Italics = lactage only

Sources: PRO SC6/935/19, SC6/1090/4, DL29/288/4719, 4720, 4734; NRO Diocesan Est/2, 2/15 and 17, Diocesan Est/10, NRS 2848 12 F1, DCN 60/25/1–3, DCN 62/1 and 7, NRS 2796 12E2; BL Add. Roll 26060, Add. Charter 15199–202

and Martham in the period 1326–27, 53 per cent of all cheeses produced were sold; this proportion rose to 66 per cent at Martham and 96 per cent at Thornham.[47] In the heartland of cheese production, in east-central and northeastern Norfolk, there is a clear implication that specialization in dairying was a response to market opportunities. Between 1307 and 1315 the Norwich Cathedral Priory manor at Attlebridge sold 59 per cent of its cheeses and 79 per cent of its butters; between 1305 and

[47]NRO DCN62/1.

1338 the sacristan of Norwich's manor at Bauburgh sold 63–90 per cent of its cheeses and 80 per cent of its butters; in the years 1296–97 the queen's manor at Cawston sold 93 per cent of its cheeses and 94 per cent of its butters (the remainder being paid as tithe); during the 1270s the royal manor at Costessey sold 94 per cent of its cheeses and butters; and during the same decade the Broadland manor of Acle sold its entire output of cheese and butter.[48]

Norfolk's dense network of over 120 markets must have provided a ready outlet for this dairy produce, much of which may eventually have found its way onto the Norwich food market. Costessey, for example, for which a series of detailed though damaged dairy accounts survive from the 1270s, was actively engaged in the large-scale commercial production of butter and cheese and possibly traded directly with Norwich.[49] It is situated in the valley of the River Wensum, four miles northwest of the city, and maintained a herd of 25 to 30 milking cows. This was a well-managed herd: on the three occasions for which there are legible figures of the number of cows kept and calves born, the calving rate was 100 per cent (although two out of three calves born were subsequently sold). Such a high fertility rate reflects both a favorable ratio of labor to animals—the herd was under the charge of a permanent cowman, while the dairy was staffed by a permanent dairymaid—and the careful culling of aged and sterile females. This high fertility rate was matched by similarly favorable milk yields, to judge from the quantities of cheese and butter produced and sold. One year the sale of cheese, butter, milk, and calves produced by the 25 cows kept on the demesne yielded an income of £6 8*s*.4*d*., another—

[48]NRO DCN61/11–13 and 16–19; PRO SC6/1090/4, SC6/933/13, SC6/929/1–7.
[49]PRO SC6/933/13.

this time from a herd of 26 cows—an income of £6 12*s*.0*d*. On both occasions this was equivalent to a gross income for its dairy produce alone of just over 5*s*. per cow (at least 1*s*. a head higher than the going rate at which cows were farmed for their lactage).

This intensive pastoral regime—producing butter, cheese, calves, pigs, bacon, fat lambs, poultry, and eggs for consumption and for sale—attained its fullest development during the third quarter of the fourteenth century (see Table 4), as resources were released by a contracting but still buoyant arable sector and before demand had been too seriously eroded by changing dietary preferences and prolonged population decline.[50] With the demise towards the end of the fourteenth century of the conditions of land shortage and labor surplus that had brought the system into being, this high-cost system of pastoral husbandry lost its competitive edge.

Spiraling wage rates and falling prices squeezed profit margins, and in certain markets Norfolk producers found themselves undercut by farmers whose more extensive methods meant they had lower production costs.[51] During the final quarter of the fourteenth century a trend towards more land-extensive and less labor-intensive forms of pastoralism is thus increasingly apparent in Norfolk.[52] It is at this point that

---

[50]Christopher Dyer has identified a significant decline in the amounts of dairy produce consumed by harvest workers at Sedgeford in Norfolk from the closing decades of the fourteenth century, as increasing quantities of meat were consumed: "Changes in Diet in the Late Middle Ages: The Case of Harvest Workers," *Agricultural History Review* 36/1 (1988): 25–28.

[51]For evidence of a pronounced swing from arable to pastoral production within the country at large during the fourteenth century, see Bruce M. S. Campbell, "Land, Labour, Livestock, and Productivity Trends in English Seignorial Agriculture, 1208–1450," pp. 144–82 in Campbell and Overton, *Land, Labour and Livestock*, pp. 153–59.

[52]For a parallel trend in arable production see Campbell, "Land, Labour, Livestock, and Productivity Trends," pp. 144–49.

TABLE 4
NORFOLK: TRENDS IN DEMESNE PASTORAL HUSBANDRY 1250–1449 (50-YEAR STAGGERED MEANS)

Years	Mean per Demesne							
	Livestock Units	Livestock Units per 100 Grain Acres	Percentage of Livestock Units					Oxen per Horse
			Draught Animals	Cattle other than Oxen	Young Cattle per 100 Adults	Sheep	Swine	
1250–1299	46	31	35	45	74	17	3	0.8
1275–1324	47	33	32	46	76	20	3	0.9
1300–1349	46	36	27	47	67	24	3	0.6
1325–1374	47	41	23	50	61	24	4	0.5
1350–1399	49	45	20	52	54	25	4	0.4
1375–1424	43	36	20	47	43	29	3	0.3
1400–1449	44	31	20	39	39	38	4	0.3

Source: Norfolk manorial accounts

Method: For calculation of livestock units see n. 16 above.

small, one-man plough teams began to emerge, dairy herds were frequently leased out, and sheep, with their lower unit costs and capacity to produce milk, meat, and wool at a time of uncertain markets, began to gain at the expense of cattle. In fact, when Chaucer penned his portrait of the Norfolk reeve the long-established character of pastoral husbandry in that county was on the threshold of a far-reaching transformation.[53]

It would be misleading to represent this intensive pastoral response to land shortage, labor abundance, and favorable market opportunities as unique to Norfolk, for it clearly was not. The cluster analysis of Norfolk pastoral-farming types may be matched against a corresponding cluster analysis of pastoral farming on a sample of 660 demesnes within the country as a whole during the period 1250–1349.[54] Six basic pastoral types emerge, as set out in Table 5 in descending order of intensity.

Of these six types it is the first, Cluster 1, that corresponds most closely to the intensive husbandry described in Norfolk, sharing the characteristics of a favorable ratio of nonworking to working animals, a significant contribution by horses to draught power, dominance of the nonworking sector by cattle (with some demographic bias towards adults), and the relegation of swine and especially sheep to subsidiary positions. Demesnes belonging to this cluster grouping, and therefore sharing in some measure these characteristics, were widely distributed and, as Figure 3 shows, examples may be found in most parts of the country. In part, this wide geographical scatter represents the lumping together of upland demesnes, which practiced large-scale cattle farming using extensive methods, with lowland demesnes whose cattle farming assumed a

[53]Overton and Campbell, "Norfolk Livestock Farming," pp. 383, 391, 393.

[54]Campbell and Power, "Mapping the Agricultural Geography."

TABLE 5

ENGLAND: CLASSIFICATION OF DEMESNE PASTORAL-FARMING TYPES 1250–1349

	Cluster	No. of Demesnes	% of Demesnes	Mean Percentage of Total Livestock Units					
				Horses	Oxen	Adult Cattle	Young Cattle	Sheep	Swine
1250–1349	1	100	15	11	18	34	29	6	1
	2	116	18	19	28	23	20	6	5
	3	132	20	11	21	18	14	33	3
	4	75	11	10	26	3	2	58	2
	5	40	6	25	37	6	3	12	17
	6	43	7	53	38	2	1	5	2
	7	154	23	15	79	2	2	2	1
	Overall	660	100	17	38	14	12	17	3
	*Norfolk**	*69*	*54*	*19*	*14*	*41*	*18*	*5*	*4*

*Norfolk** = mean characteristics of Cluster 1 demesnes in Norfolk

Source: National sample of manorial accounts and Norfolk manorial accounts

Method: Cluster analysis (relocation method), seven-cluster solution. For calculation of livestock units see n. 16 above.

different form and was underpinned by very intensive methods. Within such lowland contexts the distribution of these demesnes was strongly biased towards the south and east, and in certain specific localities they became the predominant pastoral type. Eastern and central Norfolk, of course, stand out, as do much of southern and eastern Suffolk, Huntingdonshire, the Soke of Peterborough, and the Lincolnshire fen-edge, east Hertfordshire, the immediate environs of London, southeast Kent, southern Hampshire, and south Somerset and east Devon. Some of these concentrations reflect the presence of environmental opportunities that were particularly advantageous for the development of pastoral husbandry—the fen-edge, the vicinity of Romney Marsh, and east Devon—but at least as important were demographic pressure, commercial opportunity, and institutional arrangements that allowed the evolution of more individualistic forms of pastoral husbandry.[55] Such factors certainly seem to have been the common denominators of the concentrations of these demesnes in East Anglia and the Home Counties.

In all of these respects the situation prevailing in Norfolk seems to have represented something of an extreme. Norfolk was fourteenth-century England's most populous county, contained England's largest provincial city, possessed a common-

---

[55]For the fen-edge see J. Ambrose Raftis, *The Estates of Ramsey Abbey: A Study in Economic Growth and Organization* (Toronto, 1957), pp. 129–58; Biddick, *Other Economy*. For Romney Marsh and east Kent see R. A. L. Smith, *Canterbury Cathedral Priory: A Study in Monastic Administration* (Cambridge, 1943), pp. 146–65, and P. F. Brandon, "Farming Techniques: South-Eastern England," pp. 312–25 in *Agrarian History*, vol. 2. For east Devon see N. W. Alcock, "An East Devon Manor in the Later Middle Ages. Part I: 1374–1420. The Manor Farm," *Report and Transactions of the Devonshire Association* 102 (1970): 141–87; K. Ugawa, "The Economic Development of Some Devon Manors in the Thirteenth Century," ibid. 94 (1962): 630–83; John Hatcher, "Farming Techniques: South-Western England," pp. 383–98 in Hallam, *Agrarian History* 2: 395–98; H. S. A. Fox, "Peasant Farmers, Patterns of Settlement and *pays*: Transformations in the Landscapes of Devon and Cornwall during the Later Middle Ages," pp. 41–73 in Robert Higham, ed., *Landscape and Townscape in the South West* (Exeter, 1989), pp. 57–64.

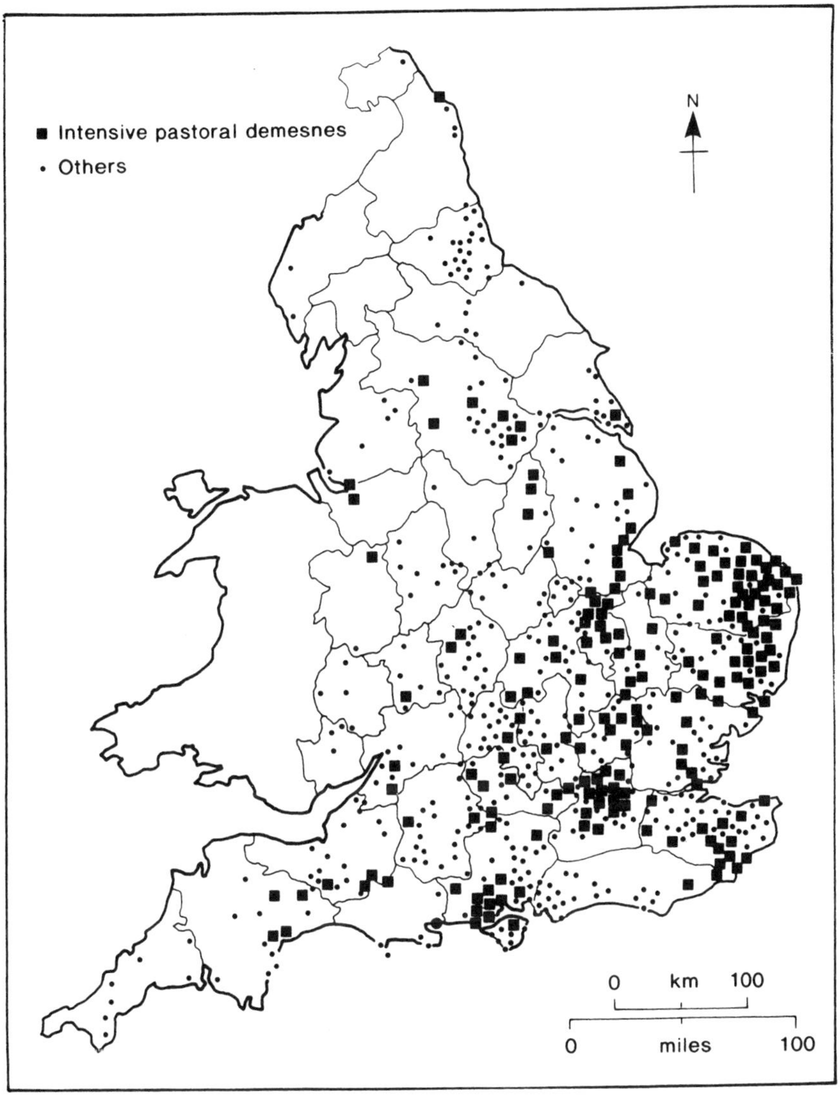

Figure 3. England: distribution of "intensive" pastoral-farming demesnes, 1250–1349 (for explanation see Cluster 1, Table 5)

field system that allowed a high degree of flexibility and freedom to the cropping of land and herding of animals, and conducted a lively trade in agricultural products both with the

Continent and other parts of the country.[56] The most progressive and productive of its arable demesnes, mostly concentrated in the east and northeast of the county, combined heavy labor inputs with the best available agricultural technology in order to obtain levels of output per unit area that were unsurpassed elsewhere in the country.[57] It can be no surprise to discover that pastoral husbandry in this area, although economically subordinate to arable production, was similarly advanced. This is borne out by a comparison of the mean characteristics of Cluster 1 demesnes in Norfolk with their mean characteristics in the country at large. As Table 5 shows, the distinctive traits of this most intensive and developed of pastoral regimes were at their most accentuated in Norfolk: here working animals comprised a smaller proportion of total livestock, horses made a greater relative contribution to draught power, and dairy cattle most eclipsed other classes of animal in their relative importance.

By medieval standards Norfolk's pastoral sector performed remarkably well. That it did so bears witness to the ability of medieval agriculture to evolve mixed-farming systems in which

[56]Alan R. H. Baker, "Changes in the Later Middle Ages," pp. 186–247 in H. C. Darby, ed., *A New Historical Geography of England* (Cambridge, 1973), pp. 190–92; Elizabeth Rutledge, "Immigration and Population Growth in Early Fourteenth-Century Norwich: Evidence from the Tithing Roll," *Urban History Yearbook* (1988): 15–30; Bruce M. S. Campbell, "The Regional Uniqueness of English Field Systems? Some Evidence from Eastern Norfolk," *Agricultural History Review* 29/1 (1981): 16–28; Anthony Saul, "Great Yarmouth in the Fourteenth Century: A Study in Trade, Politics and Society," unpub. D.Phil. thesis (Univ. of Oxford, 1975); R. A. Pelham, "Medieval Foreign Trade: Eastern Ports," pp. 298–329 in H. C. Darby, ed., *Historical Geography of England before 1800* (Cambridge, 1936), pp. 301; Vanessa Parker, *The Making of Kings Lynn: Secular Buildings from the 11th to the 17th Century* (London, 1971), pp. 3–16; Bruce M. S. Campbell, James A. Galloway, Derek Keene, and Margaret Murphy, *A Medieval Capital and Its Grain Supply: Agrarian Production and Distribution in the London Region c.1300*, Historical Geography Research Ser., no. 30 (1993), pp. 181–82.

[57]Campbell, "Agricultural Progress"; idem, "Arable Productivity in Medieval England: Some Evidence from Norfolk," *Journal of Economic History* 43/2 (1983): 379–404.

the arable and pastoral sectors were complementary rather than competitive.[58] It is particularly notable that this crucial development was taken furthest within a county more remarkable for its institutional and economic attributes than for any intrinsic environmental suitability to pastoralism. That it did not develop further and evolve into the sort of intensive mixed-farming system based more or less exclusively on fodder cropping, of the kind with which the county was to become inextricably associated during the so-called agricultural revolution of the eighteenth century, was a function of available technology and the level of market demand for livestock and their products, both of which were to advance significantly by the late seventeenth century.[59] As it was, arable and pastoral husbandry together attained a pitch of development in fourteenth-century Norfolk that was not to be matched in much of the rest of the country until several centuries later.

Herein may lie the explanation for Chaucer's choice of Norfolk as the county of origin of his reeve, an origin that he is at pains to reinforce by endowing Oswald with a number of additional Norfolk attributes—the stot upon which he rides, the

---

[58]As Biddick observes, "The simple relations between pastoral and cereal husbandry posited by Postan do not adequately account for the comparative commercialization of haulage, dairying, and wool production over the thirteenth century and the trade-offs made between producing such products and selling pastoral resources to others to produce them" (*Other Economy*, p. 130). See also Mark Bailey, *A Marginal Economy? East Anglian Breckland in the Later Middle Ages* (Cambridge, 1989), pp. 85–96.

[59]Campbell and Overton, "New Perspective," pp. 90–93, 102–03. For inadequacies in the scale of market demand as an explanation for the lack of agricultural progress in many parts of medieval England, see Bruce M. S. Campbell, "People and Land in the Middle Ages, 1066–1500," pp. 69–121 in R. A. Dodgshon and R. A. Butlin, eds., *An Historical Geography of England and Wales*, 2nd ed. (London, 1990), p. 83; Campbell, "Ecology versus Economics in Late Thirteenth- and Early Fourteenth-Century English Agriculture," in Del Sweeney, ed., *Agriculture in the Middle Ages: Technology, Practice, and Representation* (Philadelphia, 1995), pp. 76–108.

accent with which he speaks, his residence upon a heath, and, as it would now appear, the combination of livestock that he tends.[60] For as well as hailing from Norfolk, the reeve is also portrayed as a shrewd, hard-bargaining, and experienced husbandman who is fraudulent to his young master and oppressive to those socially beneath him. By associating these personal characteristics with a specific Norfolk provenance, John Matthews Manly has argued that Chaucer was alluding to an actual case of mismanagement on the estates of the earls of Pembroke (which included the manor of Foxley, in which Bawdeswell was partially situated), of which Chaucer apparently had firsthand knowledge.[61] This claim has failed to convince subsequent scholars—especially as the events in question happened almost 20 years before the General Prologue was written—and instead it is more likely that Chaucer had in mind, if not a particular person and actual events, then at least a recognizable social and regional type who would have been immediately familiar to his audience. Both Jill Mann and Alan J. Fletcher have drawn attention to a contemporary stereotype of Norfolk people as crafty, cunning, and avaricious—characteristics shared by the reeve.[62] The distinctive character of

[60]The stot, or work horse, was a quintessentially East Anglian animal; at this date stots were present in significant numbers only on demesnes in East Anglia and the southeast. Hamlets and isolated farms were a characteristic feature of Norfolk settlement, and during the later Middle Ages became especially associated with the edge of commons, heaths, and wastes: "The vast majority of medieval farms looked out over commons" (Tom Williamson, *The Origins of Norfolk* [Manchester, 1993], p. 167); David Dymond, *The Norfolk Landscape* (London, 1985), pp. 99–102. The reeve's tale makes deliberate use of northern dialect forms, blended with others that would have been current in East Anglia; see J. R. R. Tolkien, "Chaucer as Philologist: *The Reeve's Tale*," *Transactions of the Philological Society* (1934): 1–70, esp. pp. 6–7.

[61]*Some New Light*, pp. 84–94.

[62]Mann, *Chaucer and Medieval Estates Satire. The Literature of Social Classes and the General Prologue to the Canterbury Tales* (Cambridge, 1973), pp. 166; Fletcher, "Chaucer's Norfolk Reeve," *Medium Ævum* 52/1 (1983): 100–03.

Norfolk's medieval economy and society, to which its pastoral husbandry bears witness—more commercialized, technologically developed, competitive, and individualistic than that prevailing in much of the rest of the country—possibly explains how such a reputation came to be earned and why Norfolk people were seen to be so different.[63]

[63]Oswald's shrewd and hard-bargaining character was certainly shared by many of his real-life Norfolk contemporaries: Elaine Clark, "Debt Litigation in a Late Medieval English Vill," in J. Ambrose Raftis, ed., *Pathways to Medieval Peasants* (Toronto, 1981), pp. 247–79; Bruce M. S. Campbell, "Population Pressure, Inheritance and the Land Market in the Fourteenth-Century Peasant Community," in Richard M. Smith, ed., *Land, Kinship, and Life-Cycle* (Cambridge, 1984), pp. 87–134.

X

Cluster analysis and the classification of medieval demesne-farming systems

JOHN P. POWER and
BRUCE M. S. CAMPBELL

ABSTRACT
A methodology for classifying historic farming systems is developed with reference to the demesne sector of English agriculture during the period 1250–1349. Data on arable and pastoral production are combined from a sample of approximately 1000 manorial accounts representing some 388 individual demesnes. These demesnes are classified on the basis of the crops grown, livestock kept, and ratio of the latter to the former using a combination of cluster analysis and discriminant analysis.

Cluster analysis is subject to a number of limitations and the typologies it produces may be no better than the techniques employed. The method adopted here, therefore, has been to establish a stable typology through the construction of stereotypes. These are defined by those cases which are similarly classified by different clustering techniques. The stability of these 'core' groups is then assessed through the reapplication of clustering techniques and the application of discriminant analysis. On both counts the 'core' groups emerge as clearly articulated. Function coefficients obtained from discriminant analysis are then used to assign each of the remaining cases to the 'core' group to which it has the greatest probability of belonging.

The eight basic farming types thereby identified are evaluated and their character and distribution discussed. These farming types can be ranked on a scale from intensive to extensive. Overall, agriculture is shown to have been more extensive than intensive and consequently capable of considerable further development. The most developed systems are shown to have been concentrated in the east and south east where market demand, and the growth of concentrated urban demand in particular, was greatest.

INTRODUCTION

Classification of historic farming systems presents a dilemma. On one hand, all farms were unique, since in no two cases were the factor endowments of land, labour, capital, and enterprise exactly the same. On the other, the constraints of pre-industrial agricultural technology ensured that most farms shared certain features in common. The choice of crops that could be grown and livestock that could be kept was strictly limited and a dependence upon animals for draught power and manure ensured that most farms were, in effect, mixed. As a result farms differed more in terms of the relative importance and the methods of management of the same narrow range of crops and livestock than in terms of the actual crops they grew and the stock they kept.[1] The challenge in producing a classification of farming systems is therefore to devise a scheme that subsumes the many minor variations in farm enterprise while making explicit the more significant differences.

This paper is therefore concerned with developing a methodology capable of providing a national classification of farming systems from available historical evidence which is as objective as possible (although the methodology is also applicable at smaller scales). This methodology is developed within the empirical context of the demesne sector of English agriculture between 1250 and 1349. With an economy as overwhelmingly agrarian as that of the late thirteenth and early fourteenth centuries it is a matter of no small historical importance to establish the extent to which farming systems had become spatially and functionally differentiated in the crops and livestock they produced and the methods and intensity of their production.[2] Indeed, it bears directly upon the emerging debate about the extent to which the English economy experienced growth as well as expansion during the quarter millenium terminated by the Black Death.[3]

The documentation available for the late thirteenth and early fourteenth centuries is rich in both quality and quantity. It comprises the accounts rendered annually by the reeve or bailiff charged with the management of a demesne. For each farming year – usually Michaelmas to Michaelmas – income and expenditure and the crops sown and livestock kept are recorded in great detail.[4] As a basis for the classification of farming systems this evidence is superior to that available for many other periods. It is highly reliable, relates to specific working farms which – insofar as they were all seignorial demesnes – are broadly comparable in scale, and has the merit of recording livestock numbers on those farms at a more-or-less consistent time of year.[5] The limitations of this evidence are its bias towards the landlord sector and the estates of ecclesiastical landlords in particular, its uneven geographical distribution (data are particularly deficient for the remoter upland counties of the west and north), and the fact that an adequate spatial coverage is only obtainable at the sacrifice of chronological precision. Additional problems derive from the failure of some accounts to specify the acreage sown, the occasional use in others of customary measures (notably non-statute acres in the case of crops and long hundreds in the case of sheep), and a general lack of information on the precise breeds of animals and species of crops.

Although manorial accounts survive in considerable numbers analysis is here restricted to a national sample of approximately 1000, representing some 388 different demesnes. Where possible data have been extracted from several different accounts and it is the means from these that provide the basis of classification. The task of assembling this sample was considerable, for the accounts are scattered through a wide range of public and private archives.[6] Much of the data was collected by John Langdon in conjunction with his researches into the use of draught animals in medieval agriculture, but this has been supplemented by some additional information, particularly for Norfolk (which stands out as by far the best represented county in the sample).[7] As a sample it is by no means ideal, being representative of neither the geographical distribution of surviving records nor the original distribution of directly-managed seignorial demesnes, and in the long term can undoubtedly be considerably improved upon.[8] Nevertheless, it is the best currently available and is sufficient to test and demonstrate the chosen methodology and provide a provisional classification of medieval demesne-farming systems. The latter may then provide the context for further more detailed studies from which more refined classifications may emerge in due course.

The principal technique used here for deriving this agricultural classification is cluster analysis. The potential of this technique for classifying past farming systems was first recognized by Mark Overton, who pioneered its use with reference to information on crops and livestock derived from probate inventories of the sixteenth, seventeenth, and early eighteenth centuries.[9] His example has been followed by Paul Glennie, also using probate inventories. Neither, however, pays sufficient regard to the weaknesses of the technique as a tool for classification.[10] These are addressed by Campbell and Power with reference to the separate classification of medieval arable- and pasture-farming systems.[11] This paper develops that methodology in several significant respects and produces a classification of medieval demesne-farming systems based upon both crops and livestock.[12]

PROBLEMS ASSOCIATED WITH CLUSTER ANALYSIS

The appeal of cluster analysis for the classification of farming systems lies in its capacity to subdivide a dataset into groups based upon the degree of similarity or dissimilarity between cases measured across all the individual variables (e.g. the crops sown and livestock kept) of which they are composed. There is, however, an inherent danger that the groups thus defined may be as much a product of the particular clustering technique used as of any intrinsic grouping within the data. Much hinges upon the proximity measures employed in the definition of inter-case differences since it is these that form the primary input for most clustering techniques.[13] Different clustering techniques also define group membership in different ways and hence are sensitive to different types and shapes of pattern within the data.[14] Where a clear view can be obtained of the location of individual groups and their constituent cases within multi-dimensional space some evaluation of cluster solutions is possible.[15] Nevertheless, this is usually contingent upon some form of data generalization and is normally only feasible for relatively small numbers of cases. Much also depends upon the number of groups or clusters a given technique is instructed to locate. Yet *a priori* knowledge of the number of groups present within a dataset is very much the exception rather than the rule. Indeed, such knowledge presupposes a clear understanding of the definition of group types, which is precisely what cluster analysis is supposed to provide. Some clustering techniques do attempt to indicate an optimum number of groups but the measures provided are not always reliable and are usually derived by assessing the variance of the particular groups found.

It follows that the quality of any solution provided by cluster analysis is only as good as the appropriateness of the techniques applied and the choice of procedures followed. To be of utility to other researchers the method should be replicable and provide some assessment of the reliability or robustness of the solutions derived. That means applying not one but a variety of clustering techniques, since it is only by comparison of the results thereby obtained that the consistency of the solutions may be evaluated and areas of common ground identified. Evaluation should also take account of the susceptibility of the results to slight methodological variations in the form in which the variables are input to the techniques; for instance, taking different proximity measures and sub-samples from the data. Where consistent results emerge from different methods it is plausible to assume that they are less a statistitical artefact than a product of genuine configurations within the data. No two techniques will normally return identical classifications but they can provide a valuable guide to the optimum number of groups present and focus attention on the characteristics of those 'core' cases common to all solutions and therefore most typical of the groups identified. Cases outside of these 'cores', which do not fall decisively into any one category, can then be assigned probabilities of membership and classified accordingly.[16]

THE VARIABLES AND DATA DEFINITION

Farming is a complex and multi-faceted activity and farming systems are capable of classification on a variety of criteria. The choice of variables depends upon the object of the classification. With farm enterprise, for example, it is the basic production mix of crops and livestock that commands attention. Application of cluster analysis to raw figures of the acres sown and animals stocked is, however, inappropriate. Indeed, three basic problems have to be overcome before clustering techniques can be applied.

The problem of scale

Farms varied considerably in size, both in terms of the extent of their sown acreage and the numbers of animals stocked, with the result that application of cluster analysis to data expressed in absolute terms simply produces groupings based upon differences of scale rather than enterprise (a problem which is further compounded by the occasional use of customary acres). Analysis should therefore take place on the basis of relative rather than absolute measures.

The problem of units

Classification of farm enterprise on the basis of both crops and animals requires comparison of unlike units – notably, livestock with crops and one class of animal with another. The problem of different classes of animal can be resolved by means of a system of weighting; converting the raw livestock numbers into livestock units. The system employed here is based upon feed requirements and is a modified version of that used by J. A. Yelling and R. C. Allen for the early modern period.[17] In contrast, the problem of equating livestock with crops is altogether more intractable. The normal solution is to convert the raw figures of acres sown and animals stocked into some standard unit of measurement. Thus, J. T. Coppock in his *Agricultural Atlas of England and Wales* employs labour requirements expressed as standard man-days per acre, whereas Overton and Glennie in their analyses of early modern farming employ valuations.[18] Unfortunately, neither conversion method is compatible with medieval data. Twentieth-century norms of labour input are hardly applicable to farming operations undertaken under the organizational and technological conditions prevailing in the Middle Ages. Nor is there any simple and reliable method of valuing the crops and livestock recorded in the accounts. Some alternative and more straight-

X

TABLE I. *Variables used for analysis*

Variable	*Ceiling*
CROPS (acres):	
R_1 ratio of winter to spring grain	3·68
R_2 wheat as % of winter grain	
R_3 oats as % of spring grain	
R_4 ratio of legumes to total grain	none
LIVESTOCK (units):*	
R_5 ratio of horses to oxen	18·16
R_6 ratio of non-working to working animals	none
R_7 ratio of immature to adult cattle	3·72
R_8 sheep as % of non-working animals	
R_9 swine as % of livestock units	
CROPS AND LIVESTOCK:	
R_{10} total livestock units* per 100 sown acres	none

Notes: *horses × 1·0; oxen × 1·2; mature cattle × 1·2; immature cattle × 0·8; sheep × 0·1; swine × 0·1

forward method of converting the raw data must therefore be found.

The problem of closed numbers

Highly correlated variables can bias cluster analysis. Any method of standardization or comparison which employes percentages may therefore be undesirable, particularly when these percentages are derived from a common base where the use of the same denominator introduces a measure of correlation between the variables.[19] Some alternative method of expressing the relative importance of different variables should therefore be employed.

A solution to all three of these problems lies in the employment of ratios. Ratios provide a relative rather than absolute measure, obviate the need to convert crops and livestock into standardized units, and avoid the problem of closed-numbers. It is, however, essential to chose a set of ratios which embodies all the main features of the activity being classified since the relationships expressed by ratios may predispose the analysis towards certain types of solution. This presumes a knowledge and understanding of the activity under investigation. Where ratios are inappropriate and proportions have to be employed it is crucial that the proportions are calculated using independent bases. In the scheme adopted here six ratios and four independent proportions have been employed; five relating to livestock, four to crops, and one to the relationship between them (Table I).

One problem with ratios is their capacity to produce extreme and often misleading values, especially where the denominator is zero. Such cases have been resolved by imposing a maximum four standard deviations away from the mean ratio calculated on the remaining cases (four standard deviations were chosen so as to impose an extreme maximum value but limit its influence to a specified ceiling). Logarithmic ratios to base 10 have also been employed so that cases with numerators less than denominators are not restricted to the range 0·0–1·0. To keep all variables on the same scale the logarithms of variables measured as straight proportions have also been used. Finally, to remove correlations between variables and eliminate any bias arising from the uneven representation of crop and livestock variables, a principal components analysis was conducted and the scores of the first six components – which account for almost 83 per cent of total variance within the ten chosen variables – stored for use in cluster analysis (Table II).[20] These scores provide a set of similarity measures based on orthogonalized variables and help to sharpen the focus of study.[21]

THE METHOD

The method relies on the creation of 'stereotypes' through identification of cluster 'cores'.[22] Cluster 'cores' are deemed to comprise those cases which are similarly classified by different clustering techniques. In all, three techniques were applied to the data – Ward's, K-means, and Normix. Both Ward's and the K-means methods were applied to a squared euclidean-distance matrix calculated from the principal components scores. Normix uses a method of 'maximum likelihood' to estimate a multivariate mixture of cluster distributions which allocates cases to those groups to which they have the greatest probability of membership. Here, too, the principal components scores were used. The Normix technique contains a test for the optimal number of clusters which evaluates the presence of k and k-1 cluster groups (where k = number of clusters).[23] Reviews of this test by D. Wishart and Brian Everitt conclude that, although not wholly reliable, it is nonetheless one of the best available.[24]

As both the K-means and Normix methods are optimization techniques different starting configurations were supplied to reduce the risk of locating local rather than global optimum solutions.[25] Starting configurations were derived randomly and from the

TABLE II. *Principal components correlation matrix*

Variable	*PC1*	*PC2*	*PC3*	*PC4*	*PC5*	*PC6*
R_1	−0·224	−0·232	0·582	0·013	0·272	0·673
R_2	−0·242	−0·346	0·668	0·231	0·060	−0·361
R_3	−0·414	0·490	−0·027	0·421	0·386	0·030
R_4	0·456	−0·525	0·205	−0·153	−0·406	0·075
R_5	0·535	−0·349	−0·473	0·095	0·300	0·217
R_6	0·909	0·193	0·084	−0·009	0·173	0·022
R_7	0·696	0·225	0·089	0·470	−0·187	0·193
R_8	0·360	0·259	0·248	−0·707	0·415	−0·118
R_9	0·331	−0·573	−0·032	0·298	0·445	−0·325
R_{10}	0·534	0·484	0·420	0·165	−0·120	−0·148
Variance	26·1%	15·3%	13·1%	11·0%	9·3%	8·2%

Note: for definition of R_1–R_{10} see Table I

output of Ward's method using three different inter-case distance matrices. The output from one optimization technique was also used as input to the other. The optimum solutions produced by Normix for from ten to two cluster groups were then evaluated and the hypothesis of the presence of eight cluster groups accepted ($P<0{\cdot}05$). The classifications produced by each of the three methods were then compared to assess their degree of correspondence and identify cluster 'cores'.

If all three methods returned identical classifications there would be little doubt that clear clustering of cases exists. Such an outcome is, however, improbable, since individual techniques are rarely entirely successful at identifying all group structures even when these are well-defined. Conversely, where there is little coincidence between the three classifications emergent groups are more likely to be arbitrary partitions of uniform distributions. In fact, the solution obtained from the three methods employed here lies midway between these two extremes. The lowest observed two-way match was that between the Ward's and the Normix methods at 68 per cent, while, overall, there was a three-way match of 57 per cent. This level of correspondence is high enough to suggest the presence of genuine clustering within the data, but not so high as to imply a clear and consistent pattern of demarcation between the various clusters.[26]

Grouping is most clearly defined in the case of Cluster Two ('light-land intensive') and Cluster Five ('sheep-corn husbandry'), whose size and composition remain essentially the same in all three solutions. There is also a significant degree of correspondence in the definition of Cluster Three ('mixed-farming with cattle'), Cluster Seven ('extensive arable husbandry'), and Cluster Eight ('oats and cattle'). But with Cluster One ('intensive mixed-farming'), Cluster Four ('arable husbandry with swine'), and Cluster Six ('extensive mixed-farming'), the three-way match is significantly poorer. The problem is most acute in the case of Cluster Four where a search for a common 'core' yields only two demesnes similarly classified by all three solutions. Such problems of definition may arise from irregularities in the shapes of these three clusters. Alternatively, it may be a consequence of less clearly defined clustering patterns. Indeed, the dividing line between 'intensive mixed-farming' (Cluster 1) and 'extensive mixed-farming' (Cluster 6) appears to have been a fine one. Significantly, many cases are allocated to either one or the other, producing a minimum two-way match in the combined composition of the two clusters of 52 per cent and a maximum of 59 per cent. Since the match between the Normix and the K-means methods was generally better for Clusters One, Four, and Six than that of either method with Ward's, the demesnes comprising the 'cores' of these three clusters were identified less strictly according to the two-way match between these two methods.

The 261 'core' demesnes thus identified can be assumed to represent the nuclei of their respective clusters and to embody their most distinctive features. This assumption was tested through the re-application of the K-means and Ward's methods to the sub-sample of cluster 'cores': in both cases the classifications produced mirrored the original 'core' group identities. The exercise was repeated using the original ten variables in their untransformed state and the match with the 'core' classification was again

TABLE III. *Fisher's linear discriminant functions*
Classification function coefficients

	Cluster 1	2	3	4	5	6	7	8
d_1	−3·68	−4·79	0·52	−2·66	−2·84	−3·97	−5·07	−26·55
d_2	23·26	8·09	22·38	21·07	24·13	27·91	28·57	0·63
d_3	5·15	2·73	14·75	9·08	9·97	13·72	13·86	15·96
d_4	−4·25	−5·78	−8·24	−3·72	−3·76	−7·89	−6·52	−6·66
d_5	−3·18	−1·12	−4·32	−7·87	−10·58	−6·39	−8·66	−7·74
d_6	2·83	4·56	1·48	0·18	1·47	0·66	−10·56	−5·59
d_7	−2·10	−4·90	−0·78	−23·01	−24·41	−0·67	−19·92	−11·10
d_8	0·05	0·08	0·02	0·09	0·32	0·15	0·11	0·10
d_9	0·22	0·11	0·18	1·21	0·33	0·15	0·29	0·30
d_{10}	−12·04	−17·92	−18·34	−24·12	−16·92	−8·86	−17·52	−12·76
a	−18·56	−15·70	−28·78	−58·68	−65·52	−32·65	−67·04	−44·72

Notes: d_1–d_{10} = discriminant function scores for R_1–R_{10} (for definition see Table I)
a = constant

found to be high, with 95·3 per cent of the 261 cases correctly relocated. This second test confirms that the cluster 'core' identities are reasonably resilient to variable transformations. It also demonstrates that there is no serious dilution of inter-core differences as a result of the restriction of analysis to the first six principal component scores. The high degree of success with which the 'core' groups were reconstructed would also appear to confirm the possession by each of a well defined identity with clear differences between them. This should come as no surprise for the 'core' groups were created to maximize cluster differences and the readiness with which they are reconstructed indicates no more than the success of this objective. But what of the remaining 127 'peripheral' cases representing virtually a third of all demesnes?

The cluster identities of the residual demesnes excluded from the 'core' were determined using discriminant analysis. This method has particular utility for evaluating classifications and determining the membership of additional cases.[27] When applied to the 261 'core' cases it had 98·1 per cent success at reclassifying them correctly. Given this close fit the remaining 127 cases were allocated between the eight cluster groupings by substituting the R_1–R_{10} values into eight discriminant equations and accepting the group identity of the highest score:

$$\text{Max}\ (f_1, f_2, f_3, f_4, f_5, f_6, f_7, f_8)$$

where the f score for each group 1 to 8 is given by:

$$\text{Constant} + d_1(R_1) + d_2(R_2) + \ldots d_{10}(R_{10}) \qquad (1)$$

and d_1–d_{10} are the function coefficients for each group as given in Table III. These coefficients were used to assign the 'peripheral' cases based on the 'core' characteristics of each of the eight clusters as measured across the ten original variables. The resultant match between the 'core' and predicted discriminant group membership of the 'peripheral' cases is shown in Table IV. The same discriminating functions can also, of course, be used to predict group membership of any other demesnes for which data may become available.

DEMESNE-FARMING SYSTEMS 1250–1349: A CLASSIFICATION

The classification of demesne-farming systems which emerges from these techniques is most readily understood in terms of the mean crop and livestock characteristics of each of the eight cluster groupings (expressed as percentages of total sown acreage and total livestock units), together with the mean number of livestock units per 100 sown acres (which expresses the relative balance between the two sectors). Table V summarizes the picture with reference to the 261 'core' demesnes, whose distribution is shown in Figure 1. Table VI gives corresponding information for all 388 demesnes, as classified according to discriminant analysis, and their distribution is shown in Figure 2. As will be noted, it is the demesnes comprising the 'core' groups, as mapped in Figure 1, whose distribution exhibits the sharpest geographical focus. The eight farming types are discussed in order

TABLE IV. *Match between 'core' groups and predicted discriminant groups for cases outside the 'core'*

	Discriminant groups: Cluster								
	1	*2*	*3*	*4*	*5*	*6*	*7*	*8*	*Total*
'Core' groups:									
Cluster 1	26	0	0	0	0	0	0	0	26
Cluster 2	1	27	0	0	0	0	0	0	28
Cluster 3	0	0	43	0	0	0	0	0	43
Cluster 4	0	0	0	16	0	0	0	0	16
Cluster 5	0	0	0	0	30	0	0	0	30
Cluster 6	2	0	2	0	0	48	0	0	52
Cluster 7	0	0	0	0	0	0	59	0	59
Cluster 8	0	0	0	0	0	0	0	7	7
Unassigned	37	7	27	3	15	17	21	0	127
Total	66	34	72	19	45	65	80	7	388

progressing from the most intensive to the most extensive, the degree of intensity being gauged from the internal characteristics of each farming type.

Cluster One: 'intensive mixed-farming'

This farming type comprises those demesnes which lay at the intensive extreme of the mixed-farming spectrum. The three clustering techniques differ quite significantly in their definition of this farming type with the result that a 'core' of only 26 demesnes is common between them. To these a further 40 'peripheral' demesnes are added by discriminant analysis. The latter are less sharply defined – they sowed slightly larger proportions of wheat and oats and smaller proportions of barley, stocked fewer livestock, and made greater relative use of oxen – and the implication seems to be that they were less intensive in their methods.

On the arable side this intensive mixed-farming regime possessed three main distinguishing features. First, within its winter and spring cropping schedules priority was given to the higher valued grains of greatest commercial potential: wheat in preference to rye and barley and/or dredge in preference to oats.[28] Second, spring grains occupied a disproportionate share of the sown acreage, a practice consistent with the employment of relatively complex, flexible, and advanced rotations.[29] Third, legumes were sown on a substantial scale. Indeed, legumes were crucial to the viability of this entire farming system, supplying nitrogen to the soil and, along with oats, fodder to the livestock.[30]

Fodder cropping was one of the principal means by which these demesnes succeeded in maintaining a more favourable ratio of livestock to crops than was the norm in most other farming systems at this time (Tables V and VI).[31] An emphasis upon the more intensive and productive forms of pastoral husbandry also helped. Horses, with their higher costs but advantages of speed and efficiency, rivalled oxen as the principal draught animal and helped to contain the number of working animals which it was necessary to maintain.[32] In fact, non-working outnumbered working animals by roughly two to one. Among non-working animals cattle were dominant, with a herd structure demographically skewed towards adults, indicative of a specialist interest in dairying (the most food-productive form of pastoralism per unit area).[33] The presence of sheep and swine in modest but significant numbers also demonstrates the advantage taken of all available ecological opportunities.

The geographical distribution of the 'core' demesnes shows a pronounced and historically significant pattern. Notable concentrations occur in east Norfolk and east Kent and, to a lesser extent, in the Soke of Peterborough. Independent studies of agriculture in these three localities confirm the intensive and integrated character of the mixed-farming systems employed and demonstrate the exceptionally high returns per unit area thereby obtained.[34] All three localities were characterized by naturally fertile and readily cultivated loam soils, access to substantial local markets, and wider access, via navigable rivers and coastal trading ports, to major external markets.[35] In short, they were characterized by a convergence of high Ricardian and high von Thünen economic rent.[36]

The addition of the 'peripheral' demesnes reinforces and extends this pattern (Figure 2A). East

TABLE V. *'Core' group means and standard deviations*

	Cluster 1	2	3	4	5	6	7	8	*Overall*
%TOTAL SOWN ACRES:									
wheat	22·45	8·07	33·69	33·15	36·01	32·71	42·57	0·67	30·96
	9·10	*6·30*	*14·84*	*13·45*	*10·35*	*11·38*	*14·09*	*1·76*	*16·25*
rye	4·10	19·16	7·24	2·83	2·07	1·63	0·52	7·76	4·72
	6·80	*12·60*	*12·67*	*4·11*	*7·02*	*3·79*	*1·32*	*8·34*	*9·43*
winter mixtures	1·44	1·90	5·78	0·55	4·55	1·20	0·16	0·00	2·13
	3·65	*7·22*	*11·76*	*2·02*	*8·40*	*3·90*	*0·86*	*0·00*	*6·68*
barley	34·73	43·42	3·98	15·64	12·45	13·07	8·82	10·46	16·04
	18·58	*13·64*	*5·79*	*12·38*	*10·88*	*11·04*	*10·36*	*25·30*	*17·27*
oats	12·68	15·20	41·96	28·05	25·59	42·50	38·08	78·05	33·64
	6·87	*7·75*	*12·25*	*21·02*	*14·95*	*16·24*	*17·98*	*24·93*	*19·96*
spring mixtures	4·54	2·24	2·28	6·16	9·10	1·87	3·47	0·00	3·65
	9·60	*8·14*	*6·11*	*13·48*	*13·13*	*4·69*	*12·17*	*0·00*	*10·03*
legumes	20·24	10·01	4·22	13·61	9·90	4·62	4·97	3·07	7·88
	7·02	*5·48*	*5·16*	*12·34*	*9·51*	*5·37*	*7·10*	*4·06*	*8·57*
TOTAL SOWN ACRES	180·46	179·89	229·74	181·50	176·74	216·18	144·16	106·79	184·94
	93·84	*114·74*	*100·51*	*111·62*	*94·85*	*141·35*	*94·32*	*90·58*	*112·84*
%TOTAL LIVESTOCK UNITS*:									
horses	16·66	19·38	18·33	33·26	11·97	7·52	15·31	14·39	15·52
	11·39	*8·16*	*13·28*	*13·57*	*11·18*	*4·58*	*12·85*	*9·98*	*12·34*
oxen	16·27	9·43	35·95	39·30	38·76	32·40	84·41	57·66	42·50
	7·26	*11·84*	*23·81*	*22·93*	*16·44*	*13·07*	*12·95*	*23·19*	*29·60*
mature cattle	29·05	28·75	22·24	5·11	0·68	18·51	0·24	18·02	14·26
	13·00	*14·01*	*15·06*	*12·97*	*2·44*	*15·34*	*1·54*	*18·47*	*16·38*
immature cattle	18·41	12·18	14·60	0·00	0·00	11·80	0·00	7·48	8·10
	8·77	*6·69*	*10·14*	*0·00*	*0·00*	*7·15*	*0·00*	*8·23*	*9·43*
sheep	14·45	27·86	5·61	5·22	47·58	27·21	0·00	2·45	16·63
	18·43	*22·16*	*9·67*	*14·08*	*19·83*	*18·73*	*0·00*	*4·48*	*21·60*
swine	5·16	2·40	3·28	17·12	1·02	2·55	0·03	0·00	2·99
	4·41	*2·58*	*3·43*	*13·77*	*1·84*	*2·33*	*0·16*	*0·00*	*5·69*
TOTAL LIVESTOCK UNITS*	84·34	50·74	57·05	28·44	50·99	110·28	23·87	40·06	59·26
	65·76	*43·84*	*26·17*	*13·69*	*30·68*	*60·55*	*15·01*	*38·13*	*50·95*
STOCKING DENSITY‡	45·94	31·56	27·30	18·55	32·02	60·83	17·90	40·31	34·52
	22·37	*22·13*	*14·33*	*10·02*	*15·81*	*31·19*	*7·64*	*27·86*	*25·07*

Notes: means in roman; standard deviations in *italics*
*(horses × 1·0) + ([oxen + mature catte] × 1·2) + (immature cattle × 0·8) + ([sheep + swine] × 0·1)
‡total livestock units per 100 sown acres

TABLE VI. *Discriminant group means and standard deviations*

	Cluster 1	2	3	4	5	6	7	8	*Overall*
%TOTAL SOWN ACRES:									
wheat	27·43	9·97	35·91	32·89	32·97	33·81	40·59	0·67	31·68
	9·61	*7·64*	*15·47*	*12·76*	*13·55*	*11·26*	*16·94*	*1·76*	*15·97*
rye	3·81	18·41	5·62	2·38	4·99	2·15	3·40	7·76	5·20
	5·49	*12·44*	*10·48*	*3·90*	*10·94*	*4·74*	*8·23*	*8·34*	*9·47*
winter mixtures	1·63	2·51	4·29	1·62	3·80	2·13	0·48	0·00	2·27
	5·55	*7·20*	*9·96*	*5·33*	*7·79*	*6·10*	*2·37*	*0·00*	*6·72*
barley	26·44	39·55	5·34	16·08	13·59	13·73	8·59	10·46	15·58
	16·58	*14·71*	*6·45*	*12·56*	*12·76*	*11·10*	*10·30*	*25·30*	*15·87*
oats	16·87	17·42	39·29	25·03	27·06	38·67	35·51	78·05	31·21
	11·74	*10·05*	*13·43*	*21·03*	*17·89*	*16·74*	*18·54*	*24·93*	*19·18*
spring mixtures	3·89	2·68	2·23	6·99	7·49	2·20	4·35	0·00	3·79
	8·38	*8·31*	*5·25*	*12·89*	*13·58*	*5·36*	*12·21*	*0·00*	*9·49*
legumes	19·89	9·46	6·50	13·87	9·38	5·68	6·02	3·07	9·43
	8·78	*5·45*	*6·77*	*12·20*	*9·38*	*6·10*	*8·34*	*4·06*	*9·39*
TOTAL SOWN ACRES	196·89	202·25	229·59	187·62	185·23	227·70	136·96	106·79	192·80
	94·13	*149·71*	*116·81*	*113·98*	*97·62*	*145·92*	*88·19*	*90·58*	*118·66*
% TOTAL LIVESTOCK UNITS*:									
horses	15·38	18·53	17·61	32·70	12·07	7·69	16·67	14·39	15·49
	9·01	*9·16*	*11·11*	*16·24*	*10·47*	*4·84*	*12·98*	*9·98*	*11·66*
oxen	20·70	9·67	33·84	38·63	38·59	32·37	82·30	57·66	40·45
	12·24	*11·06*	*21·13*	*24·89*	*16·21*	*13·55*	*13·15*	*23·19*	*28·06*
mature cattle	27·59	28·67	23·45	5·38	2·04	16·84	0·72	18·02	15·35
	12·36	*14·35*	*14·28*	*12·49*	*5·00*	*12·96*	*2·56*	*18·47*	*15·77*
immature cattle	18·30	12·18	15·40	0·41	0·08	11·16	0·00	7·48	9·07
	8·07	*7·14*	*10·19*	*1·79*	*0·39*	*6·52*	*0·00*	*8·23*	*9·81*
sheep	12·88	28·59	6·06	6·82	45·30	29·28	0·02	2·45	16·36
	15·44	*21·04*	*9·50*	*14·21*	*18·75*	*16·79*	*0·19*	*4·48*	*20·35*
swine	5·15	2·37	3·64	16·07	1·92	2·66	0·29	0·00	3·27
	3·89	*2·64*	*3·74*	*12·87*	*3·38*	*2·64*	*1·22*	*0·00*	*5·22*
TOTAL LIVESTOCK UNITS*	83·06	55·75	62·94	29·40	51·81	113·72	23·03	40·06	62·66
	50·40	*45·74*	*32·62*	*12·86*	*27·12*	*64·39*	*14·36*	*38·13*	*50·57*
STOCKING DENSITY‡	44·93	30·86	30·61	19·41	32·07	57·96	19·33	40·31	35·46
	33·45	*20·81*	*15·97*	*11·35*	*16·45*	*29·22*	*12·64*	*27·86*	*26·00*

Notes: means in roman; standard deviations in *italics*
*(horses × 1·0) + ([oxen + mature cattle] × 1·2) + (immature cattle × 0·8) + ([sheep + swine] × 0·1)
‡total livestock units per 100 sown acres

Norfolk, east Kent, and the Soke of Peterborough emerge even more strongly and are joined by a light scatter of demesnes in the east midlands (Cambridgeshire, Leicestershire and Rutland, Lincolnshire), the upper Thames Valley, and along the south coast. Studies of agriculture in several of these areas have shown that it often exhibited a tendency towards the kinds of intensive methods which found their fullest and most productive expression in east Norfolk and east Kent.[37]

Cluster Two: 'light-land intensive'

This is one of the most distinctive and geographically most sharply focused husbandry types. The correspondence between the various clustering methods in locating this group is high with little substantive difference between them. Discriminant analysis adds only six demesnes to the initial 'core' of 28.

On the arable side its main distinguishing features were the prominence accorded rye and barley as the principal winter and spring grains: in no other farming type did these two cereals occupy such large respective shares of the sown acreage. A strong bias towards spring-sown crops implies flexible rotations, although with only moderate sowings of legumes (which performed indifferently on light soils) these are unlikely to have been as intensive as those employed by the 'intensive mixed-farming' demesnes.

Pastoral husbandry was similarly distinctive. This is the only farming type in which horses decisively outnumbered oxen, to the considerable benefit of other classes of livestock.[38] The ratio of non-working to working animals was more favourable on these demesnes than any others. Dairy cattle were prominent – the demographic bias of herds towards adults being even more pronounced than in Cluster One – but so too were sheep. The precise ratio between cattle and sheep varied a good deal, but most demesnes stocked both. Nevertheless, for all the highly developed pastoral profile of these demesnes, stocking densities were relatively low.

The prominence within this husbandry system of rye, barley, horses, and sheep is symptomatic of light-land husbandry and this is confirmed by the distribution of 'core' demesnes. These exhibit a strong regional bias towards Norfolk and a pronounced locational bias towards areas of light soil, notably the immediate hinterland of Norwich, the 'good sands' region of north-west Norfolk, the Breckland of north-west Suffolk, and the Sandlings of east Suffolk.[39] The allocation of 'peripheral' demesnes by discriminant analysis confirms the strong geographically-specific character of this farming type. Only three isolated examples are added in other parts of the country, all of which appear to be associated with localized patches of light soil.[40]

Cluster Three: 'mixed-farming with cattle'

This represents a less intensive husbandry system than either Cluster One or Cluster Two. It was also practised by a larger number of demesnes over a wider geographical area. The three clustering methods identify a 'core' of 43 demesnes, to which discriminant analysis adds a further 29 'peripheral' demesnes. The latter, however, do little to alter the overall profile of this farming type.

On the arable side this farming system bears all the hallmarks of a basically three-course system of cropping, comprising winter corn, spring corn, and fallow (although a two-course system cannot be ruled out). Thus, winter and spring grains were sown in almost exactly equal proportions, with wheat predominant within the winter course and oats within the spring. Some rye and maslin (a wheat/rye mixture) also feature within the winter course, but little barley or dredge (a barley/oats mixture) within the spring. Nor were legumes grown on such a scale as to distort this seasonal symmetry.

The strong arable focus of this farming system is reflected in the large mean size of the demesnes which practised it and a generally low ratio of livestock to crops. The emphasis upon cultivation also ensured that working animals usually accounted for over half of all livestock units. As is consonant with this farming system's generally extensive character, oxen tended to provide the bulk of draught power. Nevertheless, most demesnes stocked some horses and the ratio of horses to oxen exhibited much variation. Cattle made up the bulk of non-working animals. They were needed to breed replacement oxen, but it is also plain from their herd structure that breeding was combined with dairying. In the latter respect, the pastoral sector of this farming system was more intensive than the arable.

In Warwickshire and the east midlands on one hand, and south Somerset on the other, this farming type is associated with regular commonfield systems.[41] Demesnes in these locations exhibited the strongest reliance upon oxen for draught and kept only relatively modest supplementary herds of cattle in which the breeding function far outweighed the dairying.[42] In Suffolk and the Home Counties (Essex, Hertfordshire, Middlesex, south Buckinghamshire,

Surrey, Kent, and the lower Thames Valley), however, there was no such association with regular commonfield systems.[43] Significantly, it was here, where communal controls were weakest and metropolitan demand strongest, that horses made their greatest contribution to draught power and the dairying function of cattle herds was most fully developed.[44]

Cluster Four: 'arable husbandry with swine'

This is very much a minority farming type and is represented by a comparatively small number of demesnes. Although differences between the 'core' and discriminant groups are small, the cluster was defined very differently by the three clustering methods. Moreover, it is closely related to Cluster Seven ('extensive arable husbandry'), which it resembles in its low stocking density and the dominance of working over non-working animals, and shares some affinities in its pattern of cropping with Cluster Five ('sheep-corn husbandry'). What sets it apart is the relatively favourable ratio of horses to oxen among working animals, and, above all, the prominence of swine among non-working animals. On the arable side there was little that was distinctive except for relatively substantial sowings of legumes, which were an important source of fodder for both horses and hogs.

In distribution this farming type is more associated with those parts of the country that were agriculturally most developed – the east midlands, East Anglia, and the south-east – than with the north, west, or south-west. Individual examples are, however, widely scattered which suggests that its occurrence owed more to local than to regional circumstances. Swine herding was traditionally associated with woodland, which may help to explain the modest concentrations of examples around the edge of the Kent and Sussex Weald and in the more wooded parts of Northamptonshire.[45] But sty management based on the feeding of legumes and other fodder crops offered an intensive alternative: hence, perhaps, the incidence of cases in east Kent and the Soke of Peterborough.[46]

Cluster Five: 'sheep-corn husbandry'

This farming type is clearly identified by all three clustering techniques and only marginal differences occur between the overall breakdown of each method. Nor do substantive differences occur between the 'core' and discriminant groups, even though the latter technique increases the number of constituent demesnes from 30 to 45. Its distinctiveness principally derives from its pastoral sector, with draught animals – principally oxen – comprising approximately half of all livestock units and sheep most of the remainder. The virtual absence of cattle herds, notwithstanding a strong reliance upon oxen, is striking and implies that replacement work animals must have been obtained from elsewhere.[47] This is an intrinsically extensive pastoral regime and is associated with a strong reliance upon temporary and permanent pasture rather than produced fodder.[48] Stocking densities were correspondingly unimpressive.

Patterns of cropping, by contrast, were much more assorted and therefore far less distinctive. Wheat was the principal winter crop, and oats the spring, with moderate quantities of barley, dredge, and legumes also being grown. Although some of the barley was winter sown there was a general bias towards spring cropping. With the exception of an above average emphasis upon spring mixtures, this is a pattern of cropping which conforms relatively closely to the overall national pattern.

Demesnes practising this farming system were mostly located away from the more intensively cultivated areas of East Anglia and the Home Counties, where horses and cattle held greater sway.[49] They rarely comprise a locally or regionally dominant type but show up wherever pastoral resources and institutional circumstances favoured the large-scale development of sheep farming.[50] This included demesnes with access to extensive marshland and heathland grazings, as in East Anglia and on the south coast, but, above all, demesnes with access to the short, sweet grasslands of downlands and wolds. Hence the prominence of this farming type in the limestone country of south Yorkshire and, above all, on the chalk downlands of southern England. Here a regime commonly prevailed whereby sheep fed on permanent pastures by day were folded on the arable by night, whose soil they enriched with their treading, dung, and urine.[51]

Cluster Six: 'extensive mixed-farming'

This is a well-defined and relatively widely distributed farming type which shares certain features in common with the other 'mixed-farming' systems – Cluster One and Cluster Three. Its 52 'core' demesnes are chiefly distinguished by substantial cropped acreages and high stocking densities, indicative of the juxtaposition, if not the integration, of arable and pastoral husbandry on a relatively extensive scale.[52] On the arable side production was dominated by the cultivation of wheat and especially oats, with some barley and limited quantities of legumes. This is consistent with a basically two- or three-course system of cropping and would have been compatible with a two- or three-field commonfield regime. On the pastoral side, these demesnes carried significant numbers of livestock and, on average, exhibited higher stocking densities than any other farming types. Nevertheless, it was the most extensive forms of pastoralism that tended to prevail. Oxen were preferred to horses for draught with the result that working animals accounted on average for 40 per cent of all livestock units (in contrast to the 28–35 per cent of Clusters One and Two). Within the non-working sector, cattle husbandry and sheep farming assumed roughly equal importance. Moreover, the demographic bias of cattle herds in favour of mature adults was less pronounced than in the more intensive mixed-farming systems represented by Clusters One, Two, and Three. In part this probably reflected the need to rear up replacement oxen.

Discriminant analysis adds a further 17 demesnes to bring the total within this cluster grouping to 65. In distribution most are located within the general ambit of the midland commonfield system as defined by H. L. Gray, stretching in a great belt from the Vale of York in the north-east to Somerset in the south-west.[53] A few are, however, located to the east and the south-west of this zone in areas of more irregular field systems.[54] Nevertheless, whether associated with regular or irregular field systems, their well-developed pastoral profile is explicable in no other terms than a relatively generous endowment with temporary and permanent pasturage. Hence, no doubt, the occurrence of demesnes belonging to this cluster grouping in certain known pasture-rich locations: in the Fens, on the edge of Romney Marsh in Kent, along the Sussex coast, on the Wiltshire Downs, in south Devon, and at a wide scatter of locations in northern England. Yet, with certain notable exceptions, this is not a farming type that was typical of either East Anglia or the Home Counties.

Cluster Seven: 'extensive arable husbandry'

This is the best represented and geographically most widespread of the eight cluster types, with 59 'core' demesnes scattered through 17 counties. The most arresting feature of these demesnes is the virtual absence of non-working animals and consequently low ratio of livestock to crops. In effect, farm enterprise was exclusively concerned with arable production. Moreover, that arable production was extensive rather than intensive. An overwhelming preference for oxen rather than horses for farm work in part ensured that this was the case. So, too, did a narrow reliance upon wheat and oats as the leading crops with only very limited sowings of barley and legumes, which implies a basically two- or three-course system of cropping. Indeed, these demesnes largely conform in distribution to the western edge of the midland commonfield system and are very much complementary in distribution to the 'extensive mixed-farming demesnes' of Cluster Six.

In some cases the absence of livestock other than those essential to the cultivation of the arable may reflect institutional factors, since demesnes which for one reason or another fell into the king's hands were often stripped of all but their working animals.[55] Nevertheless, the number of demesnes is too great and their association with certain parts of the country too strong for this to apply to more than a minority of cases. As Ian Blanchard has demonstrated for Derbyshire, in 'upland' Britain a spatial division often existed between pastoral farming on upland vaccaries, bercaries, and studs, and arable farming on lowland demesnes, the latter often carrying working animals only and obtaining their replacements from the uplands.[56] Such an explanation is consistent with the strong northern, western, and south-western bias to the distribution of demesnes belonging to this farming type. As such, their failure to exploit available resources more intensively through the fuller development of mixed farming is also symptomatic of relatively low levels of economic rent borne of below average densities of population and remoteness from major markets.[57] In this context it is significant that, with the exception of a few stray examples, demesnes of this type are largely absent from East Anglia, the east midlands, and the Home Counties.

This spatial distinction is blurred somewhat by the application of discriminant analysis and allocation of an additional 21 'peripheral' demesnes to this cluster type, bringing the overall total to 80. Although the established relationship with the zone of regular commonfield systems is reinforced, other demesnes are added from south Buckinghamshire, Suffolk, and Surrey which lay well outside and to the east of that zone.[58] In their case allocation to this cluster grouping owes more to an absence of non-working animals and low stocking densities than to any basic similarity of cropping.

Cluster Eight: 'oats and cattle'

This is the smallest of the eight cluster groupings and comprises only seven demesnes. Moreover, these demesnes were themselves small as measured by both sown acreage and livestock units. Oats was overwhelmingly their principal crop, with limited quantities of rye and barley but few legumes and virtually no wheat. The dominance that attached to oats on the arable side was matched by cattle, especially oxen, on the pastoral: few horses were kept, fewer sheep, and no swine. On four of the seven demesnes oxen accounted for the majority of cattle stocked (West Derby, Lancashire; Lambeth, Surrey; and Great Sandal and Soothill, Yorkshire West Riding), but at Exminster, Devon, and, to a lesser extent, Harrow, Middlesex, matures and immatures were also stocked in substantial numbers, suggestive of a specialist interest in breeding and/or fattening.

This farming type is associated with two very specific and contrasting types of location. Demesnes were either located adjacent to major urban centres – Harrow and Lambeth near London and Hickley outside Southampton – where there was a highly concentrated demand for oats, beef, and dairy produce, or they were to be found in parts of the country – Devon, Lancashire, and the West Riding of Yorkshire – where environmental conditions placed constraints on the crops that could be cultivated and extensive grazings and rough pastures provided opportunities for cattle raising on a substantial scale.[59] These same localities were also associated with specialist cattle farms whose economies were exclusively pastoral.[60]

EVALUATING THE SIMILARITIES AND DIFFERENCES BETWEEN FARMING TYPES

As the discussion of both the method and its results has demonstrated, some of these farming types were much more distinctive and self-contained than others. Which types were most distinctive and by how much can be gauged by 'mapping' individual cases in multidimensional and geographical space. Thus, on the evidence of Andrews plots and discriminant function plots of the Normix classification, 'light-land intensive' (Cluster 2), 'sheep-corn husbandry' (Cluster 5), and 'extensive arable husbandry' (Cluster 7) emerge as the most homogeneous and discrete of the eight farming types, genuinely different in character from most others.[61] As already observed, each was clearly and consistently identified by all three clustering techniques.[62] Of these farming types, however, only 'light-land intensive' was similarly distinctive in its geographical distribution, insofar as examples were almost exclusively confined to East Anglia and to the light soils of Norfolk in particular (Figures 1A and 2A). 'Sheep-corn husbandry' and 'extensive arable husbandry', by contrast, were much less regionally specific in their distributions (Figures 1C and 2C and 1D and 2D). True, both were more typical of some parts of the country, and some types of location, than others. But they generally occurred in conjunction with other farming types and therefore represent a specific response to localized, environmental and institutional circumstances. Interestingly, the same, in part, holds true of 'arable husbandry with swine' (Cluster 4). This was the most inconsistently identified of the eight farming types and, as Andrews and discriminant function plots show, it shared certain attributes in common with 'extensive arable husbandry' (Cluster 7) – notably a shortage of non-working livestock. Nevertheless, its largely independent geographical distribution (Figures 1B and 2B) suggests there is utility in distinguishing it as a separate farming type.

Of the remaining farming types, 'intensive mixed-farming' (Cluster 1), 'mixed-farming with cattle' (Cluster 3), and 'extensive mixed-farming' (Cluster 6) all represent different stages on a mixed-farming continuum. It has already been observed that the three clustering techniques differed quite significantly in their allocation of cases between Clusters One and Six, which implies the existence of strong affinities between them. Andrews and discriminant function plots confirm that this was indeed the case and show that the constituent demesnes of all three mixed-farming types were located in relatively close proximity to each other. Clusters Three and Six ('mixed-farming with cattle' and 'extensive mixed-farming') are particularly closely located and are not clearly distinguished by shape. Cluster One ('intensive mixed-farming') is more clearly identified though some of its points are in close proximity to Cluster Six ('extensive mixed-farming'), possibly on account of the common denominator of high stocking densities. Yet, whatever similarities these three mixed-farming systems may have exhibited in nature, they were very different in location and on these grounds it is certainly valid to distinguish between them. Thus, as Figures 1A and 2A demonstrate, 'intensive mixed-farming' demesnes were largely concentrated in specific, highly-favoured localities in the extreme east and south-east, with a scatter of outliers in the east midlands. 'Mixed-farming with cattle', by contrast,

FIGURE 1. Demesne-farming systems 1250–1349: 'core' cluster groups

occupied two very different types of location. Demesnes of this type were concentrated, on the one hand, in the Home Counties, and, on the other, in areas of regular commonfield agriculture in the midlands and the south-west (Figures 1B and 2B). The latter, in part, complemented the distribution of 'extensive mixed-farming' demesnes, which exhibited a strong geographical bias towards those parts of northern England, the east midlands, and southern England that were reasonably well provided with permanent pasture (Figures 1C and 2C).

Only in parts of the east midlands and the Thames Valley was there any convergence in the distributions of these three basic mixed-farming types. In part this reflects the commercial and environmental diversity of the localities concerned, but it is also likely that the demesnes in question practised forms of mixed-farming which were transitional between the three main types and thereby present a genuine problem of classification. In this respect the discriminant functions contained in Table III are not only of great value for establishing group membership but also for

FIGURE 2. Demesne-farming systems 1250–1349: discriminant cluster groups

locating the position of individual demesne farms on the mixed-farming spectrum.[63]

Finally, the handful of demesnes practising 'oats and cattle' (Cluster 8) emerge as relatively widely scattered in both multi-dimensional and geographical space. By implication, this is the least satisfactorily defined of the eight farming types and raises as many questions as it answers. Its definition would undoubtedly benefit from a fuller and more representative coverage of data from the extreme north-west and south-west of the country where four of the seven identified examples are located.

CONCLUSION

It might be objected that the use of cluster analysis to classify medieval demesne-farming systems represents the application of an imprecise method to imperfect data and as such is hardly likely to yield a definitive solution. Yet historical geographers are

X

rarely in a position to deal with certainties and the concern must be to derive solutions and explanations that are the most consistent possible given available techniques and information. As far as cluster analysis is concerned, Everitt has observed:

> Cluster analysis is potentially a very useful technique but it requires care in its application, because of the many associated problems. In many of the applications . . . the authors have either ignored or been unaware of these problems, and consequently few results of lasting value can be pointed to.[64]

The method outlined above goes a considerable way towards resolving these problems and thereby offers a way forward for others interested in deriving classification schemes from historical evidence. Obviously, the classification scheme advanced here would carry greater conviction if it were founded upon a fuller and more representative sample of data. Record survival may ultimately limit the degree to which a good sample can be collected, but the current sample of 388 demesnes can certainly be improved upon. Nevertheless, for all its deficiencies, it does provide a unique opportunity to investigate differences in agricultural enterprise at a *national* level. Hitherto most work on past agricultural systems has been rooted in local and regional analysis, yet it is only through analysis at a national scale that the broader contrasts in the pattern of agricultural activity – with all that these imply for larger-scale spatial relationships – will be recognized and appreciated.[65]

From the provisional classification of farming types presented here a picture emerges of an agricultural sector still more extensive than intensive in its methods of production, although one increasingly exposed to forces of specialization and intensification. The principal stimulus to these forces was market expansion in general and the growth of concentrated urban demand in particular.[66] The heightened spatial differentiation to which this gave rise was most in evidence in the east and south-east, especially outside the zone of regular commonfield systems, for it was here that farm enterprise assumed its most developed and intensive forms. Norfolk, above all, stands out as dominated by intensive mixed-farming systems and in many respects was already by far the most distinctive of English farming counties.[67] Kent, too, exhibited many of the same traits but contained a wider range of farming types, a feature it shared with Northamptonshire. In fact, on the currently available evidence Kent and Northamptonshire were the two most heterogeneous medieval farming counties, each containing at least six of the eight farming types to be found within the country as a whole.

There is much here that requires detailed local investigation and verification. With the national picture established in outline the challenge must now be to extend and refine this classification. For instance, additional demesnes can be incorporated using the function coefficients contained in Table III. But analysis ought to go further than this. Other variables need to be included, particularly those which bear upon the productivity of arable and pastoral husbandry, and the exercise needs to be repeated at a more detailed scale in order to bring into sharper focus the local as well as the regional variations by which medieval farming systems were so plainly characterized.[68] Nevertheless, there are limits to the insights which even the most refined classifications can provide into past farming systems, since the reconstruction of pattern is no substitute for the investigation of process. Thus, the classification presented here begs the essential historical question whether formal similarities in the crops that were grown and livestock that were kept reflect functional similarities in the environmental, institutional, economic, and managerial factors which determined the enterprise of individual demesne farms. In that sense classification *per se* represents an exploratory technique whose true utility lies in its capacity to help formulate and focus further research questions.

NOTES

1. THIRSK, J. (ed.) (1967) *The agrarian history of England and Wales, IV, 1500–1640* (Cambridge University Press, Cambridge) pp. 1–112
2. CAMPBELL, B. M. S. (1990) 'People and land in the Middle Ages, 1066–1500', in DODGSHON, R. A. and BUTLIN, R. A. (eds) *An historical geography of England and Wales* (Academic Press, London, 2nd edn) pp. 70–102
3. ASTON, T. H. and PHILPIN, C. H. E. (eds) (1985) *The Brenner debate: agrarian class structure and economic development in pre-industrial Europe* (Cambridge University Press, Cambridge); PERSSON, K. G. (1988) *Pre-industrial economic growth, social organization and technological progress in Europe* (Basil Blackwell, Oxford); BRITNELL, R. H. (1989) 'England and northern Italy in the early fourteenth century: the economic contrasts', *Trans. R. Hist. Soc.*, 5th ser. 39: 167–83; SNOOKS, G. (1990) 'Economic growth during the last millenium: a

quantitative perspective for the British Industrial Revolution' (The Australian National Univ., Working Papers in Economic History, 140); CAMPBELL, B. M. S. (ed.) (1991a) *Before the Black Death: studies in the 'crisis' of the early fourteenth century* (Manchester University Press, Manchester)

4. STITT, F. B. (1953) 'The medieval minister's account', *Soc. Local Archivists' Bull.* 11: 2–8; HARVEY, P. D. A. (1976) 'Introduction, Part II, accounts and other manorial records', in *idem* (ed.) *Manorial records of Cuxham, Oxfordshire, circa 1200–1359* (Oxfordshire Record Soc., 50, London) pp. 12–71
5. This is a particular problem for historians of the early modern period reliant upon probate inventories: OVERTON, M. (1984) 'Probate inventories and the reconstruction of agrarian landscapes', in REED, M. (ed.) *Discovering past landscapes* (Croom Helm, London) pp. 167–94; SYMONS, L. (1970) *Agricultural geography* (Bell and Sons, London) pp. 209–15
6. A summary register of manorial records may be consulted at the National Register of Archives, Quality House, Quality Court, Chancery Lane, London
7. The bulk of these accounts are listed in LANGDON, J. L. (1983) 'Horses, oxen and technological innovation: the use of draught animals in English farming from 1066 to 1500', unpubl. Ph.D. thesis, Dept., of Hist., Univ of Birmingham, Appendix C. We are grateful to Dr Langdon for making these data available to us. The Norfolk demesnes are a selection of the best documented examples: for this county alone over 120 demesnes are documented by almost 1200 accounts during the period 1250–1349
8. For instance, a comprehensive database comprising some 461 accounts and representing some 204 demesnes from ten counties around London during the period 1290–1315 has recently been compiled by the 'Feeding the City I' project at the Centre for Metropolitan History, Institute of Historical Research, London
9. OVERTON, M. (1981) 'Agricultural change in Norfolk and Suffolk, 1500–1740', unpubl. Ph.D. thesis, Dept. of Geogr., Univ of Cambridge; *Idem* (1985) 'Agricultural regions in early modern England: an example from East Anglia' (Univ of Newcastle, Dept. of Geogr., seminar paper 42)
10. GLENNIE, P. (1988) 'Continuity and change in Hertfordshire agriculture 1550–1700: I – patterns of agricultural production', *Agric. Hist. Rev.* 36: 55–75
11. CAMPBELL, B. M. S. and POWER, J. P. (1989) 'Mapping the agricultural geography of medieval England', *J. Hist. Geogr.* 15: 24–39
12. *Ibid.*, pp. 25–6
13. GATRELL, A. C. (1983) *Distance and space: a geographical perspective* (Clarendon Press, Oxford) pp. 8–43
14. Some clustering techniques are particularly sensitive to intermediary points between groups: EVERITT, B. S. (1980) *Cluster analysis* (SSRC Reviews of Current Research, 11, Heinemann, London, 2nd edn) pp. 67–8
15. See below pp. 33–6
16. Such an approach is similar to that utilized for the handling of poorly defined sets within regional geography: GATRELL *op cit.*, pp. 11–13
17. YELLING, J. A. (1977) *Common field and enclosure in England, 1450–1850* (Macmillan, London) p. 159; ALLEN, R. C. (1991) 'The two English agricultural revolutions, 1459–1850', in CAMPBELL, B. M. S. and OVERTON, M. (eds) *Land, labour and livestock: historical studies in European agricultural productivity* (Manchester University Press, Manchester) pp. 245–6. The weightings are given in Table I: for a discussion see CAMPBELL, B. M. S. (1991b) 'Land, labour, livestock, and productivity trends in English seignorial agriculture, 1208–1450', in *IDEM* and OVERTON *op cit.*, pp. 156–7
18. COPPOCK, J. T. (1964) *An agricultural atlas of England and Wales* (Faber, London) p. 213; OVERTON (1981 and 1985) *op cit.*; GLENNIE *op. cit.*
19. KING, L. J. (1969) *Statistical analysis in geography* (Prentice-Hall, Englewood Cliffs, N.J.) p. 179
20. POWER, J. P. (1991) 'A provisional classification of English medieval seignorial agriculture (1250–1349): a geographical perspective', unpubl. diploma thesis, School of Social Stud., Univ of Essex, p. 13
21. JOHNSTON, R. J. (1978) *Multivariate statistical analysis in geography: a primer on the general linear model* (Longman, London) p. 146
22. For a fuller exposition see POWER *op. cit.*
23. WISHART, D. (1987) *Clustan user manual* (Computing Laboratory, Univ of St Andrews, 4th edn)
24. *Ibid.*, pp. 229–31; EVERITT *op. cit.*, pp. 93–5
25. POWER *op. cit.*, p. 19
26. The two-way matches between cluster methods used to identify 'core' groups are given in Tables 4 and 5 of POWER *op. cit.*, pp. 20 and 24
27. HIE, N. H., HULL, C. H., JENKINS, J. G., STEINBRENNER, K. and BENT, D. (1975) *Statistical package for the social sciences* (McGraw Hill, London) pp. 434–67; JOHNSTON *op. cit.*, pp. 224–52
28. For the relative values of the principal grains see CAMPBELL (1991b) *op. cit.*, pp. 168–9
29. For the connection between extensive spring cropping and flexible/intensive rotations see CAMPBELL, B. M. S. (1981) 'The regional uniqueness of English field systems? Some evidence from eastern Norfolk', *Agric. Hist. Rev.* 29: 21–2
30. CHORLEY, G. P. H. (1981) 'The agricultural revolution in northern Europe, 1750–1880: nitrogen, legumes and crop productivity', *Econ. Hist. Rev.* 2nd ser. 34: 71–93; SHIEL, R. S. (1991) 'Improving soil productivity in the pre-fertiliser era', in CAMPBELL and OVERTON *op cit.*, pp. 51–77

31. Fodder cropping encouraged the integration of arable and pastoral production; for its productivity benefits see OVERTON, M. and CAMPBELL, B. M. S. (1991) 'Productivity change in European agricultural development', in CAMPBELL and OVERTON *op. cit.*, pp. 42–4
32. CAMPBELL, B. M. S. (1988) 'Towards an agricultural geography of medieval England', *Agric. Hist. Rev.* 36: 24–39; *IDEM* (forthcoming [a]) 'Intensive pastoral husbandry in medieval England: a Norfolk perspective', in DEWINDT, E. B. (ed.) *Festschrift for Professor J. A. Raftis* (Medieval Institute, Western Michigan Univ., Kalamazoo)
33. CAMPBELL, B. M. S. (forthcoming [b]) 'Commercial dairy production on medieval English demesnes: the case of Norfolk', in GRANT, A. (ed.) *Animals and their products in trade and exchange, Anthropozoologica, Quatrième Numéro Spécial* (Paris)
34. CAMPBELL, B. M. S. (1983) 'Agricultural progress in medieval England: some evidence from eastern Norfolk', *Econ. Hist. Rev.* 2nd ser. 36: 26–46; SMITH, R. A. L. (1943) *Canterbury Cathedral Priory: a study in monastic administration* (Cambridge University Press, Cambridge) pp. 128–65; MATE, M. (1985) 'Medieval agrarian practices: the determining factors?', *Agric. Hist. Rev.* 33: 22–31; BIDDICK, K. (1989) *The other economy: pastoral husbandry on a medieval estate* (University of California Press, Berkeley and Los Angeles) pp. 50–77
35. EDWARDS, J. F. and HINDLE, B. P. (1991) 'The transportation system of medieval England and Wales', *J. Hist. Geogr.* 17: 123–34
36. GRIGG, D. B. (1982) *The dynamics of agricultural change: the historical experience* (Hutchinson, London) pp. 50–1; CHISHOLM, M. (1962) *Rural settlement and land-use: an essay on location* (Hutchinson, London) pp. 20–32; THÜNEN, J. H. VON (1966) *Der isolierte staat*, trans. by WARTENBERG, C. M. as *Von Thünen's isolated state*, HALL, P. (ed.) (Pergamon Press, Oxford)
37. RAVENSDALE, J. R. (1974) *Liable to floods: village landscape on the edge of the Fens AD 450–1850* (Cambridge University Press, Cambridge) pp. 116–20; HARVEY, P. D. A. (1965) *A medieval Oxfordshire village: Cuxham 1240 to 1400* (Oxford University Press, Oxford) pp. 39–65; BRANDON, P. F. (1971) 'Demesne arable farming in coastal Sussex during the later Middle Ages', *Agric. Hist. Rev.* 19: 113–42; HALLAM, H. E. (ed.) (1988) *The agrarian history of England and Wales*, II, *1042–1350* (Cambridge University Press, Cambridge) pp. 318–24
38. LANGDON, J. L. (1986) *Horses, oxen and technological innovation: the use of draught animals in English farming from 1066–1500* (Cambridge University Press, Cambridge) pp. 101–5; CAMPBELL (1989) *op. cit.*, pp. 91–3
39. BAILEY, M. (1989) *A marginal economy? East Anglian Breckland in the later Middle Ages* (Cambridge University Press, Cambridge); CAMPBELL, B. M. S. (forthcoming [c]) 'The Middle Ages: arable and pastoral husbandry', in WADE-MARTINS, P. (ed.) *A Norfolk historical atlas* (Norfolk County Council, Norwich)
40. Fulstow, Lincolnshire is the only outlying 'core' demesne. Wootton, Hampshire, Kettering, Northamptonshire, and Bromham, Wiltshire are added by discriminant analysis
41. GRAY, H. L. (1915) *English field systems* (Harvard University Press, Cambridge, Mass.); CAMPBELL, B. M. S. (1981) 'Commonfield origins – the regional dimension', in ROWLEY, T. (ed.) *The origins of open field agriculture* (Croom Helm, London) pp. 112–29; HALLAM *op. cit.*, pp. 341–5, 369–75
42. CAMPBELL and POWER *op. cit.*, p. 36; HALLAM *op. cit.*, pp. 375–7
43. GRAY *op. cit.*, pp. 272–402; BAKER, A. R. H. and BUTLIN, R. A. (eds) (1973) *Studies of field systems in the British Isles* (Cambridge University Press, Cambridge) pp. 281–429; BRITNELL, R. H. (1983) 'Agriculture in a region of ancient enclosure, 1185–1500', *Nottingham Medieval Stud.* 27: 37–55; RODEN, D. (1969) 'Demesne farming in the Chiltern Hills', *Agric. Hist. Rev.* 17: 9–23; HALLAM *op. cit.*, pp. 272–81
44. CAMPBELL (1988) *op. cit.*, pp. 95–7; *IDEM* and POWER *op. cit.*, pp. 36–7; CAMPBELL (forthcoming [b]) *op. cit.*; KEENE, D. J. (1989) 'Medieval London and its region', *London J.* 14: 99–111; GALLOWAY, J. A. and MURPHY, M. M. (1991) 'Feeding the city: London and its agrarian hinterland', *London J.* 16: 1–12
45. DARBY, H. C. (1977) *Domesday England* (Cambridge University Press, Cambridge) pp. 172–8, 190–4; HALLAM *op. cit.*, pp. 315–16
46. BIDDICK, K. (1984) 'Pig husbandry on the Peterborough Abbey estate', in GRIGSON, C. and CLUTTON-BROCK, J. (eds) *Animals and archaeology: 4. Husbandry in Europe* (British Archaeological Reports International ser. 227, Oxford) pp. 161–77; *IDEM* (1989) *op. cit.*, pp. 121–5
47. For inter-manorial transfers of draught animals on the estate of Peterborough Abbey see: BIDDICK (1989) *op. cit.*, pp. 81–91, 116–21
48. SIMMONS, I. G. (1974) *The ecology of natural resources* (Edward Arnold, London) pp. 201–6
49. Cf. THIRSK *op. cit., passim*
50. Cf. OVERTON (1985) *op. cit.*, pp. 5–8
51. SCOTT, R. (1959) 'Medieval agriculture', in PUGH, R. B. (ed.) *Victoria history of the County of Wiltshire*, IV (Oxford University Press, London) pp. 19–21; HARE, J. N. (1981) 'Change and continuity in Wiltshire agriculture in the later Middle Ages', in MINCHINTON, W. (ed.) *Agricultural improvement: medieval and modern* (Exeter papers in economic history, 14) pp. 4–9; HALLAM *op. cit.*, p. 323
52. OVERTON and CAMPBELL *op. cit.*, pp. 34–5; BIDDICK (1991) 'Agrarian productivity on the estates of the Bishopric of Winchester in the early thirteenth

century: a managerial perspective', in CAMPBELL and OVERTON *op. cit.*, p. 115

53. GRAY *op. cit.*
54. BAKER, A. R. H. (1973) 'Field systems of southeast England', in *IDEM* and BUTLIN *op. cit.*, pp. 377–429; FOX, H. S. A. (1972) 'Field systems of east and south Devon. Part I: east Devon', *Trans. Devonshire Ass.* 104: 81–135; *IDEM* (1990) 'Peasant farmers, patterns of settlement and *pays*: transformations in the landscapes of Devon and Cornwall during the later Middle Ages, in HIGHAM, R. (ed.) *Landscape and townscape in the south-west* (University of Exeter publications in history, Exeter) pp. 41–73
55. TITOW, J. Z. (1962) 'Land and population on the Bishop of Winchester's estates 1209–1350', unpubl. Ph.D. thesis, Univ of Cambridge, pp. 44–7; BIDDICK (1991) *op. cit.*, pp.98–104
56. BLANCHARD, I. S. W. (1967) 'Economic change in Derbyshire in the late Middle Ages, 1272–1540', unpubl. Ph.D. thesis, Univ of London, pp. 168–74
57. CAMPBELL (1990) *op. cit.*, pp. 81–92; *IDEM* (forthcoming [d]) 'Ecology versus economics in late thirteenth- and early fourteenth-century English agriculture', in SWEENEY, D. (ed.) *Reality and image in medieval agriculture* (Pennsylvania University Press, State College, Pa.)
58. For the ecological constraints of commonfield husbandry see: FOX, H. S. A. (1984) 'Some ecological dimensions of medieval field systems', in BIDDICK, K. (ed.) *Archaeological approaches to medieval Europe'* (Medieval Institute, Western Michigan Univ., Kalamazoo) pp. 119–58
59. MILLER, E. (1975) 'Farming in northern England during the twelfth and thirteenth centuries', *Northern Hist.* 11: 9–10; HALLAM *op. cit.*, pp. 405–8
60. For case studies of upland stock farms see: TUPLING, G. H. (1927) *The economic history of Rossendale* (Chetham Society, Manchester) pp. 17–41; SHAW, R. C. (1956) *The royal forest of Lancaster* (Guardian Press, Preston) pp. 353–91; DAVIES, R. R. (1978) *Lordship and society in the March of Wales 1282–1400* (Clarendon Press, Oxford) pp. 115–16; DONKIN, R. A. (1978) *The Cistercians: studies in the geography of medieval England and Wales* (Pontifical Inst. of Medieval Stud., Studies and Texts 38, Toronto) pp. 75–82; HALLAM *op. cit.*, pp. 408–11
61. POWER *op. cit.*, pp. 21–5 (where the clusters are numbered 1[1]; 2[2]; 3[3]; 4[5]; 5[4]; 6[7]; 7[6]; [8] – current numbers in square brackets)
62. Above, pp. 15–16
63. HIE, *et al. op. cit.*; JOHNSTON *op. cit.*
64. EVERITT *op. cit.*, p. 106
65. KUSSMAUL, A. (1990) *A general view of the rural economy of England, 1538–1840* (Cambridge University Press, Cambridge) pp. 6–7
66. CAMPBELL (1990) *op. cit.*, pp. 81–3; CAMPBELL and POWER *op. cit.*, pp. 37–8; OVERTON and CAMPBELL *op. cit.*, pp. 20–2; GALLOWAY and MURPHY *op. cit.*
67. CAMPBELL (1983) *op. cit.*; CAMPBELL (forthcoming [a]) *op. cit.*; *IDEM* and OVERTON, M. (forthcoming) 'A new perspective on medieval and early modern agriculture: six centuries of Norfolk farming, *c.* 1250–*c.* 1850', *Past & Present*
68. For example, CAMPBELL (forthcoming [c]) *op. cit.*

XI

Economic rent and the intensification of English agriculture, 1086–1350

The eleventh, twelfth, and thirteenth centuries witnessed a European-wide expansion in population and economic activity.[1] Although the trend is unmistakable, in England alone is it possible to quantify the scale on which landuse was transformed. Thus, on the evidence of the number of ploughteams recorded by Domesday Book, the arable area may be estimated at approximately 8.5 million acres in 1086.[2] By 1300, after two centuries of active reclamation and colonization, it is unlikely to have exceeded the 11.5 million acres (4.7 million ha) attained at the height of the ploughing-up campaign of the Napoleonic Wars.[3] By dint of much hard effort – assarting, draining and reclaiming, and making good the devastation wrought by William I in the north – the cultivated area was extended by approximately a third.[4] Yet over the same period the population grew from an estimated 1.5–2.5 million in 1086 to an estimated 3.8–7.2 million *c.* 1300.[5] Although there is much debate over which are the most acceptable of these various global estimates, few would dispute that the population at least doubled and may even have trebled.[6] As a result the amount of arable land *per caput* roughly

[1] M. M. Postan, ed., *The Cambridge economic history of Europe*, I, *The agrarian life of the Middle Ages* (2nd edn., Cambridge, 1966), 291–659.

[2] The estimate of ploughteams and total arable area is based on that given by R. V. Lennard, *Rural England 1086–1135* (Oxford, 1959), 393, correcting for the fact that his estimate omits eleven of the English counties.

[3] For the arable area *c.* 1300 (and the population which it was capable of supporting) see B. M. S. Campbell, J. A. Galloway, D. Keene, and M. Murphy, *A medieval capital and its grain supply: agrarian production and distribution in the London region c. 1300*, Historical Geography Research Series 30 (1993), 44–5.

[4] For a survey of colonization and reclamation in this period see, R. A. Donkin, "Changes in the early Middle Ages", 73–135 in H. C. Darby, ed., *A new historical geography of England* (Cambridge, 1973), 98–106; also, H. C. Darby, R. E. Glasscock, J. Sheail, and G. R. Versey, "The changing geographical distribution of wealth in England 1086–1334–1524", *Journal of Historical Geography* 5 (1979), 249–56.

[5] R. M. Smith, "Human resources", 188–212 in G. Astill and A. Grant, eds., *The countryside of medieval England* (Oxford, 1988), 189–91; R. M. Smith, "Demographic developments in rural England 1300–1348", 25–77 in B. M. S. Campbell, ed., *Before the Black Death: studies in the 'crisis' of the early-fourteenth century* (Manchester, 1991), 47–50.

[6] For a dissentient view see A. R. Bridbury, "The Domesday valuation of manorial income", 111–32 in *The English economy from Bede to the Reformation* (Woodbridge, 1992), 121–5.

halved over this 200–year period. The decline in the *per caput* supply of grassland and woodland is likely to have been even greater.

Such a substantial reduction in the ratio of land to people lends substance to those who have argued that the twelfth and thirteenth centuries exemplify the classic Malthusian scenario of population outstripping available food supplies. Indeed, M. M. Postan and J. Z. Titow both maintain that once reserves of colonizable land were depleted continued population growth was largely sustained at the price of a serious erosion of peasant living standards.[7] Nor does Postan concede that the adoption of more intensive methods of production, especially among the peasantry who made up the mass of cultivators, did much to compensate for the declining supply of land. Yet some intensification there must have been; after all, it was inherent to the very process of land-reclamation. The progressive upgrading of land from marsh, to pasture, and eventually meadow, which H. S. A. Fox has described at Podimore in Somerset, for instance, was only achieved through the expenditure of much labour.[8] Its reward was a significant addition to that township's limited stock of pastoral resources. The same kind of thing was going on in countless other townships, repeating in miniature what was being undertaken on a large scale in the Somerset levels, Romney Marsh, the Fens of East Anglia, and, most spectacularly, across the North Sea in the polder lands of the Low Countries. In every case, labour, capital, and enterprise (in the form of organization) were being lavished upon land in order to make it more productive.[9] Nor was this a one-off investment; converting land from pastoral to arable use may have delivered a higher food yield per unit area but it also incurred a permanent increase in unit labour costs.[10] It was not, however, intensification through the expansion of agricultural/arable land that Postan doubted; rather, he questioned whether a given unit of arable land itself could be made more productive. If not, as Titow points out, "the quantity of food produced per head of population must have been declining".[11]

Raising the productivity per unit of arable land necessarily entailed both the improvement of existing techniques (involution) and some measure of technological change (innovation). In the absence of innovation some degree of involution is almost always possible. Ricardo's landuse model implies that the intensity of production rises with the demand for land, as higher levels of economic rent justify the expenditure

---

[7] J. Z. Titow, *English rural society 1200–1350* (London, 1969), 64; M. M. Postan, *The medieval economy and society: an economic history of Britain in the Middle Ages* (London, 1972), 44.

[8] H. S. A. Fox, "The alleged transformation from two-field to three-field systems in medieval England", *Economic History Review*, 2nd series XXXIX (1986), 544–5.

[9] Compare E. Thoen, "The birth of 'the Flemish husbandry': agricultural technology in medieval Flanders" and P. Hoppenbrouwers, "Agricultural production and technology in the Netherlands, *c.* 1000–1500", 69–88 and 89–114 in G. Astill and J. Langdon, eds., *Medieval farming and technology: the impact of agricultural change in northwest Europe* (Leiden, 1997). On the often quite complex organization and collaboration involved in the reclamation of the Fens see H. E. Hallam, *Settlement and society: a study of the early agrarian history of south Lincolnshire* (Cambridge, 1965), 16–22, 218–20.

[10] G. Clark, "Labour productivity in English agriculture, 1300–1860", 211–35 in B. M. S. Campbell and M. Overton, eds., *Land, labour and livestock: historical studies in European agricultural productivity* (Manchester, 1991), 230–31.

[11] Titow, *English rural society*, 72.

of more labour.[12] This chimes with Boserup's empirical observation that at a given level of technology there is a positive correlation between population density, the intensity of agriculture, and agricultural output per unit area.[13] Yet Postan and Titow deny that any such technological advance took place over this period.[14] Whether this was because of a genuine paucity of new technical ideas, the communal organization of agriculture, or the manorial regime's stultifying effect upon investment at all levels, Postan is undecided.[15] Robert Brenner, however, is less diffident:

> the inability of the serf-based agrarian economy to innovate in agriculture ... is understandable in view of the interrelated facts, first, of heavy surplus extraction by the lord from the peasant and, second, the barriers to mobility of men and land which were themselves part and parcel of the unfree surplus-extraction relationship. ... At the same time, given his unfree peasants, the lord's most obvious mode of increasing income from his lands was not through capital investment and the introduction of new techniques, but through squeezing the peasants, by increasing either money rents or labour services.[16]

Such a verdict, of course, presumes that medieval farmers did indeed fail to match at least in part rising population with rising output per unit area. There are grounds now for doubting whether this was in fact the case.

In an exclusively organic age, higher land productivity was contingent upon the evolution of increasingly intensive, self-sustaining, mixed-farming systems. In the seventeenth and eighteenth centuries the development of such systems is associated with a degree of technological novelty: new crops, new breeds, novel implements and methods, and the physical and tenurial reorganization of farms and fields.[17] Medieval farmers enjoyed fewer such options. The establishment of rabbit warrens, erection of windmills, and substitution of horses for oxen were the main technological novelties of the age.[18] More productive husbandry systems therefore had mostly to be fashioned from the existing range of crops and animals. Much could nevertheless be achieved. Rather than "a few technological leaps" the process of agricultural change during the

---

12 On Ricardo, see M. Blaug, *Economic theory in retrospect* (3rd edn., Cambridge, 1978), 91–112.

13 E. Boserup, *Population and technology* (Oxford, 1981), 15–28.

14 "... the inertia of medieval agricultural technology is unmistakable": Postan, *Medieval economy and society*, 44. See also Titow, *English rural society*, 72.

15 Postan, *Medieval economy and society*, 42–4.

16 R. Brenner, "Agrarian class structure and economic development in pre-industrial Europe", (*Past and Present* 70 [1976], 30–75) 10–63 in T. H. Aston and C. H. E. Philpin, eds., *The Brenner debate: agrarian class structure and economic development in pre-industrial Europe* (Cambridge, 1987), 31.

17 For the post-medieval development of English agriculture see *The agrarian history of England and Wales, IV–VI* (Cambridge, 1967–89).

18 J. Sheail, *Rabbits and their history* (Newton Abbot, 1971); M. Bailey, "The rabbit and the medieval East Anglian economy", *Agricultural History Review* 36 (1988), 1–20; R. Holt, *The mills of medieval England* (Oxford, 1988), 20–21; J. L. Langdon, *Horses, oxen and technological innovation: the use of draught animals in English farming from 1066–1500* (Cambridge, 1986); J. Langdon, "Was England a technological backwater in the Middle Ages?", 275–92 in Astill and Langdon, eds., *Medieval farming and technology*.

twelfth and thirteenth centuries consisted of "a long chain of small improvements".[19] It assumed seven main forms:

1) A refocusing of production upon those agricultural food chains which were most productive of food and energy per unit area, namely (a) crops rather than animals and animal products, (b) pastoral regimes based upon a combination of grazing and fodder cropping rather than grazing alone, (c) coppiced timber rather than natural woodland. This was reflected in the expansion of arable at the expense of grassland, coupled with a greater emphasis upon fodder cropping, hay meadows, and coppiced woodland, all of which required higher factor inputs per unit area.[20]

2) The substitution of crops and animals of higher financial, food, and/or energy yield for those of lower: industrial and horticultural crops for grain crops, food grains for drink grains, pottage grains for bread grains,[21] legumes for bare fallows, draught horses for draught oxen, dairy animals for meat animals, cattle for sheep, and sty-fed for pannage-fed pigs. In this way the aggregate value of agricultural output was raised per unit area.[22]

3) The closer integration of arable and pastoral husbandry to create mixed-farming systems in which the two sectors were complementary rather than competitive in their respective landuse requirements. This was crucial if the resultant arable-based mixed-farming systems were to prove ecologically sustainable in more than the short-term. It was achieved via better control of fallow grazing and folding, the development of convertible-farming systems in which land alternated between arable and temporary pasture, a greater reliance upon fodder crops (principally legumes and oats) in conjunction with the stall and sty feeding of animals, especially in winter, and a more systematic recycling of nitrogen via the collection and application of animal wastes to the arable fields.[23]

---

[19] K. G. Persson, *Pre-industrial economic growth, social organization and technological progress in Europe* (Oxford, 1988), 28.

[20] On agricultural food chains see D. G. Grigg, *The dynamics of agricultural change* (London, 1982), 68–80; on woodland management see O. Rackham, *Ancient woodland: its history, vegetation and uses in England* (London, 1980).

[21] Consuming grains, like oats, as pottage, rather than, as in the case of wheat, grinding them into flour and processing the flour into bread, maximizes the kilocalorie extraction rate: that is, provided that wheat and oats yield equally well, more people per unit area can be supported on a diet of porridge than upon a diet of bread.

[22] Campbell and others, *Medieval capital*, 41–2; I. G. Simmons, *The ecology of natural resources* (London, 1974), 20–22, 170–72.

[23] M. Bailey, "Sand into gold: the evolution of the foldcourse system in west Suffolk, 1200–1600", *Agricultural History Review* 38 (1990), 40–57; E. Searle, *Lordship and community: Battle Abbey and its banlieu, 1066–1538*, Pontifical Institute of Mediaeval Studies, Studies and Texts XXVI (Toronto, 1974), 272–91; B. M. S. Campbell and M. Overton, "A new perspective on medieval and early modern agriculture: six centuries of Norfolk farming c.1250–c.1850", *Past and Present* 141 (1993), 61–2; B. M. S. Campbell, "The diffusion of vetches in medieval England", *Economic History Review*, 2nd series XLI (1988), 193–208;

4) The diversification of rotations (and, where necessary, modification of field systems through the creation of extra field divisions), reduction of fallows, and increase in the frequency of cropping. As a corollary it became necessary to pay greater attention to (a) the preparation of the seed-bed via repeated ploughings, (b) the quality and quantity of seed sown, (c) the maintenance and improvement of soil structure and fertility through the cultivation of nitrogen-fixing legumes, systematic folding of sheep, and applications of farmyard manure, marl, lime, night-soil and the like, and (d) weed control, via systematic weeding, heavier seeding rates, and multiple summer ploughings of bare fallows.[24]

5) The intensification and rationalization of labour processes to achieve higher standards of arable and pastoral management. On seigneurial demesnes this tended to comprise increased labour inputs, more careful management and supervision of workers, greater specialization of labour, and the partial or complete substitution of waged for servile labour to improve work motivation. It was facilitated by the widespread adoption of a system of annual accounting.[25]

6) Improvements to tools and implements and fuller investment in farm-buildings intended for the storage of harvests and protection of stock and livestock, including barns, stables, byres, sties, and sheepcotes. Also, investment in equipment and machinery intended for the processing of agricultural products.[26]

7) Exploitation of the opportunities afforded by market expansion to specialize according to comparative advantage, thereby securing the benefits of a greater spatial division of labour.[27]

K. Biddick, *The other economy: pastoral husbandry on a medieval estate* (London, 1989), 116–25; B. M. S. Campbell, "Agricultural progress in medieval England: some evidence from eastern Norfolk", *Economic History Review*, 2nd series XXXVI (1983), 26–46; R. A. L. Smith, *Canterbury Cathedral Priory: a study in monastic administration* (Cambridge, 1943), 128–65; H. E. Hallam, ed., *The agrarian history of England and Wales*, II, *1042–1350* (Cambridge, 1988), 272–496.

[24] For example, Campbell, "Agricultural progress"; M. Mate, "Medieval agrarian practices: the determining factors?", *Agricultural History Review* 33 (1985), 22–31; P. F. Brandon, "Cereal yields on the Sussex estates of Battle Abbey during the later Middle Ages", *Economic History Review*, 2nd series XXV (1972), 403–20; D. Postles, "Cleaning the medieval arable", *Agricultural History Review* 37 (1989), 130–43.

[25] For example, C. Thornton, "The determinants of land productivity on the bishop of Winchester's demesne of Rimpton, 1208 to 1403", 183–210 in Campbell and Overton, eds., *Land, labour and livestock*, 201–7; for the spread of manorial accounting see P. D. A. Harvey, ed., *Manorial records of Cuxham, Oxfordshire, circa 1200–1359*, Oxfordshire Record Society 50 and Royal Commission on Historical Manuscripts joint publication 23 (London, 1976), 12–71.

[26] J. G. Hurst, "Rural building in England and Wales: England", 854–965 in Hallam, ed., *Agrarian History, II,* 859, 867–8, 888–98; Holt, *Mills of medieval England*; B. M. S. Campbell, "Commercial dairy production on medieval English demesnes: the case of Norfolk", 107–18 in E. Grant, ed., *Animals and their products in trade and exchange*: *Anthropozoologica* 16 (1992), 112.

[27] M. Overton and B. M. S. Campbell, "Productivity change in European agricultural development", 1–50 in Campbell and Overton, eds., *Land, labour and livestock*, 19–22; Persson, *Pre-industrial economic growth*, 10–12, 31.

All of these methods and strategies can be documented as having been employed to some extent in some part of England by the beginning of the fourteenth century. Their selective adoption is reflected in a greater differentiation of mixed-farming types, as exemplified by the eight basic systems which have been identified as in operation on seigneurial demesnes at this time (Table 1). The range of farming systems on peasant holdings was undoubtedly wider. With higher labour to land ratios than most demesnes, peasant holdings were potentially far more intensively cultivated. Peasants, for instance, outpaced landlords in the replacement of the ox with the horse and, on the evidence of the *Nonarum inquisitiones* of 1342 (which for some counties detail the value of the small tithes on flax, hemp, and cider), were significant producers of industrial crops.[28]

Although the peasant sector is the larger and more crucial it is the demesne sector that is the better documented thanks to the survival of manorial accounts in large numbers. Table 1 is based upon a national sample of such accounts from 388 demesnes.[29] On its evidence the more intensive seigneurial mixed-farming systems – "intensive mixed-farming" and "light-land intensive" – remained outnumbered roughly two to one by such relatively extensive systems as "sheep-corn husbandry", "extensive mixed-farming", "extensive arable husbandry" and "oats and cattle". Features diagnostic of intensification are most conspicuous on the pastoral side. For instance, intensive demesnes tended to make greater use of the faster, stronger, but more expensive horse than the slower, cheaper ox. The dividend was the far higher proportion of non-working animals which this enabled them to stock, particularly dairy cattle, but also sheep (especially on "light-land intensive demesnes") and some swine. The livestock profiles of intensive demesnes were therefore particularly well developed, to the extent that this often translated into an above average ratio of livestock to crops. This had obvious benefits for the recycling of nitrogen and maintenance of soil fertility. Although these were the farming systems which made greatest use of the horse they were also the systems which devoted the smallest proportions of their cropped acreage to oats (thus refuting the widely canvassed notion that greater use of the horse promoted an increase in the oats acreage).[30] Enough oats were grown to satisfy the requirements of traction and haulage and no more. Oats had the lowest cash value of any of the grains hence pride of place within the spring-sown schedule was allocated to barley, which commanded a substantially higher relative price and was preferred for

---

[28] Langdon, *Horses, oxen and technological innovation*, 172–253; G. Vanderzee, ed., *Nonarum Inquisitiones in Curia Scaccarii*, Record Commissioners (London, 1807); A. Sutton, "The early linen and worsted industry of Norfolk and the evolution of the London Mercers' Company", *Norfolk Archaeology* XL (1989), 201–25; N. Evans, *The East Anglian linen industry: rural industry and local economy 1500–1850*, Pasold studies in textile history 5 (Aldershot, 1985), 41–6.

[29] Omitted were solely pastoral regimes: for example, M. A. Atkin, "Land use and management in the upland demesne of the de Lacy estate of Blackburnshire", *Agricultural History Review* 42 (1994), 1–19.

[30] For example, C. Parain, "The evolution of agricultural technique", 125–79 in Postan, ed., *Cambridge economic history of Europe*, 162; Persson, *Pre-industrial economic growth*, 30.

Table 1. The principal English seigneurial mixed-farming systems 1250–1349 (mean characteristics)

	Farming type								All
	1	2	3	4	5	6	7	8	
% total sown acres:									
wheat	22	8	34	33	36	33	43	1	31
rye	4	19	7	3	2	2	1	8	5
winter mixtures	1	2	6	1	5	1	<1	0	2
barley	35	43	4	16	12	13	9	10	16
oats	13	15	42	28	26	43	38	78	34
spring mixtures	5	2	2	6	9	2	3	0	4
legumes	20	10	4	14	10	5	5	3	8
Total sown acres	180	180	230	182	177	216	144	107	185
% total livestock units:*									
horses	17	19	18	33	12	8	15	14	16
oxen	16	9	36	39	39	32	84	58	43
mature cattle	29	29	22	5	1	19	<1	18	14
immature cattle	18	12	15	0	0	12	0	7	8
sheep	14	28	6	5	48	27	0	2	17
swine	5	2	3	17	1	3	<1	0	3
Total livestock units*	84	51	57	28	51	110	24	40	59
Stocking density+	46	32	27	19	32	61	18	40	35

Notes:

1 intensive mixed-farming (66 demesnes)
2 light-land intensive (34 demesnes)
3 mixed-farming with cattle (72 demesnes)
4 arable husbandry with swine (19 demesnes)
5 sheep-corn husbandry (45 demesnes)
6 extensive mixed-farming (65 demesnes)
7 extensive arable husbandry (80 demesnes)
8 oats and cattle (7 demesnes)

* (horses x 1.0) + ([oxen + mature cattle] x 1.2) + (immature cattle x 0.8) + ([sheep + swine] x 0.1)

\+ total livestock units per 100 sown acres.

Source: J. P. Power and B. M. S. Campbell, "Cluster analysis and the classification of medieval demesne-farming systems", *Transactions of the Institute of British Geographers*, new series 17 (1992), 234.

brewing.[31] The emphasis upon barley was part-and-parcel of a general bias towards spring-sown crops, which often occupied well over two-thirds of the cropped acreage. This bias provides a clue to the generally intensive and flexible character of rotations, especially in the case of "intensive mixed-farming". Demesnes practising this most exacting form of husbandry devoted the lion's share of the winter course to wheat, the most demanding crop of all, and partly to replenish soil nitrogen, partly for fodder, and partly for food, grew legumes on a larger scale than in any other farming system. This is consistent with virtually continuous cropping of the arable and the near elimination of fallows which independent investigation reveals to have been the case on some of the most intensively cultivated of this group of demesnes.[32] In contrast, "light-land intensive" demesnes favoured rye rather than wheat and grew smaller acreages of legumes. Their soils would not support such demanding rotations hence they practised a variety of irregular rotations, including the periodic alternation of land between crops and temporary pasture.

Typically, the more extensive the farming system the less closely integrated the arable and pastoral sectors, which were sometimes conducted as virtually separate enterprizes.[33] Even when stocking densities were relatively high, as in the case of "extensive mixed farming", this usually owed more to an abundance of permanent grassland than fodder cropping or the operation of some kind of convertible regime. Again, the composition of the pastoral sector is revealing. Usually grass-fed oxen outnumbered the grass- and fodder-fed horse by over four to one. Working animals made up practically 40 per cent of total livestock units, and a significant proportion of the remainder were devoted to rearing replacement draught animals (hence the more balanced ratio between mature and immature cattle than that maintained on more intensively-managed demesnes where the prime function of cattle herds was milk production). Among the non-working animals, sheep – the most grassland dependent animals of all – were particularly prominent, especially on "sheep-corn" demesnes, with ewes often supplanting cows as the principal dairy animal.[34] "Extensive arable" demesnes, however, effectively stocked working animals alone (again primarily oxen). For replacement draught animals they either relied upon transfers from other manors or purchase. Irrespective of the size of their pastoral sectors, most of these demesnes often operated some version of two- or three-course cropping, with wheat the predominant winter

[31] Nationally, the relative price per bushel of the principal grains *c.* 1300 was: wheat 1.00, rye 0.75; barley 0.70, and oats 0.41 (B. M. S. Campbell, "Land, labour, livestock, and productivity trends in English seigniorial agriculture", 144–82 in Campbell and Overton, eds., *Land, labour and livestock*, 169). See also G. Comet, "Technology and agricultural expansion in the Middle Ages: the example of France north of the Loire", 11–40 in Astill and Langdon, eds., *Medieval farming and technology.*

[32] Campbell, "Agricultural progress", 28–36; B. M. S. Campbell, "Arable productivity in medieval England: some evidence from Norfolk", *Journal of Economic History* XLIII (1983), 390–94.

[33] K. Biddick with C. Bijleveld, "Agrarian productivity on the estates of the bishopric of Winchester in the early thirteenth century: a managerial perspective", 95–123 in Campbell and Overton, eds., *Land, labour and livestock*, 115.

[34] On ewe-dairying see H. E. Hallam, *Rural England 1066–1348* (London, 1981), 129–30, 248; Biddick with Bijleveld, "Agrarian productivity", 115–18.

crop and oats the predominant spring. Some legumes were grown, but on too small a scale to make much contribution to available supplies of soil nitrogen. Regular fallowing must therefore have played a key role in the maintenance of soil fertility, with cropping restricted to a maximum of two consecutive courses.[35]

A wide productivity gulf separated the most from the least intensive of these farming systems, as exemplified by arable rental values of 12–36 pence an acre on the most intensive mixed-farming demesnes *c.* 1300, compared with valuations of 4 pence, 3 pence, or even as little as 2 pence an acre on demesnes operating extensive mixed-farming systems.[36] In Norfolk at this time "intensive mixed-farming" demesnes commonly obtained mean gross yields per acre for all the principal grain crops in the range 15–25 bushels per acre (1,350–2,250 litres per hectare), rising to 20–30 bushels per acre (1,800–2,700 lit/ha) on the most productive of all. The latter constitute the highest recorded English medieval yields, comparable with yield levels more usually associated with the era of the agricultural revolution in the late eighteenth and early nineteenth centuries and almost on a par with the exceptionally high yields documented by J. Derville in parts of northern France at the close of the thirteenth century.[37] Yields on more extensively farmed demesnes were generally much lower, the unimpressive and, sometimes, dismal yields obtained on many of the midland demesnes of Westminster Abbey and southern demesnes of the bishopric of Winchester, St Swithin's Priory, Winchester, and Glastonbury Abbey, being largely responsible for medieval agriculture's reputation for low productivity.[38] For instance, aggregate output per arable acre on the bishop of Winchester's demesne of Rimpton in Somerset was only a third that prevailing on the prior of Norwich's intensively farmed demesne of Martham in eastern Norfolk.[39] Moreover, whereas it has been claimed that yields tended to fall on many of the Winchester demesnes as pressure upon the land mounted during the second half of the thirteenth century, mean yields actually rose in intensively-cropped Norfolk over the same period.[40] In other words, there is a strong positive association between the intensity and the sustainability of husbandry.[41]

[35] R. S. Shiel, "Improving soil fertility in the pre-fertiliser era", 51–77 in Campbell and Overton, eds., *Land, labour and livestock*, 70–73.

[36] These figures are derived from an unpublished analysis of the extents attached to *inquisitiones post mortem* held at the Public Record Office, London.

[37] As reported by Thoen, "Birth of 'the Flemish husbandry'". Corresponding yield *ratios* were nevertheless well below those obtained on these French farms.

[38] R. Lennard, "The alleged exhaustion of the soil in medieval England", *Economic Journal* XXXII (1922), 12–27; M. M. Postan, "Medieval agrarian society in its prime", 548–632 in Postan, ed., *Cambridge economic history of Europe*, 556-9; J. Z. Titow, *Winchester yields: a study in medieval agricultural productivity* (Cambridge, 1972); D. L. Farmer, "Grain yields on Westminster Abbey Manors, 1271–1410", *Canadian Journal of History* XVIII (1983), 331–47; N. Hybel, *Crisis or change. The concept of crisis in the light of agrarian structural reorganization in late medieval England*, trans. J. Manley (Aarhus, 1989).

[39] Thornton, "Determinants of land productivity", 191–3.

[40] Titow, *Winchester yields*, 12–33; Campbell, "Land, labour, livestock, and productivity trends", 159–74.

[41] Campbell, "Land, labour, livestock, and productivity trends", 144–6.

Why was this productivity gap so wide? Certainly, by the end of the eighteenth century it had closed dramatically, and nowhere did mean yields per county deviate far from the national mean.[42] And why, if food was in such short supply, had so few demesnes intensified, innovated, and raised their productivity to the maximum sustainable given available technology? Patently, medieval cultivators did not lack the technological means to create sustainable systems which produced more from the land. As Postan mused, "the real problem of medieval technology is not why new technological knowledge was not forthcoming, but why the methods, or even the implements, known to medieval men were not employed, or not employed earlier or more widely than they in fact were".[43] In eastern Norfolk and northeast Kent medieval husbandmen had solved the central dilemma of how to reconcile the conflicting landuse requirements of the arable and pastoral sectors without jeopardising the fragile ecological basis of reproduction. They did this by integrating crop and livestock production into a mutually reinforcing mixed-farming regime in which fodder cropping, and especially the cultivation of nitrogen-fixing legumes, played a crucial role. Half a millenium later, rapid and widespread diffusion of an improved version of this mixed-farming system was to be one of the cornerstones of the so-called agricultural revolution.[44] The late thirteenth and early fourteenth centuries, however, experienced no such agricultural revolution, and seigneurial agriculture at least remained more extensive than intensive, with low rather than high productivity the norm. The explanations most commonly advanced to account for this are either Whiggish or Marxist, and stress either the conservatism of cultivators and their resistance to change or the disincentives to investment and innovation provided by the feudal system and the licence it gave lords to raise revenues instead by raising feudal exactions.[45] Unfortunately, such explanations do not sit easily with the clear empirical evidence that medieval cultivators – both peasants and lords – could and did innovate and adopt new practices when it suited them. Rather than a failure of supply, the problem may have lain more with demand and a lack of sufficient incentives to invest, innovate, and intensify.

Demand is translated into landuse and farming systems via the medium of economic rent. Economic rent is the return due for the use of the land alone as a factor of production. It represents that part of a farmer's revenue above production costs, but excluding remuneration derived from the three other main factors of production, namely labour, capital, and enterprise. So long as self-sufficiency was the predominant objective of most cultivators, economic rent – as Ricardo recognized – was largely a function of land quality and the demand for land (that is population density mediated via institutional controls upon rent levels and access to land). Where competition for land for subsistence was strongest and that land was inherently most productive, rents

[42] See the yield figures summarized in R. C. Allen and C. Ó Gråda, *On the road again with Arthur Young: English, Irish, and French Agriculture during the Industrial Revolution*, Discussion paper 86–38, Department of Economics, University of British Columbia (1986), 42–4.

[43] Postan, *Medieval economy and society*, 42.

[44] Campbell and Overton, "A new perspective", 88–95.

[45] For a summary see, Persson, *Pre-industrial economic growth*, 3–7, 63–4; also, Langdon, "Was England a technological backwater?".

would be highest. But once markets developed and agriculture became more commercialized so, as J. H. von Thünen demonstrated, economic rent was increasingly determined by the cost of transporting goods to market, rents rising with proximity to the market.[46] In both cases, the higher the economic rent the greater the incentive and justification for raising inputs of labour, capital, and enterprise. Where, close to major cities, von Thünen rents generally exceeded Ricardian rents it was the market which determined the character and intensity of production, whether for consumption or exchange. At a greater distance, however, the rent for subsistence production (with no effective transport costs to bear) may have exceeded that for commercial production, resulting in different types and intensities of production between the subsistent and commercial sectors. In a medieval context this meant that in areas of low von Thünen economic rent peasants producing for consumption may have been more intensive in their methods than demesnes producing for exchange. Areas of high economic rent – be it because of good soils, strong demand for land, and/or favourable access to markets – should therefore have been characterized by more intensive and productive husbandry systems than areas of low economic rent. A test of whether this was in fact the case is provided by the geographical distribution of the principal demesne-farming systems.

As will be seen from Figures 1 and 2, the two most intensive farming systems – "intensive mixed-farming" and " light-land intensive" – were both comparatively limited in distribution. Notable concentrations of "intensive mixed-farming" demesnes occurred in eastern Norfolk and eastern Kent and, to a lesser extent, in the Soke of Peterborough and on the better soils of the lower Thames valley just upstream from London. Independent studies of agriculture in these localities confirm the intensive and integrated nature of the mixed-farming systems employed and the high returns per unit area thereby obtained, which were also reflected in exceptionally high per acre valuations.[47] All four localities were characterized by naturally fertile and readily cultivated loam soils, access to substantial local markets, and wider access, via navigable rivers and coastal trading ports, to major external markets.[48] East Norfolk and northeast Kent were also weakly manorialized and exceptionally densely populated. In short, they were characterized by an extreme coincidence of high Ricardian and von Thünen economic rent with the result that it was here that English medieval agriculture attained its greatest peak of intensity and productivity (the superior intensity and productivity of husbandry in parts of Flanders and northern France at this date implying even higher levels of economic rent, as was consistent with their more

[46] M. Chisholm, *Rural settlement and land-use: an essay on location* (London, 1962), 20–32; Grigg, *The dynamics of agricultural change*, 135–40; M. Bailey, "The concept of the margin in the medieval English economy", *Economic History Review*, 2nd series XLII (1989), 1–17.

[47] Campbell, "Agricultural progress"; Smith, *Canterbury Cathedral Priory*, 128–65; Mate, "Medieval agrarian practices", 22–31; Biddick, *The other economy*, 50–77; Hallam, ed., *Agrarian History*, II, 320; Campbell and others, *Medieval capital*, 128–44.

[48] J. Langdon, "Inland water transport in medieval England", *Journal of Historical Geography* 19 (1993), 1–11.

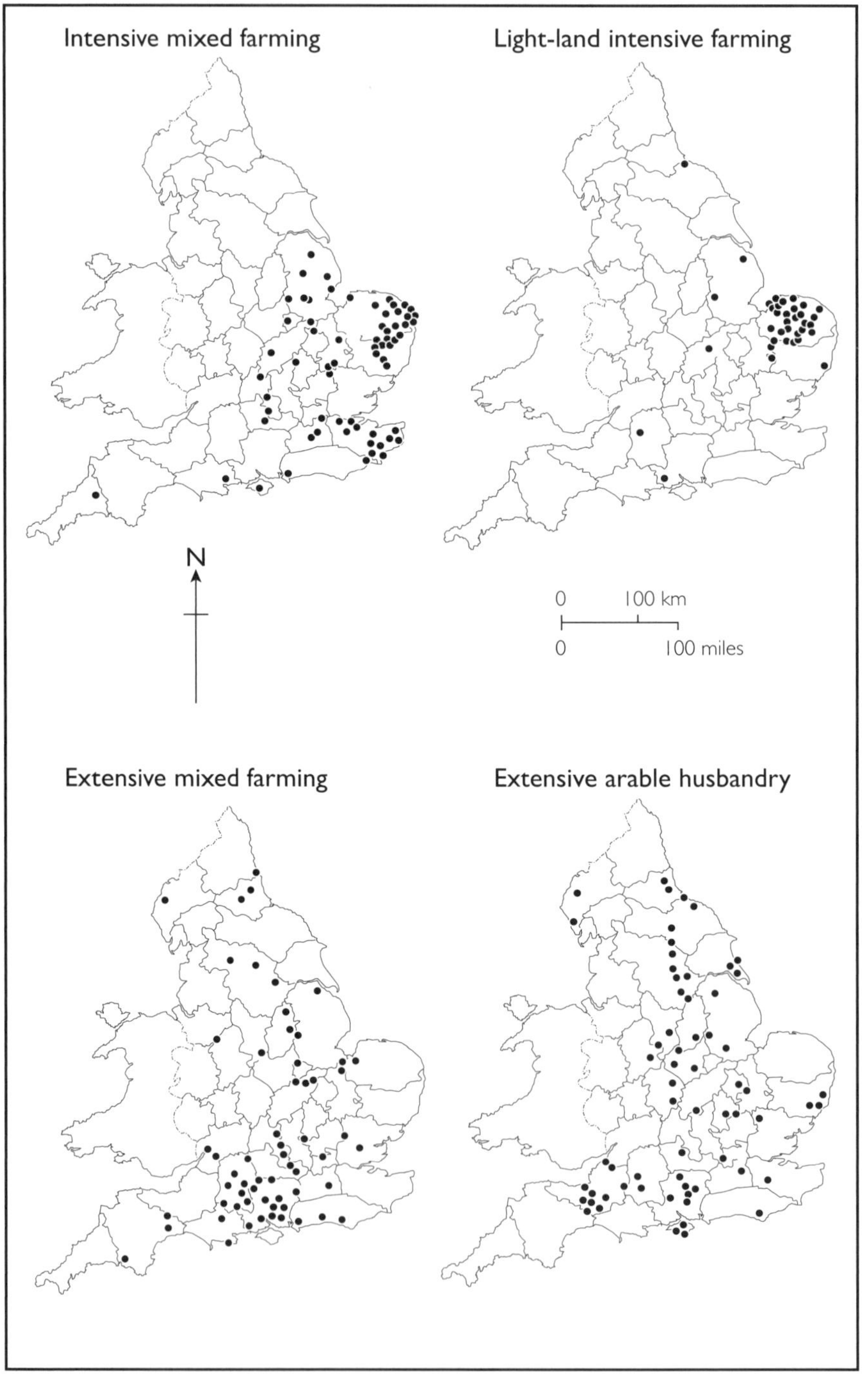

Figure 1. Intensive and extensive seigneurial mixed-farming systems in England, 1250–1349.

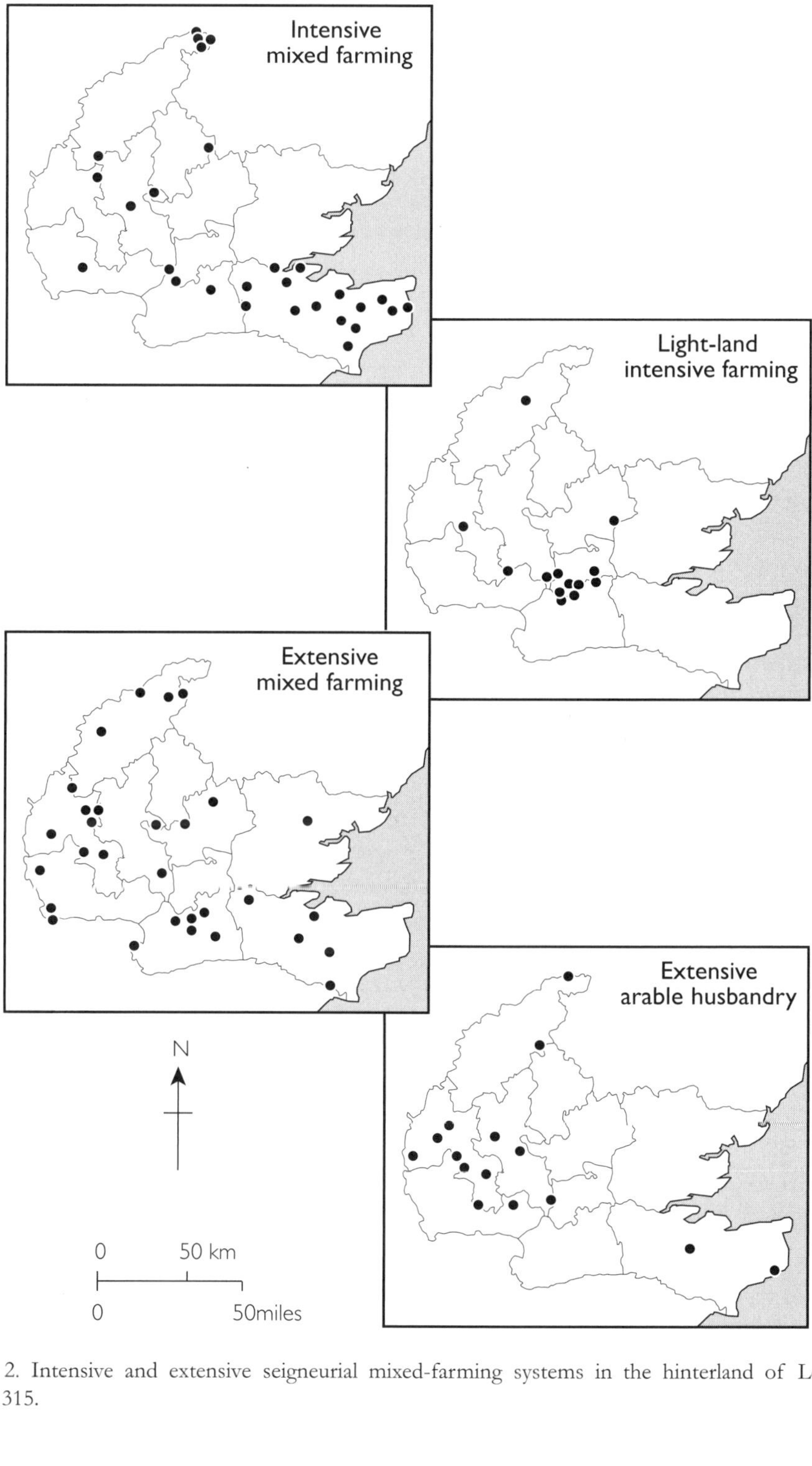

Figure 2. Intensive and extensive seigneurial mixed-farming systems in the hinterland of London, 1288–1315.

specialized and urbanized economies).[49] Outside of these localities "intensive mixed-farming" demesnes were to be found at a scatter of locations in the east midlands (Lincolnshire, Leicestershire, Rutland, and Cambridgeshire), the upper Thames valley, and along the south coast. Studies of agriculture in several of these areas have shown that it often exhibited a tendency towards the kinds of intensive methods which found their fullest and most productive expression in eastern Norfolk and eastern Kent.[50] "Intensive mixed-farming" demesnes were, however, conspicuously absent from the rest of the country.

"Light-land intensive" demesnes were, if anything, even more specific in distribution. First and foremost, this was a husbandry system associated with Norfolk, where it was especially characteristic of the county's lighter soils (a distribution which spills over into the Breckland of north-west, and the Sandlings of east Suffolk). The concentration of "light-land intensive" demesnes in the immediate vicinity of Norwich – the second city after London with a population of perhaps 25,000 – is especially notable and is probably to be explained by the coexistence of light soils and strong urban demand for the rye and barley which were the principal crops grown (neither of which was as capable as wheat of bearing the cost of carriage from a distance).[51] The need to produce relatively bulky, low-value crops close to consuming centres – as von Thünen's landuse model predicts – no doubt helps to explain the corresponding cluster of "light-land intensive" demesnes which occurs in the immediate vicinity of London (Figure 2), where rye and barley were also grown in quantity and there was a premium to be gained from maximising the ratio of non-working to working animals.[52] Holywell, an outlying "light-land intensive" demesne on the outskirts of Oxford, shares the same close association with a major urban centre. The few other isolated outliers – in Lincolnshire, Northamptonshire, and Wiltshire – are mostly explicable in environmental and/or institutional terms.

Both the most intensive farming systems therefore display a close association with areas of high economic rent, especially where this was borne of access/proximity to major urban markets at home and overseas. Significantly, the very areas where these two systems were most developed were those from which "extensive mixed-farming" and "extensive arable husbandry" were most conspicuously absent. These more extensive farming systems are symptomatic of lower levels of economic rent and, since they are comparatively widespread (Figure 1), there is a clear implication that moderate to low economic rent was the norm throughout the greater part of the country. Examples of "extensive mixed farming" are to be found from the far north to the

---

49 Compare Thoen, "Birth of 'the Flemish husbandry'".

50 J. R. Ravensdale, *Liable to floods: village landscape on the edge of the Fens AD 450–1850* (Cambridge, 1974), 116–20; Brandon, "Demesne arable farming"; Hallam, ed., *Agrarian History*, II, 318–24.

51 E. Rutledge, "Immigration and population growth in early fourteenth-century Norwich: evidence from the tithing roll", *Urban History Yearbook 1988*, 15–30.

52 Campbell and others, *Medieval capital*, 111–25. London's demand for oats and fat cattle may also account for the presence of "oats and cattle" demesnes (no. 8 in the farming-type classifications: see Table 1) at Harrow, Middlesex, and Esher and Lambeth, Surrey (a farming type otherwise more characteristic of moist, upland areas).

extreme south of the country, wherever demesnes were well endowed with both arable and grassland. More numerous and as widespread are demesnes practising "extensive arable husbandry", which display a particular bias towards Somerset, the midlands, and the north of England. These areas were remote from major concentrations of demand and characterized, for the most part, by below average population densities, circumstances which evidently encouraged neither the fuller exploitation of pastoral resources nor the closer integration of crop and livestock production.

On this analysis, extensive farming systems of one sort or another were very much the norm throughout much of the country, and intensive farming systems the exception, confined to a few favoured areas. By implication, therefore, low to moderate levels of economic rent were far more typical than high. This is consistent with what is known about the number, size, and location of the leading urban centres and the extent of their provisioning zones at the climax of medieval demographic, urban, and commercial expansion *c.* 1300. By that date there were probably at least sixteen English cities with 10,000 or more inhabitants.[53] With a population of perhaps 80–100,000, London was by far the largest English city and second only to Paris among the cities north of the Alps.[54] It owed its impressive size to its precocious economic and political primacy and growing centrality.[55] Already it was the focus of the country's road network and was well served by river and coastal communications. These rendered it the country's busiest port and enabled it to exercise a magnetic pull upon a wide area for the supply of foodstuffs and raw materials.

In years of normal harvest London regularly drew on an area of 4,000 square miles in extent for its grain supplies. This area was irregular in shape, reflecting the availability of water transport, and included at its furthest extent several ports on the south and east coasts which were 100 miles or more distant from the city.[56] Livestock, which were capable of walking to market, and the higher-valued livestock products were probably drawn in from further afield. The city's supply lines were well developed and it was served by a sophisticated and well-articulated marketing network. In years of abnormal harvest this extended outwards to tap a wider area. In 1317, for instance, at the height of the worst medieval harvest failure on record, the king ordered his sheriffs to procure essential provisions for the royal household at Westminster. Hay, one of the bulkiest of commodities, was to be obtained from the counties closest to London: Middlesex, Essex, Hertfordshire, Surrey, and Sussex. Grain, better able to withstand the costs of carriage, was to come from a much wider geographical area, comprising Kent, Surrey, Sussex, Hertfordshire, Essex, Suffolk, Norfolk, Cambridge-

53 Campbell and others, *Medieval capital*, 9–11.

54 D. Keene, "A new study of London before the Great Fire", *Urban History Yearbook 1984*, 11–21; D. Keene, *Cheapside before the Great Fire* (London, 1985). But see also P. Nightingale, "The growth of London in the medieval English economy', 89–106 in R. Britnell and J. Hatcher, eds., *Progress and problems in medieval England: essays in honour of Edward Miller* (Cambridge, 1996), 95–6, who argues that the city's medieval population never exceeded 60,000.

55 D. Keene, "Medieval London and its region", *London Journal* 14 (1989), 99–111; again, for a contrary view, see Nightingale, "Growth of London", 95–6.

56 Campbell and others, *Medieval capital*, 46–77.

shire, and Huntingdonshire (in the last two cases presumably shipped to London via the major grain entrepot of Kings Lynn). Finally, livestock were to be procured from a wide scatter of inland counties from as far afield as Gloucestershire, and Somerset to the west and Cambridgeshire and Huntingdonshire to the north, and thence driven overland to Westminster.[57]

This example illustrates how London drew upon different areas at different distances for different commodities in much the way that von Thünen's model predicts. It also demonstrates that even in one of the worst years on record London's provisioning zone remained confined to certain very specific counties. In normal years its hinterland was even more circumscribed, embracing – for all commodities – perhaps a fifth of the country's total land area. Of this, less than half was engaged in the regular supply of grain to the city. Much of the country, therefore, lay beyond the stimulus of the capital's influence. Nor did the needs of other urban centres provide adequate compensation since they were few and far between and their provisioning hinterlands were even smaller. Winchester, for example, lacking a navigable river and situated in a region of below average productivity, was supplied with grain from within a radius of about 12 miles.[58] Exeter, smaller than Winchester but situated within an even less productive hinterland, regularly drew its supplies from up to 20 miles or more away.[59] In a world where, even on the most generous estimates, a maximum of 417,200 people and 10.4 per cent of the country's population lived in towns of at least 10,000 inhabitants, it was inevitable that the agricultural impact of urban markets remained restricted and selective.[60] Under these circumstances, as David Farmer has emphasized, "the local markets and the communities around them were the more important outlets for the produce of the countryside".[61] Such markets were, however, incapable of stimulating economic rent and agricultural intensification to the same extent as major urban concentrations of demand, with the result that across much of the country extensive methods of production remained the most rational form of landuse, especially on the extensive demesne holdings of lords.

Nor, with certain notable exceptions such as the Cornish stannaries, does it appear that areas beyond the reach of major markets had yet hit upon the solution of producing manufactured goods and marketing these at a distance (thereby turning their low land values and cheap food to advantage).[62] This was not to occur until the close of the

---

[57] *Calendar of Close Rolls, 1313–1318* (HMSO, London, 1893), 513–14. I am grateful to Dr Derek Keene for this reference.

[58] D. Keene, *Survey of medieval Winchester*, Winchester Studies 2 (Oxford, 1985), 251–5.

[59] M. Kowaleski, "The grain trade in fourteenth-century Exeter", 1–53 in E. B. Dewindt, ed., *The salt of common life: individuality and choice in the medieval town, countryside and church. Essays presented to J. Ambrose Raftis on the occasion of his 70th birthday* (Kalamazoo, 1995), 28–31.

[60] Campbell and others, *Medieval capital*, 10–11; R. H. Britnell, "Commercialisation and economic development in England, 1000–1300", 7–26 in R. H. Britnell and B. M. S. Campbell, eds., *A commercialising economy: England 1086–1300* (Manchester, 1995), 9–12.

[61] D. L. Farmer, "Marketing the produce of the countryside, 1200–1500", 324–430 in E. Miller, ed., *The agrarian history of England and Wales*, III, *1348–1500* (Cambridge, 1991), 329.

[62] J. Hatcher, *Rural economy and society in the Duchy of Cornwall 1300–1500* (Cambridge, 1970).

Middle Ages when it was to transform the fortunes of many hitherto under-developed areas in the midlands, north and west of the country. Indeed, John Langton and Göran Hoppe have stressed the mutually beneficial tripartite relationship which subsequently developed during the early modern period between expanding metropolitan demand, the evolution of capitalist agriculture, and the growth of proto-industrialization.[63] But in the Middle Ages, the corresponding relationship remained dual rather than tripartite, since proto-industrialization existed only in embryo, and the capacity of urban demand alone to stimulate agrarian development was further qualified by the lesser scale of the metropolis and other urban centres. The rural-urban nexus was consequently a less powerful agent of change in the Middle Ages than it was to be in later centuries.[64] It was in Flanders – northern Europe's most urbanized region – rather than England that its impact was greatest in this period.[65]

Moreover, even within provisioning range of cities high economic rent was the exception rather than the rule. According to the von Thünen model intensive landuse systems of one sort or another occupied only the inner 25 per cent of a city's hinterland, systems of lesser intensity occupying the remainder.[66] Thus, in the case of London *c.* 1300 only 25 per cent of documented demesnes within provisioning range of the capital practised either "intensive mixed-farming" or "light-land intensive" husbandry, all other demesnes operated more extensive systems. Most cities therefore relied more on the extensive than the intensive production of grain and other provisions. Beyond their provisioning range, wool – in many respects the most extensive agricultural commodity of all – was practically the only product capable of being marketed at a distance and was therefore a commercial lifeline for many rural producers otherwise largely dependent upon the dispersed demand of local markets. A. R. Bridbury has estimated that by the opening of the fourteenth century the wool of approximately eight million sheep was being exported in one form or another, the volume of this trade testifying to the low economic rent that prevailed across so much of the country.[67]

On this evidence, it required cities a great deal larger than the largest medieval cities to raise economic rent over a sufficiently wide area to encourage agricultural intensification by more than a minority of farmers in a few favoured localities. When appropriate incentives existed medieval cultivators were not backward in adopting more intensive and productive methods, but across much of the country the impulse

---

[63] J. Langton and G. Hoppe, *Town and country in the development of early modern Western Europe*, Historical Geography Research Series 11 (Norwich, 1983).

[64] E. A. Wrigley, "A simple model of London's importance in changing English society and economy, 1650–1750", *Past and Present* 37 (1967), 44–70; E. A. Wrigley, "Urban growth and agricultural change: England and the Continent in the early modern period", *Journal of Interdisciplinary History* 15 (1985), 683–728; Overton and Campbell, "Productivity change", 35–44; Campbell and Overton, "A new perspective", 88–105.

[65] Persson, *Pre-industrial economic growth*, 73–6.

[66] Campbell and others, *Medieval capital*, 5–6.

[67] A. R. Bridbury, "Before the Black Death" (*Economic History Review*, 2nd series XXX [1973], 393–410), 180–99 in *From Bede to the Reformation*, 185–6.

to change was weak with the result that limited technological development and low productivity remained the order of the day on the majority of demesnes. Had London and the leading provincial cities been larger there can be little doubt that their correspondingly enlarged hinterlands would have experienced little difficulty in producing and supplying the additional foodstuffs and raw materials demanded, for where market signals were strong specialization and intensification almost invariably resulted and marketing links and commercial institutions were forthcoming.[68] At the culmination of medieval economic expansion *c.* 1300 this was most conspicuously the case in the east and southeast of England, and it can be no coincidence that it is here that adoption of the various new technological innovations of the age – windmills, rabbits, horse traction and haulage, vetches and legumes – made greatest progress. It was also here that medieval farmers achieved their greatest technological breakthrough of all by evolving an integrated mixed-farming system capable of a sustained high level of production. Intriguingly, however, it was not in the immediate vicinity of London or even of Norwich that this breakthrough was made, but in eastern Norfolk and eastern Kent, localities accessible to, but at some remove from both those cities.

What these two localities offered over the more immediate environs of these two major cities were good soils, a relatively free and enterprising rural population, and riverine and maritime access to a variety of different markets at home and overseas. They serve as a salutary reminder that environmental, institutional, and cultural factors also exercised an important influence upon the course of agricultural development. But these were also localities dominated by a numerous small-holding peasantry and it is possible that here, as in the Low Countries, it was they rather than the lords who pioneered adoption of more intensive methods.[69] Peasants generally had a more favourable ratio of labour to land than the larger demesne holdings (whose distinguishing feature is more likely to have been a superior productivity of labour than of land) and fewer problems of work motivation. As pressure mounted to raise agricultural output in line with population they were therefore in a stronger position to substitute labour for land. Moreover, it was upon their holdings that they largely had to rely for their income, be that in the form of food produced for direct consumption or cash received from marketed goods (except where this could be augmented by the sale of either surplus labour for wages or craft goods produced with that labour). In this sense most peasants were primarily subsistence rather than commercial producers. This distinction is important for there were undoubtedly many parts of the country where the economic rent for subsistence production of a commodity was superior to that for its commercial production, thus inducing and sustaining more intensive methods of

---

68 Compare the situation in relatively highly urbanised Flanders: Thoen, "Birth of 'the Flemish husbandry'".

69 A. Verhulst, "L'intensification et la commercialisation de l'agriculture dans les Pays-Bas méridionaux au XIIIe siècle", 89–100 in *La Belgique rurale. Mélanges offerts à J. J. Hoebanx* (Brussels, 1985); A. Verhulst, "The 'agricultural revolution' of the Middle Ages reconsidered", 17–28 in B. S. Bachrach and D. Nicholas, eds., *Law, custom and the social fabric in medieval Europe: essays in honor of Bryce Lyon* (Kalamazoo, 1990), 25.

production.[70] The impression of medieval agriculture formed from the evidence of demesnes may therefore be unduly weighted towards extensive methods of production and low per unit area levels of productivity.

It may have been peasants rather than lords who were most energetic and successful at reconciling the widening gap between the expanding population to be fed and the shrinking area *per caput* from which to feed it. At late thirteenth century demesne yield levels, and with the same basic product mix, there would have been little difficulty in feeding a population in 1086 of 2.5 million from the estimated arable area of approximately 8.5 million acres.[71] By *c.* 1300, however, the same yields and production mix would have had difficulty in feeding a population of more than 4.0 million from an arable area which is unlikely to have exceeded 11.5 million acres and may well have been less.[72] Yet most recent estimates of the population *c.* 1300 favour a figure in the region of 6.0 million.[73] Clearly, something in the equation does not fit; the estimates of either the total population, the total arable area, or the output per unit area are wrong. Of course, none of the estimates is robust, but the most plausible explanation of the inconsistency may well be the demesne sector's unrepresentativeness of the productivity of agriculture as a whole. If, for instance, it is assumed that intensification and specialization during the twelfth and thirteenth centuries not only raised demesne productivity but raised peasant productivity further, to a level significantly above that of the demesnes, then the estimates of total population could more easily be squared with those of the arable area available to support it. Only further research can resolve this crucial enigma.

Acknowledgements

I am grateful to Ken Bartley, Jim Galloway, and Margaret Murphy for research assistance and to Derek Keene for advice. Part of the data upon which this paper is based were collected as part of the Feeding the City Project, funded by the Leverhulme Trust and based at the Centre for Metropolitan History, Institute of Historical Research, University of London. Above all, The Queen's University of Belfast has provided encouragement and support over a long period.

70 Bailey, "The concept of the margin", 4.

71 A. R. Bridbury has, in fact, suggested that the Domesday population may have been significantly higher: "Domesday valuation of manorial income", 121–5.

72 Campbell and others, *Medieval capital*, 44–5.

73 For example, Smith, "Demographic developments", 48–9.

Bibliography

The agrarian history of England and Wales, IV–VI (Cambridge, 1967–89).

R. C. Allen and C. Ó Gråda, *On the road again with Arthur Young: English, Irish, and French Agriculture during the Industrial Revolution*, Discussion paper 86–38, Department of Economics, University of British Columbia (1986).

G. Astill and J. Langdon, eds., *Medieval farming and technology: the impact of agricultural change in northwest Europe* (Leiden, 1997).

M. A. Atkin, "Land use and management in the upland demesne of the de Lacy estate of Blackburnshire", *Agricultural History Review* 42 (1994), 1–19.

M. Bailey, "The rabbit and the medieval East Anglian economy", *Agricultural History Review* 36 (1988), 1–20.

M. Bailey, "The concept of the margin in the medieval English economy", *Economic History Review*, 2nd series XLII (1989), 1–17.

M. Bailey, "Sand into gold: the evolution of the foldcourse system in west Suffolk, 1200–1600", *Agricultural History Review* 38 (1990), 40–57.

K. Biddick, *The other economy: pastoral husbandry on a medieval estate* (London, 1989).

K. Biddick with C. Bijleveld, "Agrarian productivity on the estates of the bishopric of Winchester in the early thirteenth century: a managerial perspective", 95–123 in Campbell and Overton, eds., *Land, labour and livestock.*

M. Blaug, *Economic theory in retrospect* (3rd edn., Cambridge, 1978).

E. Boserup, *Population and technology* (Oxford, 1981), 15–28.

P. F. Brandon, "Cereal yields on the Sussex estates of Battle Abbey during the later Middle Ages", *Economic History Review*, 2nd series XXV (1972), 403–20.

R. Brenner, "Agrarian class structure and economic development in pre-industrial Europe" (*Past and Present* 70 [1976], 30–75), 10–63 in T. H. Aston and C. H. E. Philpin, eds., *The Brenner debate: agrarian class structure and economic development in pre-industrial Europe* (Cambridge, 1987).

A. R. Bridbury, "Before the Black Death" (*Economic History Review*, 2nd series XXX [1973], 393–410), 180–99 in *From Bede to the Reformation.*

A. R. Bridbury, *The English economy from Bede to the Reformation* (Woodbridge, 1992).

A. R. Bridbury, "The Domesday valuation of manorial income", 111–32 in *From Bede to the Reformation.*

R. H. Britnell, "Commercialisation and economic development in England, 1000–1300", 7–26 in R. H. Britnell and B. M. S. Campbell, eds., *A commercialising economy: England 1086–1300* (Manchester, 1995).

Calendar of Close Rolls, 1313–1318 (HMSO, London, 1893).

B. M. S. Campbell, "Arable productivity in medieval England: some evidence from Norfolk", *Journal of Economic History* XLIII (1983), 379–404.

B. M. S. Campbell, "Agricultural progress in medieval England: some evidence from eastern Norfolk", *Economic History Review*, 2nd series XXXVI (1983), 26–46.

B. M. S. Campbell, "The diffusion of vetches in medieval England", *Economic History Review*, 2nd series XLI (1988), 193–208.

B. M. S. Campbell, "Land, labour, livestock, and productivity trends in English seigniorial agriculture", 144–82 in Campbell and Overton, eds., *Land, labour and livestock.*

B. M. S. Campbell, "Commercial dairy production on medieval English demesnes: the case of Norfolk", in A. Grant, ed., *Animals and their products in trade and exchange: Anthropozoologica* 16 (1992), 107–18.

B. M. S. Campbell, J. A. Galloway, D. Keene, and M. Murphy, *A medieval capital and its grain supply: agrarian production and distribution in the London region c. 1300*, Historical Geography Research Series 30 (1993).

B. M. S. Campbell and M. Overton, "A new perspective on medieval and early modern agriculture: six centuries of Norfolk farming *c.* 1250–*c.* 1850", *Past and Present* 141 (1993), 38–105.

B. M. S. Campbell and M. Overton, eds., *Land, labour and livestock: historical studies in European agricultural productivity* (Manchester, 1991).

M. Chisholm, *Rural settlement and land-use: an essay on location* (London, 1962).

G. Clark, "Labour productivity in English agriculture, 1300–1860", 211–35 in Campbell and Overton, eds., *Land, labour and livestock.*

G. Comet, "Technology and agricultural expansion in the Middle Ages: the example of France north of the Loire", 11–40 in Astill and Langdon, eds., *Medieval farming and technology.*

H.C. Darby, R. E. Glasscock, J. Sheail, and G. R. Versey, "The changing geographical distribution of wealth in England 1086–1334–1524", *Journal of Historical Geography* 5 (1979), 247–62.

R. A. Donkin, "Changes in the early Middle Ages", 73–135 in H. C. Darby, ed., *A new historical geography of England* (Cambridge, 1973).

N. Evans, *The East Anglian linen industry: rural industry and local economy 1500–1850*, Pasold studies in textile history 5 (Aldershot, 1985).

D. L. Farmer, "Grain yields on Westminster Abbey Manors, 1271–1410", *Canadian Journal of History* XVIII (1983), 331–47.

D. L. Farmer, "Marketing the produce of the countryside, 1200–1500", 324–430 in E. Miller, ed., *The agrarian history of England and Wales*, III, *1348–1500* (Cambridge, 1991).

H. S. A. Fox, "The alleged transformation from two-field to three-field systems in medieval England", *Economic History Review*, 2nd series XXXIX (1986), 526–48.

D. G. Grigg, *The dynamics of agricultural change* (London, 1982).

H. E. Hallam, *Settlement and society: a study of the early agrarian history of south Lincolnshire* (Cambridge, 1965).

H. E. Hallam, *Rural England 1066–1348* (London, 1981).

H. E. Hallam, ed., *The agrarian history of England and Wales*, II, *1042–1350* (Cambridge, 1988).

P. D. A. Harvey, ed., *Manorial records of Cuxham, Oxfordshire, circa 1200–1359*, Oxfordshire Record Society 50 and Royal Commission on Historical Manuscripts joint publication 23 (London, 1976).

J. Hatcher, *Rural economy and society in the Duchy of Cornwall 1300–1500* (Cambridge, 1970).

R. Holt, *The mills of medieval England* (Oxford, 1988).

P. Hoppenbrouwers, "Agricultural production and technology in the Netherlands, *c.* 1000–1500", 89–114 in Astill and Langdon, eds., *Medieval farming and technology.*

J. G. Hurst, "Rural building in England and Wales: England", 854–965 in Hallam, ed., *Agrarian History*, II.

N. Hybel, *Crisis or change. The concept of crisis in the light of agrarian structural reorganization in late medieval England*, trans. J. Manley (Aarhus, 1989).

D. Keene, "A new study of London before the Great Fire", *Urban History Yearbook 1984*, 11–21.

D. Keene, *Cheapside before the Great Fire* (London, 1985).

D. Keene, *Survey of medieval Winchester*, Winchester Studies 2 (Oxford, 1985).

D. Keene, "Medieval London and its region", *London Journal* 14 (1989), 99–111.

M. Kowaleski, "The grain trade in fourteenth-century Exeter", 1–53 in E. B. Dewindt, ed., *The salt of common life: individuality and choice in the medieval town, countryside and church. Essays presented to J. Ambrose Raftis on the occasion of his 70th birthday* (Kalamazoo, 1995).

J. L. Langdon, *Horses, oxen and technological innovation: the use of draught animals in English farming from 1066–1500* (Cambridge, 1986).

J. Langdon, "Inland water transport in medieval England", *Journal of Historical Geography* 19 (1993), 1–11.

J. Langdon, "Was England a technological backwater in the Middle Ages?", 275–92 in Astill and Langdon, eds., *Medieval farming and technology.*

J. Langton and G. Hoppe, *Town and country in the development of early modern Western Europe*, Historical Geography Research Series 11 (Norwich, 1983).

R. V. Lennard, "The alleged exhaustion of the soil in medieval England", *Economic Journal* XXXII (1922), 12-27.

R. V. Lennard, *Rural England 1086–1135* (Oxford, 1959).

M. Mate, "Medieval agrarian practices: the determining factors?", *Agricultural History Review* 33 (1985), 22–31.

P. Nightingale, "The growth of London in the medieval English economy', 89–106 in R. Britnell and J. Hatcher, eds., *Progress and problems in medieval England: essays in honour of Edward Miller* (Cambridge, 1996).

M. Overton and B. M. S. Campbell, "Productivity change in European agricultural development", 1–50 in Campbell and Overton, eds., *Land, labour and livestock.*

C. Parain, "The evolution of agricultural technique", 125–79 in Postan, ed., *Cambridge Economic History of Europe.*

K. G. Persson, *Pre-industrial economic growth, social organization and technological progress in Europe* (Oxford, 1988).

M. M. Postan, ed., *The Cambridge economic history of Europe*, I, *The agrarian life of the Middle Ages* (2nd edn., Cambridge, 1966).

M. M. Postan, "Medieval agrarian society in its prime", 548–632 in Postan, ed., *Cambridge Economic History of Europe.*

M. M. Postan, *The medieval economy and society: an economic history of Britain in the Middle Ages* (London, 1972).

D. Postles, "Cleaning the medieval arable", *Agricultural History Review* 37 (1989), 130–43.

O. Rackham, *Ancient woodland: its history, vegetation and uses in England* (London, 1980).

J. R. Ravensdale, *Liable to floods: village landscape on the edge of the Fens AD 450–1850* (Cambridge, 1974).

E. Rutledge, "Immigration and population growth in early fourteenth-century Norwich: evidence from the tithing roll", *Urban History Yearbook 1988*, 15–30.

E. Searle, *Lordship and community: Battle Abbey and its banlieu, 1066–1538*, Pontifical Institute of Mediaeval Studies, Studies and Texts XXVI (Toronto, 1974).

J. Sheail, *Rabbits and their history* (Newton Abbot, 1971).

R. S. Shiel, "Improving soil fertility in the pre-fertiliser era", 51–77 in Campbell and Overton, eds., *Land, labour and livestock.*

I. G. Simmons, *The ecology of natural resources* (London, 1974).

R. A. L. Smith, *Canterbury Cathedral Priory: a study in monastic administration* (Cambridge, 1943).

R. M. Smith, "Human resources", 188–212 in G. Astill and A. Grant, eds., *The countryside of medieval England* (Oxford, 1988).

R. M. Smith, "Demographic developments in rural England 1300–1348", 25–77 in B. M. S. Campbell, ed., *Before the Black Death: studies in the 'crisis' of the early-fourteenth century* (Manchester, 1991).

A. Sutton, "The early linen and worsted industry of Norfolk and the evolution of the London Mercers' Company", *Norfolk Archaeology* XL (1989), 201–25.

E. Thoen, "The birth of 'the Flemish husbandry': agricultural technology in medieval Flanders" 69–88 in Astill and Langdon, eds., *Medieval farming and technology.*

C. Thornton, "The determinants of land productivity on the bishop of Winchester's demesne of Rimpton, 1208 to 1403", 183–210 in Campbell and Overton, eds., *Land, labour and livestock.*

J. Z. Titow, *English rural society 1200–1350* (London, 1969).

J. Z. Titow, *Winchester yields: a study in medieval agricultural productivity* (Cambridge, 1972).

G. Vanderzee, ed., *Nonarum Inquisitiones in Curia Scaccarii*, Record Commissioners (London, 1807).

A. Verhulst, "L'intensification et la commercialisation de l'agriculture dans les Pays-Bas méridionaux au XIIIe siècle", 89–100 in *La Belgique rurale. Mélanges offerts à J. J. Hoebanx* (Brussels, 1985).

A. Verhulst, "The 'agricultural revolution' of the Middle Ages reconsidered", 17–28 in B. S. Bachrach and D. Nicholas, eds., *Law, custom and the social fabric in medieval Europe: essays in honor of Bryce Lyon* (Kalamazoo, 1990).

E. A. Wrigley, "A simple model of London's importance in changing English society and economy, 1650-1750", *Past and Present* 37 (1967), 44–70.

E. A. Wrigley, "Urban growth and agricultural change: England and the Continent in the early modern period", *Journal of Interdisciplinary History* 15 (1985), 683–728.

XII

The Demesne-Farming Systems of Post-Black Death England: A Classification[1]

By BRUCE M S CAMPBELL, KENNETH C BARTLEY, and JOHN P POWER

Abstract
What was the character of English demesne-farming systems in the half century or so after the Black Death and how does this compare with their character before? Data from three major samples of accounts (representing Norfolk, a ten-county area around London, and the country as a whole) are analysed in an attempt to answer this question. To clarify developments demesnes are classified into seven basic types, replicating the methodology used to develop an equivalent typology for the earlier period. The same methodology is also used to test the relative merits of regionally- versus nationally-derived classifications, with the latter being shown to possess significant advantages over the former. Each of the resultant seven national farming types is both mapped and described and the paper concludes with a consideration of what their configuration reveals about the changing agricultural geography of England in this post-plague era of population decline and economic contraction.

IN few periods have farmers had to contend with demand shifts as dramatic as those of the fourteenth century.[2] Between 1315 and 1375 the country's population, along with that of much of north-west Europe, was reduced by at least half.[3] Suddenly there were significantly fewer mouths to be fed and, relatively, more land from which to feed them. Concomitant shifts in factor prices rendered land far cheaper than it had been at the peak of rural congestion at the beginning of the century, and labour much dearer (although the immediate post-Black Death labour laws endeavoured to restrain the inflationary wage rise).[4] Higher land to labour ratios and rising wage rates meant improved living standards and better diets for the majority. Christopher Dyer's analysis of harvest diets, for instance, demonstrates that by the close of the century workers were eating less bread and consuming more meat and ale.[5] Moreover, the food and drink were of a higher quality: wheaten

[1] The research for this paper was undertaken in conjunction with the project 'The Geography of Seignorial Land-ownership and Land-use, 1270–1349', funded by the Leverhulme Trust and based in the Department of Economic and Social History, The Queen's University of Belfast. We are also grateful for assistance and cooperation received from Dr James Galloway, Dr Derek Keene, Professor John Langdon, Dr Margaret Murphy, Miss Olwen Myhil, Miss Jenitha Orr, and Professor Mark Overton. Responsibility for any errors, of course, remains our own.

[2] The much discussed late seventeenth-century demand shift was much smaller: see E L Jones, 'Agriculture and economic growth in England, 1660–1750: agricultural change', *J Econ Hist*, XXV, 1965, pp 1–18 (reprinted in E L Jones, *Agriculture and the Industrial Revolution*, Oxford, 1974, 1–18); A Kussmaul, 'Agrarian change in seventeenth-century England: the economic historian as paleontologist', *J Econ Hist*, XLV, 1985, pp 1–30; H J Habakkuk, 'The agrarian history of England and Wales: regional farming systems and agrarian change, 1640–1750', *Econ Hist Rev*, 2nd series, XL, 1987, pp 281–96. Not until the revolution in world food markets of the late nineteenth century was a comparable shock experienced, although on that occasion its origin lay with supply rather than demand.

[3] Evidence of the English demographic experience is reviewed in R M Smith, 'Human resources', in G Astill and A Grant, eds, *The Countryside of Medieval England*, Oxford, 1988, pp 188–212; R M Smith, 'Demographic developments in rural England, 1300–48: a survey', in B M S Campbell, ed, *Before the Black Death: Studies in the 'Crisis' of the Early Fourteenth Century*, Manchester, 1991, pp 25–77.

[4] A crude comparison of demesne land values in a ten-county area around London for the periods 1270–1339 and 1375–1400 (using valuations given in extents attached to *Inquisitiones post mortem*) indicates a 27 per cent decline in the unit value of arable, a 19 per cent decline in the unit value of meadow, and an 18 per cent decline in the unit value of pasture (Feeding the City I, IPM database; Feeding the City II, IPM database). Over the same period the real wage rate of a building craftsman rose by 33 per cent (E H Phelps Brown and Sheila V Hopkins, 'Seven centuries of the prices of consumables, compared with builders' wage-rates', *Economica*, XXII, 1956, pp 296–314, Appendix B (reprinted in E M Carus-Wilson, ed, *Essays in Economic History*, 1962, pp 179–96). For rural wage rates and the impact of the Statutes of Labourers, see D L Farmer, 'Prices and wages, 1350–1500', in E Miller, ed, *The Agrarian History of England and Wales, III, 1348–1500*, (hereafter *Ag Hist III*) 1991, pp 467–90.

[5] C Dyer, 'Changes in diet in the late Middle Ages: the case of harvest workers', *AHR*, 36, 1988, pp 21–37; *idem*, *Standards of Living in the Later Middle Ages: Social Change in England c 1200–1520*, 1989.

bread replaced that baked from rye and barley, beef displaced bacon, and ale supplanted cider. These demographic and dietary changes imply an agriculture in which animals assumed an enhanced importance and in which the calorifically extravagant brewing-grains occupied an enlarged share of a reduced cropped acreage as the imperative to maximize the area devoted to the vital bread grains was relaxed.[6] It also implies an agriculture in which farmers traded down to less labour intensive and more land extensive systems of production, a development which should further have favoured livestock over crops and will have had important implications for land and labour productivity.[7] To what extent are these agricultural developments borne out by the evidence?

Systematic, quantifiable data of agricultural production are solely available for the demesne sector. At the opening of the fourteenth century, when direct demesne management was at its height, this sector probably comprised just under a third of the total arable area, but by the close of the century that share had undoubtedly shrunk as narrowing profit margins and greater economic uncertainty encouraged growing numbers of landlords to lease out their demesnes.[8] It is the accounts rendered annually for those manors whose demesnes landlords kept in hand which are medieval historians' principal source of agricultural information.[9] Patterns of documentary creation and survival mean that a wider cross-section of estates is documented at the end of the fourteenth century than the beginning, although after 1375 the actual number of documented demesnes progressively diminishes as direct management more and more became the exception rather than the rule, increasingly confined to the home farms of lay and ecclesiastical households. Reconstructing the composition of demesne production from a national sample of accounts must therefore, perforce, draw upon a smaller potential sample for the post-Black Death period than is available before, in the heyday of direct demesne management.

Two national samples of accounts, one comprising 389 documented demesnes for the period 1250–1349 the other comprising 297 demesnes for the period 1350–1449, provide a basis for quantifying the main changes in demesne production which occurred between the opening and close of the fourteenth century (Table 1).[10]

[6] The conversion of raw barley to ale resulted in a loss in available kilocalories of approximately 70 per cent: B M S Campbell, J A Galloway, D Keene, and M Murphy, *A Medieval Capital and its Grain Supply: Agrarian Production and Distribution in the London Region c 1300*, (hereafter *A Medieval Capital*) Historical Geography Research Series, 30, 1993, p 34.

[7] Livestock and their products required less labour per unit of output than crops. By 1851 Gregory Clark reckons that 'output per worker in animal husbandry was about 80 per cent greater than in arable cultivation': 'Labour productivity in English agriculture, 1300–1860', in B M S Campbell and M Overton, eds, *Land, Labour and Livestock: Historical Studies in European Agricultural Productivity*, Manchester, 1991, p 231. For the relationship between land, labour, and other forms of productivity, see M Overton and B M S Campbell, 'Productivity change in European agricultural development', in Campbell and Overton, *Land, Labour and Livestock*, pp 1–50.

[8] The proportion of a third is based on the evidence of the surviving 1279 Hundred Rolls: E A Kosminsky, *Studies in the Agrarian History of England in the Thirteenth Century*, 1956, pp 87–95. The priors of Norwich, for instance, slashed by half the area which they cultivated directly between the 1340s and 1390s, and by 1430 had abandoned direct management entirely: Bodleian Library, Oxford, Ms Rolls, Norfolk 20–47; Norfolk Record Office (hereafter NRO), DCN 1/1, 40/13, 60/4, 60/8, 60/10, 60/13, 60/14, 60/15, 60/18, 60/20, 60/23, 60/26, 60/28, 60/29, 60/33, 60/35, 60/37, 61/35–6, 62/1–2; DCN R233 B 4626, L'Estrange IB 1/4, 3/4, 4/4, NNAS 5890–918 20 D1–3; Raynham Hall, Norfolk, Townshend Mss. On the farming out of manors see, B F Harvey, 'The leasing of the Abbot of Westminster's demesnes in the later Middle Ages', *Econ Hist Rev*, 2nd series, XXII, 1969, pp 17–27; R A Lomas, 'The Priory of Durham and its demesnes in the fourteenth and fifteenth centuries', *Econ Hist Rev*, 2nd series, XXXI, 1978, pp 339–53; R R Davies, *Lordship and Society in the March of Wales 1282–1400*, 1978, p 113; J N Hare, 'The demesne lessees of fifteenth-century Wiltshire', *AHR*, XXIX, 1981, pp 1–15; M Mate, 'The farming out of manors: a new look at the evidence from Canterbury Cathedral Priory', *J Med Hist*, 9, 1983, pp 331–44.

[9] R H Hilton, 'The contents and sources of English agrarian history before 1500', *AHR*, III, 1955, 3–19; P D A Harvey, ed, *Manorial Records of Cuxham, Oxfordshire*, Oxfordshire Record Society, 50, and Royal Commission on Historical Manuscripts joint publication, 23, 1976; *idem*, *Manorial Records*, Archives and the User, 5, 1984.

[10] The national samples of manorial accounts were generously made available by Professor John Langdon of the Department of History, University of Edmonton. The bulk of these accounts are listed in Appendix C, part 2, of his unpublished PhD thesis, 'Horses, oxen, and technological innovation: the use of draught animals in English farming from 1066 to 1500', University of Birmingham, 1983. Data for some additional demesnes have also been incorporated. The crop and livestock profiles of each demesne, which provide the building blocks of these aggregate estimates, comprise the

Comparison of these two samples confirms that demesnes curtailed the scale of their arable operations and devoted more of their resources to pastoral husbandry. The mean cropped acreages of demesnes retained in hand were reduced by approximately a fifth. This was achieved partly by withdrawing land from cultivation and converting it to grass, partly by lengthening or increasing fallows and thereby cropping land less frequently, and partly by leasing portions of demesne arable to tenants.[11] Within the arable sector demesnes maintained the relative share of the cropped acreage devoted to wheat – the premier bread grain regularly consumed by a growing proportion of the population – but cut back on that devoted to the cheaper and coarser bread grains, notably rye and the various winter mixtures which commonly included rye. Oats, too, declined in both relative and absolute importance, possibly because of the withdrawal of cultivation from poorer soils and the shortening of some of the more intensive rotations (in which oats had often been the final course) but also because of its substitution with other spring-sown grains.[12] Barley and dredge, for instance, both gained in their relative and absolute shares of the cropped acreage. Although oats had long been used for brewing, ale manufactured from barley and dredge was increasingly favoured by a population whose thirst for quantity combined with quality was steadily rising.[13] If the cultivation of brewing grains was on the increase so too was that of legumes, although here the connection with rising living standards and improving diets is less obvious.

mean of each individual variable calculated across the sampled accounts for that demesne. The relationships between variables have been calculated using these meaned values.

[11] All three strategies were employed on the bishop of Ely's demesne at Brandon, Suffolk, where the accounts record the areas *friscus* (ie unsown for more than one year), farmed out, and sown: Chicago University Library, Bacon Roll 643, 650–60.

[12] M M Postan, *The Medieval Economy and Society: An Economic History of Britain in the Middle Ages*, 1972, p 52.

[13] J A Galloway, D J Keene, M Murphy, and B M S Campbell, 'Changes in grain production and distribution in the London region over the fourteenth century', unpublished manuscript.

Legumes – collectively beans, peas and vetches – almost doubled their share of the sown acreage; a greater relative gain than that of any other crop. Hitherto large-scale legume cultivation had been mainly restricted to the most intensive of husbandry systems, where they provided a partial substitute for fallows, helped restore nitrogen levels within the soil, and provided a nutritious source of human and animal food.[14] Nevertheless, healthy as they were, legumes were not held in high dietary esteem due to their strong association with the pottages and coarse breads of the poor and fodder for animals.[15] Higher living standards are, therefore, hardly likely to have encouraged greater per capita consumption of legumes. Nor can their wider cultivation be attributed to a general adoption of more intensive methods of cultivation, for, on the whole, the opposite was the case.[16] Rather, legumes benefited from the expansion of pastoral farming (and especially heightened demesne investment in cart horses) with its correspondingly greater demand for fodder crops.[17] In effect, sowing land with legumes represented an alternative to converting it to grass and was especially attractive to those who wished to keep their arable options open. Indeed, in areas of entrenched commonfield agriculture, where there were institutional obstacles to the conversion of

[14] B M S Campbell, 'Agricultural progress in medieval England: some evidence from eastern Norfolk', *Econ Hist Rev*, 2nd series, XXXVI, 1983, pp 26–46; *idem*, 'The diffusion of vetches in medieval England', *Econ Hist Rev*, 2nd series, XLI, 1988, pp 193–208.

[15] '... what had been baked for Bayard was boon to many hungry, And many a beggar for beans obediently laboured, And every poor man was well pleased to have peas for his wages': *William Langland: Will's Vision of Piers Plowman*, trans E Talbot Donaldson, ed by E D Kirk and J H Anderson, New York, 1990, lines 193–5, pp 65–6.

[16] See below, p 148.

[17] On the connection between legumes and the use of horses see, Campbell, 'Diffusion of vetches', pp 205–6; L W Hepple and A M Doggett, *The Chilterns*, Chichester, 1992, pp 94–5; *Ag Hist III*, p 271. On the considerable fodder requirements of cart horses see K Biddick, *The Other Economy: Pastoral Husbandry on a Medieval Estate*, Berkeley and Los Angeles, 1989, pp 116–21. On the increasing use of horses post 1349 see, J L Langdon, *Horses, Oxen and Technological Innovation: The Use of Draught Animals in English Farming, 1066–1500*, 1986, pp 95–7; B M S Campbell, 'Towards an agricultural geography of medieval England', *AHR*, 36, 1988, pp 91–3.

TABLE I

The changing composition of seigneurial agricultural production *c* 1300–*c* 1400

Mean per demesne	*National samples of demesnes*						*Feeding the City samples of demesnes*[a]		
	Core demesnes only			*All demesnes*					
	1250–1349	*1350–1449*	*% change*	*1250–1349*	*1350–1449*	*% change*	*1288–1315*	*1375–1400*	*% change*
Percentage of sown acreage:									
wheat	31.0	30.4	−2	31.7	30.9	−3	32.8	32.3	−2
	16.3	*14.9*		*16.0*	*14.5*		*14.2*	*14.0*	
rye	4.7	2.0	−57	5.2	2.9	−44	6.4	2.2	−66
	9.4	*5.5*		*9.5*	*7.5*		*11.3*	*6.5*	
winter mixtures	2.1	1.3	−38	2.3	1.9	−17	3.9	2.1	−46
	6.7	*4.1*		*6.7*	*5.4*		*8.8*	*5.4*	
barley	16.0	23.0	+44	15.7	22.2	+41	11.2	18.3	+63
	17.3	*19.4*		*16.0*	*18.0*		*11.1*	*14.8*	
oats	33.6	23.8	−29	31.1	22.8	−27	30.0	22.6	−25
	20.0	*17.3*		*19.2*	*16.7*		*15.7*	*17.1*	
spring mixtures	3.7	5.0	+35	3.8	4.9	+29	6.6	8.2	+24
	10.0	*11.6*		*9.5*	*11.3*		*10.6*	*14.8*	
legumes	7.9	13.9	+76	9.5	13.9	+46	9.2	14.4	+57
	8.6	*12.0*		*9.4*	*11.5*		*8.9*	*10.5*	
Sown acreage	184.9	147.8	−20	193.1	151.7	−21	224.9	171.0	−24
	112.9	*82.5*		*118.5*	*85.9*		*127.7*	*81.5*	
Percentage of livestock units[b]:									
horses	15.5	14.3	−8	15.5	13.1	−15	18.3	16.1	−12
	12.3	*17.6*		*11.6*	*14.7*		*13.1*	*13.9*	
oxen	42.5	27.3	−36	40.4	24.9	−38	28.3	17.5	−38
	29.6	*26.2*		*28.0*	*23.4*		*18.8*	*15.5*	
adult cattle	14.3	16.6	+16	15.4	20.4	+32	19.8	23.9	+20
	16.4	*19.6*		*15.8*	*18.6*		*15.6*	*21.0*	
immature cattle	8.1	7.5	−7	9.1	9.1	±0	8.6	6.7	−22
	9.4	*9.3*		*9.8*	*9.4*		*7.6*	*7.6*	
sheep	16.6	31.1	+87	16.3	29.2	+79	20.8	31.2	+50
	21.6	*28.5*		*20.3*	*25.1*		*19.2*	*21.2*	
swine	3.0	3.1	+3	3.3	3.3	±0	4.2	4.6	+10
	5.7	*4.7*		*5.2*	*4.4*		*4.6*	*5.8*	
Livestock units[b]	59.3	73.1	+23	62.7	79.2	+26	66.8	80.3	+20
	51.0	*54.6*		*50.4*	*56.8*		*47.7*	*50.8*	
Stocking density[c]	34.5	56.2	+63	35.4	61.2	+73	35.3	57.9	+64
	25.1	*56.5*		*25.9*	*58.8*		*35.1*	*49.0*	
Number of demesnes	261	182	−30	389	297	−24	183	125	−32

Means in roman; standard deviations in italic.

[a] Covering Beds, Berks, Bucks, Essex, Herts, Kent, Middx, N'hants, Oxon, and Surrey.

[b] [horses × 1.0] + [(oxen + adult cattle) × 1.2] + [immature cattle × 0.8] + [(sheep + swine) × 0.1].

[c] livestock units per 100 sown acres.

Source: National samples of accounts, 1250–1349 and 1350–1449; Feeding the City I, accounts database; Feeding the City II, accounts database.

arable to grassland, increased fodder cropping offered the readiest means of raising livestock numbers.[18]

Whereas arable cultivation contracted in scale and shifted in focus pastoral husbandry underwent a wholesale expansion. On the face of it the 25 per cent rise in the number of livestock units stocked per demesne was roughly commensurate with the 20 per cent contraction in cropped acreage, but the magnitude of this rise is almost certainly understated, for the growing practice of leasing out dairy herds masks many of the younger cattle from view, in much the same way that sheep disappear from accounts on those estates – and the number was growing – which managed their flocks on an intermanorial basis and accounted for them separately.[19] Since with less land under the plough draught requirements were if anything contracting (that contraction being more pronounced in the case of oxen than of horses), the gain in livestock units principally represented an increase in the number of non-working animals being stocked.[20] The latter's share of total livestock units rose from 42 per cent to 58 per cent, with the bulk of that gain being accounted for by a marked rise in the number and size of sheep flocks. Whereas non-working cattle and swine more-or-less maintained their existing importance, sheep – notwithstanding serious outbreaks of murrain in the 1360s and depressed wool prices from the late 1370s – almost doubled their share of livestock units.[21] Sheep, significantly, were managed by predominantly extensive methods and their rising numbers probably reflect a corresponding expansion in temporary and permanent pasture.[22]

The net effect of these developments was to render the pastoral sector less subservient to the arable and ensure that demesne agriculture in general became more mixed. By the end of the fourteenth century stocking densities, as measured by the number of livestock units per sown acre, were at least 60 per cent higher than they had been at the beginning.[23]

How reliable and representative are these trends? Spatial and institutional bias in the creation and preservation of manorial accounts means that a truly random sample of all demesnes is unattainable. Nor are the samples which underlie Table 1 as large, comprehensive, and sharply focused as might have been possible given unlimited research time and resources and adequate

[18] B M S Campbell, 'A fair field once full of folk: agrarian change in an era of population decline, 1348–1500', *AHR*, 41, 1993, p 63; *Ag Hist III*, pp 216, 229. The post 1349 rise in fodder cropping is analogous to that which occurred post 1640, B M S Campbell and M Overton, 'A new perspective on medieval and early modern agriculture: six centuries of Norfolk farming *c* 1240–*c*. 1850', *Past and Present*, 141, 1993, pp 59–60, 87–8, 90–2.

[19] The livestock units employed are as follows: horses, 1.0; oxen, bulls, cows, 1.2; immature cattle, 0.8; sheep, 0.1; swine, 0.1 (see B M S Campbell, 'Land, labour, livestock, and productivity trends in English seignorial agriculture, 1208–1450', in Campbell and Overton, *Land, Labour and Livestock*, pp 156–7). The two Feeding the City samples of accounts (see below note 28) indicate a rise in the proportion of demesnes farming dairies from 13 per cent in the period 1288–1315 to 34 per cent in the period 1375–1400. For examples of centralized sheep accounting see F M Page, '"Bidentes Hoylandie": a medieval sheep farm', *Econ Hist*, 1, 1929, pp 603–5; R A L Smith, 'The estates of Pershore Abbey', unpublished MA thesis, University of London, 1939, pp 215–16; R H Hilton, 'Winchcombe Abbey and the manor of Sherborne', *Univ Birmingham Hist Jnl*, 2, 1949–50, pp 50–2; Davies, *Lordship and Society*, p 119; NRO, L'Estrange Collection IB 3/4.

[20] On the ratio of oxen to horses post 1350 see Langdon, *Horses, Oxen and Technological Innovation*, pp 95–7; Campbell, 'Towards an agricultural geography', pp 91–3.

[21] T H Lloyd, 'The movement of wool prices in medieval England', *Econ Hist Rev, Supplement VI*, 1973, pp 19–20. For associated changes in the relative price of sheep see Farmer, 'Prices and wages', pp 458, 508–12.

[22] B M S Campbell, 'The livestock of Chaucer's reeve: fact or fiction?', in E B Dewindt, ed, *The Salt of Common Life: Individuality and Choice in the Medieval Town, Countryside and Church. Essays presented to J Ambrose Raftis on the Occasion of his 70th Birthday*, Kalamazoo, 1995, pp 271–305. For direct evidence of the expansion of pasture see E Miller, ed, 'The occupation of the land', in *Ag Hist III*, pp 34–174.

[23] The rise in stocking densities was to be sustained until well into the fifteenth century: Campbell, 'Land, labour, livestock, and productivity trends', pp 153–8; *Ag Hist III*, p 314; M Overton and B M S Campbell, 'Norfolk livestock farming 1250–1740: a comparative study of manorial accounts and probate inventories', *J Hist Geog*, 18, 1992, pp 386–92.

calendaring of manorial records.[24] As it is, a broad geographical coverage was in both cases only achieved by casting a wide chronological net. This poses greatest problems in the later period, since the agricultural changes which occurred between 1350 and 1449 were more profound than those between 1250 and 1349. It was after 1375 in particular that changing factor and commodity prices really began to bite, accelerating the landlords' retreat both from the more intensive forms of production and, ultimately, from direct management. From this date, therefore, it becomes increasingly difficult to find usable accounts for many parts of the country. A third of sampled demesnes have a mean date of account from before this critical economic watershed and two-thirds from after, although most of the latter bunch into the final quarter of the fourteenth century. After 1400 there is a significant tailing off in documented demesnes as direct management increasingly became the exception rather than the rule. The sampled demesnes are therefore chronologically weighted towards the late fourteenth century with the 1380s both the modal and median decade. The corresponding modal and median decades in the earlier sample are the 1280s and 1290s. In their average characteristics the two samples therefore broadly encapsulate conditions roughly a century apart. The changes that emerge from comparison of these two samples are consequently those which took place primarily between the late thirteenth and the late fourteenth centuries.

Even with such a wide chronological net the geographical coverage of both samples remains patchy, and would have become unacceptably so had comparison been restricted to demesnes common to both periods. Not only do these amount to less than half the total but almost all belonged to perpetual institutions and are consequently especially atypical of the demesne sector at large.[25] To insist upon continuity of documentation would exclude much of the west and north-west of the country, where direct demesne management was never as widely and firmly established as in the south and east. As it is, the many demesne vaccaries and bercaries of these upland margins are largely masked from view and it is the mixed-husbandry of the more favoured lowland areas that tends to be best documented.[26] Nevertheless, even within the arable and lowland south and east, and especially after 1349, there are some remarkable inconsistencies in the pattern of account roll survival. Norfolk, for instance, is exceptionally well covered (as it is in the earlier period) whereas neighbouring Lincolnshire is barely represented at all. In many counties – Huntingdonshire, Middlesex, Somerset, Suffolk, Wiltshire, and Worcestershire – the sampled manors are decidedly bunched according to the location of estates with extant documentation, but in Devon and Cornwall, the Welsh borders, and the whole of the north-west such estates are few and far between and much less representative both of the broad spectrum of ownership types and the general character of husbandry in these regions.[27] Some of these deficiencies might be rectified by a more systematic archival search for extant

[24] Tracking down manorial accounts for demesnes still in hand is no simple task, even within a single repository. The starting point for any search is the National Register of Archives, Quality House, Quality Court, Chancery Lane, London. Listings of accounts by geographical area are especially rare. For a recent list of all known pre-1350 Kentish manorial accounts, see J A Galloway, M Murphy, and O Myhill, *Kentish Demesne Accounts up to 1350: A Catalogue*, 1993.

[25] See, for example, B M S Campbell, 'Measuring the commercialisation of seigneurial agriculture *c* 1300', in R H Britnell and B M S Campbell, eds, *A Commercialising Economy: England 1086–1300*, Manchester, 1994, pp 132–93.

[26] M A Atkin, 'Land use and management in the upland demesne of the De Lacy estate of Blackburnshire, *c* 1300', *AHR*, 42, 1994, pp 1–19; H P R Finberg, *Tavistock Abbey: a Study in the Social and Economic History of Devon*, 1951; I Kershaw, *Bolton Priory: the Economy of a Northern Monastery 1286–1325*, 1973, pp 30–112.

[27] H S A Fox has tracked down most of the post-1349 grange accounts for Devon and Cornwall: for his list see *Ag Hist III*, p 305.

accounts but many are irremediable due to the inherent patchiness of account survival.

Considerable comfort may therefore be taken in the evidence of two independent and chronologically and geographically more sharply focused samples of demesnes for the ten counties of Bedfordshire, Berkshire, Buckinghamshire, Essex, Hertfordshire, Kent, Middlesex, Northamptonshire, Oxfordshire, and Surrey, which confirm the nature and magnitude of virtually all the changes that have been described (Table 1). The earlier of these two samples embraces some 183 demesnes and relates to the period 1288–1315 at the climax of thirteenth-century demographic and commercial expansion; the later sample (hereafter referred to as the Feeding the City II sample) comprises some 125 demesnes and focuses on the years 1375–1400 when demesne producers were contending with price deflation and wage inflation. Both samples are the best that surviving documentation allows, with the individual manor means being based upon up to three annual accounts whenever possible.[28] Particular confidence can therefore be placed in the aggregate trends that they reveal, the more so as these are borne out by a comparison of 57 demesnes common to both samples. Overall, demesnes within these 'metropolitan counties' cropped 24 per cent less arable, stocked 20 per cent more livestock units, and registered a 64 per cent improvement in stocking densities (proportions remarkable close to those obtained from the national samples – Table 1) at the end of the fourteenth century than at the beginning.[29] Within the arable sector – as was the case nationally – wheat maintained its share of the cropped acreage while the respective shares of rye and rye mixtures fell and those of barley and dredge rose. Legumes again expanded in importance, increasing their share of the cropped acreage by over half. Their expansion is probably linked to the greater relative importance attached to horse power in these counties after 1349. Nevertheless, it was the non-working animals that were the principal beneficiaries of the general growth in flocks and herds. Their share of livestock units rose from 53 to 66 per cent. The bulk of this gain came from the increase in sheep numbers. Whereas cattle and swine registered virtually no change in importance sheep expanded from 21 to 31 per cent of livestock units.

It is improbable that such far-reaching changes affected all demesnes equally. What effect did they have upon the types and distributions of farming systems?

I

There is no simple or perfect method of classifying temperate agricultural systems, where the enterprise of individual farms is subject to almost infinite variation within the narrow bounds set by the limited available range of crops and animals. Yet without classification generalization is impossible and the pattern of farming systems too kaleidoscopic to comprehend. These inherent problems are greatly com-

[28] The two Feeding the City accounts databases were created by Dr James A Galloway and Dr Margaret Murphy (with some assistance from Miss Olwen Myhil) and derive from the projects 'Feeding the City (I)', funded by the Leverhulme Trust, and 'Feeding the City II', funded by the Economic and Social Research Council (Award No R000233157). Both were collaborative projects between the Centre for Metropolitan History, Institute of Historical Research, University of London and the Department of Economic and Social History, The Queen's University of Belfast, the former codirected by Dr Derek Keene and Dr Bruce Campbell, the latter by Drs Keene, Campbell, Galloway, and Murphy. The Feeding the City I accounts are listed in Appendix I of *A Medieval Capital*; a handlist of the Feeding the City II accounts is available from the Centre for Metropolitan History. Both accounts databases have been incorporated within the Pre-Black Death England Database being created at the Department of Economic and Social History, The Queen's University of Belfast. The Feeding the City II accounts database has been deposited at the ESRC Data Archive, Essex University.

[29] An alternative system of livestock units, graded to take account of the age and sex of animals and based on the internal evidence of sale and purchase prices, indicates an 18 per cent gain in livestock units and 63 per cent improvement in stocking densities. Limiting comparison to the 57 demesnes common to both Feeding the City databases yields more modest gains of 7 per cent and 48 per cent respectively.

pounded when the variables upon which classification is to be based are constrained by available historical evidence. Hence the ingenuity and effort displayed by British agricultural historians over the past thirty years, commencing with the major early modern contributions of Joan Thirsk and Eric Kerridge, in their endeavour to identify and describe farming types and farming regions.[30] Hitherto the problem has mostly been tackled at a county or regional level, these different regional solutions then being aggregated to provide an overall national picture, as in the case of volumes II, III, IV, and V of the *Agrarian History of England and Wales*.[31] This has obvious limitations, for national systems of classification should ideally be based upon the systematic application of a consistent set of criteria at a national scale. Moreover, the units of classification should be the fundamental units of agricultural production – that is, farms not farming regions – and the resultant typologies should reflect properties inherent to those farms, independent of such external forces as soils, climate, or distance from the market, which may have influenced the type of farming.[32] Whether or not specific farming types were associated with, or exclusive to, particular regions can then be established by mapping the resultant typology. That is the method adopted here, building upon the procedure already developed and employed to classify the pre-1350 national sample of demesnes.[33]

Using a common method to derive independent classifications from similarly structured sets of data is the best way to analyse the changing geography of agriculture. Even so, comparison between the two periods 1250–1349 and 1350–1449 is far from unproblematic since, for the reasons already given, the data on which the classifications rest are not the product of random sampling. Indeed, such sampling is inappropriate where it is known that the available documentation is shot through with geographical, institutional and temporal bias.[34] Tests of statistical significance have therefore been rejected as a valid comparative technique in favour of the more intuitive and exploratory approach elaborated in this paper.[35] Used in this way, the results do nevertheless point to the more important changes which probably occurred and offer a simplified typology of the principal farming systems in operation at the time.

The literature on the derivation of agricultural typologies is considerable.[36] The range of criteria which ought ideally to be included comprise the crop and livestock association, the methods used to grow the crops and produce the stock, the intensity of labour, capital, and organizational inputs and associated rates of output, and whether or not production was for consumption or

[30] Systematic regional description of past farming systems really begins in 1967 with the simultaneous publication of J Thirsk, 'The farming regions of England', in J Thirsk, ed, *The Agrarian History of England and Wales, IV, 1500–1640*, 1967, pp 1–112, and E Kerridge, *The Agricultural Revolution*, 1967, pp 41–180.

[31] H E Hallam, ed, *The Agrarian History of England and Wales, II, 1042–1350*, 1988, pp 272–496; *Ag Hist III*, pp 175–323; Thirsk, *Agrarian History, IV*, pp xxi, 1–112; J Thirsk, ed, *The Agrarian History of England and Wales, VI, 1640–1750: Regional Farming Systems*, 1984. For critiques see, respectively, B M S Campbell, 'Laying foundations: the Agrarian History of England and Wales, 1042–1350', *AHR*, 37, 1989, pp 190–1; Campbell, 'A fair field', pp 62–3; E L Jones, 'The condition of English agriculture, 1500–1640', *Econ Hist Rev*, 2nd series, XXI, 1968, pp 615–16; M Overton, 'Depression or revolution? English agriculture, 1640–1750', *J Brit Studies*, 25, 1986, pp 345–7.

[32] D B Grigg, *The Agricultural Systems of the World: An Evolutionary Approach*, 1974, pp 2–3.

[33] J P Power and B M S Campbell, 'Cluster analysis and the classification of medieval demesne-farming systems', *Trans Inst Brit Geogr*, new series, 17, 1992, pp 227–45. The classifications yielded by this method represent an improvement upon those offered in B M S Campbell and J P Power, 'Mapping the agricultural geography of medieval England', *J Hist Geog*, 15, 1989, pp 24–39. Using the pre-1350 classification to classify post-1349 farming systems is obviously inappropriate given the absolute and relative changes in farm enterprise which had taken place.

[34] For the mismatch between the social distribution of landed incomes and the institutional distribution of documented demesnes in the Feeding the City I accounts database, see Campbell, 'Measuring the commercialisation of seigneurial agriculture', p 140.

[35] On the precondition of randomness for the application of significance tests see, D Ebdon, *Statistics in Geography: a Practical Approach*, 1977, pp 12–19.

[36] Grigg, *Agricultural Systems*, pp 2–4.

exchange. Here, however, classification is based exclusively upon the crops grown, the livestock reared, and the ratio of the latter to the former since these are the most readily available quantifiable criteria for the majority of documented demesnes. They also provide an indirect index of the character and intensity of production and convey some impression of the nature and degree of economic specialization. In this context it would have been a great advantage to be able to include some index of the proportion of the total arable area that was actually cropped, but too few accounts contain this information.[37] Better recorded are seeding rates, plough type, and size of plough team; had they been included a more subtly differentiated classification might have resulted.[38]

Since the aim is to produce a classification based upon demesne enterprise rather than demesne size it is necessary to express each of the four crop and five livestock variables in relative rather than absolute terms, taking care to calculate these relative measures in such a way that closed numbers are not a problem.[39] (Percentages have merely been used in Tables 2 and 3 to summarize the final results.) This also has the merit of controlling for customary acres. Only the tenth variable – the number of livestock units per 100 cropped acres – is susceptible to variations in acre size. As a further safeguard, a principal components analysis was conducted to remove correlations between variables and eliminate any bias arising from the uneven representation of crop and livestock variables. It is the scores derived from that that have been employed as the basis of classification. The choice of the number of principal component scores used proved to be critical. In the case of the pre-1350 classification it was the scores of the first six components that were used; these accounted for 83 per cent of the variance across the ten variables and yielded a 'stable' classification comprising eight basic farming types.[40] Post-1349, however, using the first six component scores (which account for 82 per cent of variance) yielded a 'stable' classification containing only three basic farming types, with the vast majority of all demesnes falling into a single mixed-farming category. Yet, while this result broadly supports the general observation that in this later period farming systems were less differentiated than they had been before, from a historical point of view a fuller break down of farming types is plainly desirable. The decision was therefore taken to eliminate the sixth and least significant principal component score and to use only the first five components, accounting for 74 per cent of total variance and yielding a 'stable' classification comprising seven basic farming types. The classification itself was derived using the statistical technique known as cluster analysis and, in fact, it has long been observed of cluster analysis that its operation can be ineffective if variables of relatively small significance (in this case the sixth component score) generate intermediate points between cluster groupings.[41] To ensure that the resultant classification reflected real differences within the data rather than the type of clustering technique chosen three separate clustering techniques were applied – Ward's,

[37] *A Medieval Capital*, pp 129–33, 139–41.

[38] On seeding rates, see *A Medieval Capital*, pp 131, 136–8. On ploughs, plough teams, and farm vehicles, see Langdon, *Horses, Oxen and Technological Innovation*.

[39] All crops are specified in acres; all livestock are specified in livestock units (see above note 19). The variables are: R1 = Log (winter grains/spring grains); R2 = Proportion (wheat/winter grains); R3 = Proportion (oats/spring grains); R4 = Log (legumes/total grain); R5 = Log (horses/oxen); R6 = Log (non-working animals [horses + oxen]/working animals); R7 = Log (immature cattle/adult cattle [omitting oxen]); R8 = Percentage (sheep/non-working animals); R9 = Percentage (swine/livestock units); R10 = Log (livestock units/total sown). All zero values have been replaced with 0.01. Note, R10 is incorrectly specified as Log(livestock units × 100/sown) in Power and Campbell, 'Cluster analysis', p 230.

[40] Power and Campbell, 'Cluster analysis', pp 230–2.

[41] B S Everitt, *Cluster Analysis*, SSRC Reviews of Current Research, 11, 2nd ed, 1980.

TABLE 2
National farming types 1350–1449: mean characteristics of core demesnes

Mean per demesne	*Farming type/cluster group*							*All core*
	1	*2*	*3*	*4*	*5*	*6*	*7*	*demesnes*
Percentage of sown acreage:								
wheat	12.1	2.8	30.7	30.1	34.3	39.0	33.7	30.4
	5.6	*5.2*	*8.5*	*9.2*	*15.5*	*10.9*	*14.4*	*14.9*
rye	4.2	18.3	0.6	0.1	0.5	1.3	2.1	2.0
	5.2	*13.8*	*2.1*	*0.4*	*2.3*	*3.5*	*5.1*	*5.5*
winter mixtures	0.3	5.1	1.0	4.0	1.3	1.1	0.7	1.3
	0.7	*10.5*	*2.6*	*8.1*	*3.2*	*3.3*	*3.5*	*4.1*
barley	56.8	38.7	28.6	19.8	18.5	4.8	16.6	23.0
	9.3	*19.7*	*13.7*	*12.1*	*10.4*	*9.7*	*17.0*	*19.4*
oats	12.7	17.4	12.4	7.1	29.9	42.0	24.6	23.8
	5.0	*9.1*	*9.6*	*7.1*	*16.9*	*11.2*	*19.3*	*17.3*
spring mixtures	0.1	2.0	5.3	12.4	6.8	1.3	7.8	5.0
	0.4	*4.6*	*11.0*	*14.5*	*13.5*	*3.1*	*16.0*	*11.6*
legumes	13.7	1.5	21.2	26.5	8.8	10.5	14.5	13.9
	4.5	*1.2*	*13.2*	*17.2*	*7.7*	*8.5*	*12.8*	*12.0*
Sown acreage	146.0	109.4	186.1	134.8	123.0	157.7	136.3	147.8
	86.1	*105.5*	*81.2*	*72.3*	*74.0*	*80.8*	*69.2*	*82.5*
Percentage of livestock units[a]								
horses	16.5	11.9	7.7	32.4	9.4	10.1	29.9	14.3
	12.0	*4.5*	*5.8*	*26.7*	*8.8*	*5.7*	*31.7*	*17.6*
oxen	2.6	1.6	20.1	36.3	24.4	24.2	68.3	27.3
	2.8	*3.9*	*13.2*	*28.7*	*12.2*	*17.2*	*31.5*	*26.2*
adult cattle	48.1	29.9	12.3	6.0	2.9	35.0	0.7	16.6
	16.0	*22.6*	*7.4*	*11.1*	*8.4*	*12.5*	*2.4*	*19.6*
immature cattle	15.3	5.0	12.2	0.4	0.0	16.2	0.0	7.5
	6.7	*5.2*	*9.0*	*0.7*	*0.0*	*8.4*	*0.0*	*9.3*
sheep	13.6	51.4	44.3	7.0	62.0	12.2	0.0	31.1
	17.7	*29.5*	*15.7*	*14.1*	*17.5*	*14.8*	*0.0*	*28.5*
swine	3.9	0.3	3.5	17.8	1.3	2.3	1.1	3.1
	3.2	*0.6*	*2.7*	*7.9*	*1.8*	*2.2*	*1.9*	*4.7*
Livestock units[a]	51.3	71.2	118.4	26.4	73.9	85.9	22.5	73.1
	33.9	*58.4*	*64.6*	*19.0*	*45.5*	*35.0*	*12.1*	*54.6*
Stocking density[b]	38.3	59.9	69.3	20.0	77.2	63.7	18.3	56.2
	16.7	*42.5*	*36.6*	*7.8*	*92.3*	*32.5*	*9.1*	*56.5*
Number of demesnes	20	7	40	10	45	33	27	182

Means in roman; standard deviations in italic.
1 = intensive mixed-farming; 2 = light-land intensive; 3 = mixed-farming with sheep; 4 = arable husbandry with swine; 5 = sheep-corn husbandry; 6 = extensive mixed-farming; 7 = extensive arable husbandry.
[a] [horses × 1.0] + [(oxen + adult cattle) × 1.2] + [immature cattle × 0.8] + [(sheep + swine) × 0.1].
[b] livestock units per 100 sown acres.
Source: National sample of accounts, 1350–1449.

K-means, and Normix. Each farming system was then defined on the basis of those demesnes identically classified by all three cluster techniques (Table 2). In effect, this identified an unambiguous *core* of 182 clearly differentiated demesnes. This is a smaller proportion of the total sample (61.3 per cent compared with 67.1 per cent) than the core of 261 pre-1350 demesnes and is consistent with the smaller number of principal

TABLE 3

National farming types 1350–1449: mean characteristics of core and peripheral demesnes

Mean per demesne	*Farming type/cluster group* 1	2	3	4	5	6	7	*All demesnes*
Percentage of sown acreage:								
wheat	15.1	6.3	32.2	32.0	34.8	36.6	33.7	30.9
	7.1	*6.6*	*9.5*	*10.6*	*15.0*	*13.3*	*14.4*	*14.5*
rye	3.9	19.3	1.3	0.1	1.0	3.3	2.0	2.9
	6.0	*15.6*	*4.4*	*0.4*	*3.5*	*6.9*	*5.1*	*7.1*
winter mixtures	1.8	6.8	1.6	3.3	1.6	1.9	0.7	1.9
	4.5	*13.1*	*4.5*	*7.9*	*3.9*	*4.9*	*3.5*	*5.4*
barley	50.9	27.8	25.5	19.3	19.3	10.8	16.6	22.2
	12.3	*25.9*	*14.0*	*11.6*	*11.3*	*12.4*	*17.0*	*18.0*
oats	12.2	25.1	15.2	7.2	27.0	33.1	24.4	22.8
	6.5	*18.7*	*11.8*	*7.5*	*17.1*	*15.6*	*19.4*	*16.7*
spring mixtures	2.1	2.6	5.5	12.5	6.1	2.6	7.5	4.9
	6.4	*6.9*	*12.1*	*14.9*	*12.6*	*6.9*	*16.0*	*11.3*
legumes	13.9	3.4	18.6	25.6	10.3	11.5	15.1	13.9
	5.4	*6.3*	*12.5*	*16.5*	*8.8*	*10.0*	*13.2*	*11.5*
Sown acreage	155.4	120.2	185.1	132.3	129.5	147.6	135.6	151.7
	88.1	*87.4*	*96.7*	*70.9*	*73.9*	*81.1*	*69.3*	*85.9*
Percentage of livestock units[a]:								
horses	16.5	10.1	7.5	29.6	10.4	11.1	29.9	13.1
	10.5	*4.1*	*5.2*	*26.2*	*9.0*	*7.1*	*31.7*	*14.7*
oxen	7.3	5.6	19.0	39.4	23.9	24.3	68.1	24.9
	10.1	*8.2*	*12.6*	*29.5*	*12.2*	*19.2*	*31.5*	*23.4*
adult cattle	43.3	29.0	18.4	5.0	5.7	31.5	0.7	20.4
	16.1	*18.0*	*11.4*	*10.9*	*13.6*	*20.0*	*2.4*	*18.6*
immature cattle	16.1	8.0	12.3	0.3	<0.1	14.5	0.0	9.1
	7.7	*7.7*	*8.4*	*0.7*	*0.2*	*8.5*	*0.0*	*9.4*
sheep	11.6	45.9	39.3	9.0	57.8	16.6	0.0	29.2
	16.0	*25.8*	*15.8*	*16.3*	*18.9*	*16.1*	*0.0*	*25.1*
swine	5.3	1.3	3.6	16.6	2.1	2.1	1.4	3.3
	4.3	*2.2*	*3.0*	*8.0*	*2.9*	*2.2*	*2.4*	*4.4*
Livestock units[a]	50.2	88.2	119.2	27.4	72.6	82.7	23.0	79.2
	30.7	*64.1*	*68.8*	*14.9*	*43.1*	*40.7*	*12.3*	*56.8*
Stocking density[b]	36.4	72.5	76.8	21.3	71.7	68.0	18.7	61.2
	18.1	*40.3*	*58.6*	*8.0*	*84.8*	*52.3*	*9.4*	*58.8*
Number of demesnes	32	14	77	12	56	78	28	297

Means in roman; standard deviations in italic.

1 = intensive mixed-farming; 2 = light-land intensive; 3 = mixed-farming with sheep; 4 = arable husbandry with swine; 5 = sheep-corn husbandry; 6 = extensive mixed-farming; 7 = extensive arable husbandry.

[a] [horses × 1.0] + [(oxen + adult cattle) × 1.2] + [immature cattle × 0.8] + [(sheep + swine) × 0.1].

[b] livestock units per 100 sown acres.

Source: National sample of accounts, 1350–1449.

component scores employed and greater difficulty of distinguishing real differences in farm enterprise.

The common denominator of the residuum of 115 non-core or *peripheral* demesnes is that there is a measure of disagreement between the three cluster techniques over their precise cluster membership.[42] This disagreement is most effectively resolved using discriminant functions calculated from the core demesnes to determine which core group each peripheral demesne has the highest probability of membership (Table 3).[43] The discriminant functions serve three further useful purposes. First, they allow the second-choice classification of each demesne to be determined, in terms of the farming type to which it bears the next closest resemblance (Figs 6–12). This is helpful in determining the degree of similarity or dissimilarity between farming types. Second, direct comparison can be made between the pre-1350 and post-1349 classifications, using the pre-1350 discriminant functions to determine which pre-1350 farming type each post-1349 farming type most closely resembled, and vice versa (Table 4).[44] Third, they may be used to allocate additional demesnes to their appropriate cluster grouping, as more data become available.[45] Here they have been used to classify the independent samples of 125 demesnes within a ten-county area around London and 106 Norfolk demesnes (Table 5 and Figs 3 and 4), thereby lending a much sharper spatial focus to the analysis of variations in farming type. Independent classification of the Norfolk demesnes, replicating at county level the methodology employed at a national scale (Table 6 and Fig 5), provides a further insight into the appropriateness of the national typology.

II

Notwithstanding the significant shifts in the general emphasis of farming practice between 1250–1349 and 1350–1449, no fundamentally new demesne-farming systems came into being.[46] Rather, the existing range of farming systems was modified and developed. Eight basic farming types may be identified in the period 1250–1349 and these may be matched by seven in the period 1350–1449. How closely these two sets of farming system resembled one another can be established, as outlined above, by applying their respective discriminant functions to the mean characteristics of each farming type (Table 4). On this basis *arable husbandry with swine, sheep-corn husbandry*, and *extensive arable husbandry* emerge as farming systems common to both periods, with similar if not identical attributes. With the three pre-1350 and three post-1349 mixed-farming systems, however, the correspondence is less exact. The discriminant functions, for instance, reveal the *intensive mixed-farming* system of post-1349 to bear a closer resemblance to the *light-land intensive* system of pre-1350 than the *intensive mixed-farming* system of the same period. In fact, the *intensive mixed-farming* and *light-land intensive* systems of pre-1350 seem to coalesce into a single intensive mixed-farming system after 1349. Insofar as a *light-land intensive* system persisted as a separate system in its own right it was very much as a minority farming type, specific to certain environmental and economic contexts and retaining the more extreme characteristics of the *light-land intensive* system of before 1350. Similarly, the *mixed-farming with cattle* of before 1350

[42] In 182 cases (the core demesnes) there is agreement between three clustering methods, in 108 cases there is agreement between two, and in seven cases there is no agreement at all.

[43] Power and Campbell, 'Cluster analysis', p 232. When applied to the 182 core demesnes the discriminant functions were 100 per cent successful at reclassifying them correctly.

[44] In fact, the discriminant functions published in Power and Campbell, 'Cluster analysis', p 232, for the period 1250–1349 have been revised marginally on account of the addition of Hemsby, Norfolk (the most intensively cultivated and highest yielding of all known medieval English demesnes) to the original sample of 388 demesnes. Needless to say, Hemsby emerges as a core *intensive mixed-farming* demesne.

[45] For the relevant procedure and program see, K C Bartley, 'Classifying the past: discriminant analysis and its application to medieval farming systems', *History and Computing*, 8.1, 1996, pp. 1–10.

[46] Stock-only demesnes may have developed in certain fresh lowland contexts during the fifteenth century (eg N W Alcock, *Warwickshire Grazier and London Skinner 1532–155: The Account Book of Peter Temple and Thomas Heritage*, British Academy, Records of Economic and Social History, new series IV, 1981), following the wholesale abandonment of arable cultivation, but there are plenty of upland and marshland precedents for such exclusively pastoral enterprises.

TABLE 4

Match between national farming types 1250–1349 and 1350–1449 (established by applying discriminant functions to the mean characteristics of each farming type)

Farming type 1250–1349	*Nearest equivalent farming type 1350–1449*[a]	*Farming type 1350–1449*	*Nearest equivalent farming type 1250–1349*[b]
CORE DEMESNES ONLY			
intensive mixed-farming	intensive mixed-farming	intensive mixed-farming	light-land intensive
light-land intensive	intensive mixed-farming	light-land intensive	light-land intensive
mixed-farming with cattle	extensive mixed-farming	mixed-farming with sheep	extensive mixed-farming
arable husbandry with swine	arable husbandry with swine	arable husbandry with swine	arable husbandry with swine
sheep-corn husbandry	sheep-corn husbandry	sheep-corn husbandry	sheep-corn husbandry
extensive mixed-farming	extensive mixed farming	extensive mixed-farming	extensive mixed-farming
extensive arable husbandry	extensive arable husbandry	extensive arable husbandry	extensive arable husbandry
oats and cattle	light-land intensive		
ALL DEMESNES			
Farming type 1250–1349	*Nearest equivalent farming type 1350–1449*[a]	*Farming type 1350–1449*	*Nearest equivalent farming type 1250–1349*[b]
intensive mixed-farming	extensive mixed-farming	intensive mixed-farming	intensive mixed-farming
light-land intensive	intensive mixed-farming	light-land intensive	light-land intensive
mixed-farming with cattle	extensive mixed-farming	mixed-farming with sheep	intensive mixed-farming
arable husbandry with swine	arable husbandry with swine	arable husbandry with swine	arable husbandry with swine
sheep-corn husbandry	sheep-corn husbandry	sheep-corn husbandry	sheep-corn husbandry
extensive mixed-farming	extensive mixed-farming	extensive mixed-farming	extensive mixed-frming
extensive arable husbandry	extensive arable husbandry	extensive arable husbandry	extensive arable husbandry
oats and cattle	light-land intensive		

[a] using post-1349 discriminant functions; [b] using pre-1350 discriminant functions.
Source: J P Power and B M S Campbell, 'Cluster analysis and the classification of medieval demesne-farming systems', *Trans Inst Brit Geogr*, new series, 17, 1992, Tables III, V, and VI; K C Bartley, 'Classifying the past: discriminant analysis and its application to medieval farming systems', *History and Computing*, 8.1, 1996, Tables 2 and 3.

finds its closest counterpart in the *extensive mixed-farming* of post-1349 rather than any more intermediate system. After 1349 it is *mixed-farming with sheep* which seems to fall somewhere between the intensive and extensive extremes of the mixed-farming spectrum since the discriminant functions reveal it to have combined many of the features of *extensive mixed-farming* with some of those of *intensive mixed-farming*. In fact, the discriminant functions identify a closer coincidence between the *extensive mixed-farming* of post 1349 and the *extensive mixed-farming* of pre-1350 than the *mixed-farming with cattle* with which the prominence of cattle within its pastoral profile might have suggested a closer match. Significantly, the only farming system to disappear entirely from view after 1349 is the smallest and least satisfactorily defined pre-1350, namely *oats with cattle*[47].

Cluster One: 'intensive mixed-farming'

This is at once the most intensive and most arable of the three main mixed-farming systems which may be recognized post 1349. Its intensity is apparent from the choice of animals stocked and crops grown, while its strong arable emphasis is manifest in a mean stocking density below the national average and well below the stocking densities of the other mixed-farming systems (Table 2). It is also the most clearly defined of the three mixed-farming systems, insofar as there is a higher degree of

[47] Power and Campbell, 'Cluster analysis', pp 239, 241.

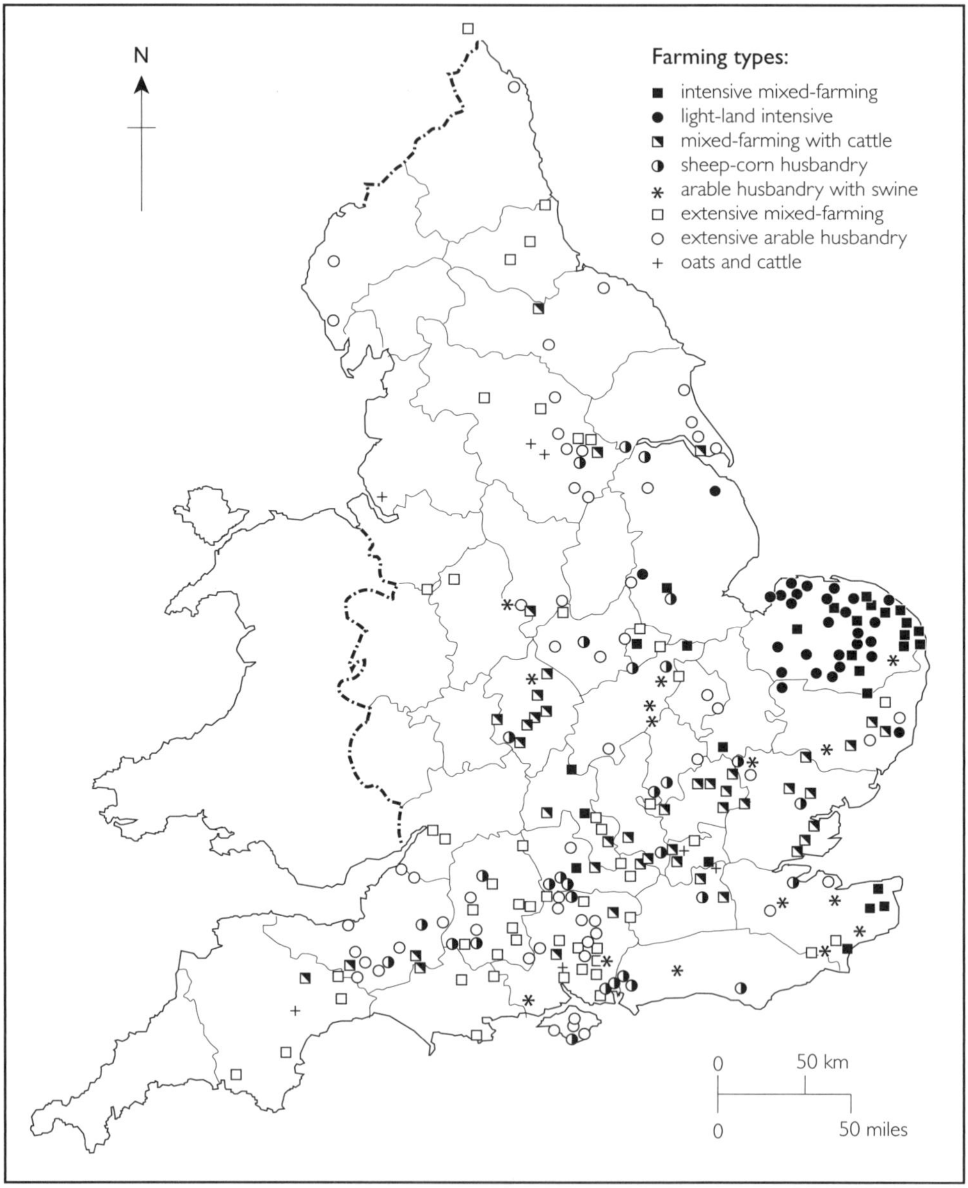

FIGURE I
National farming types, 1250–1349: core demesnes.

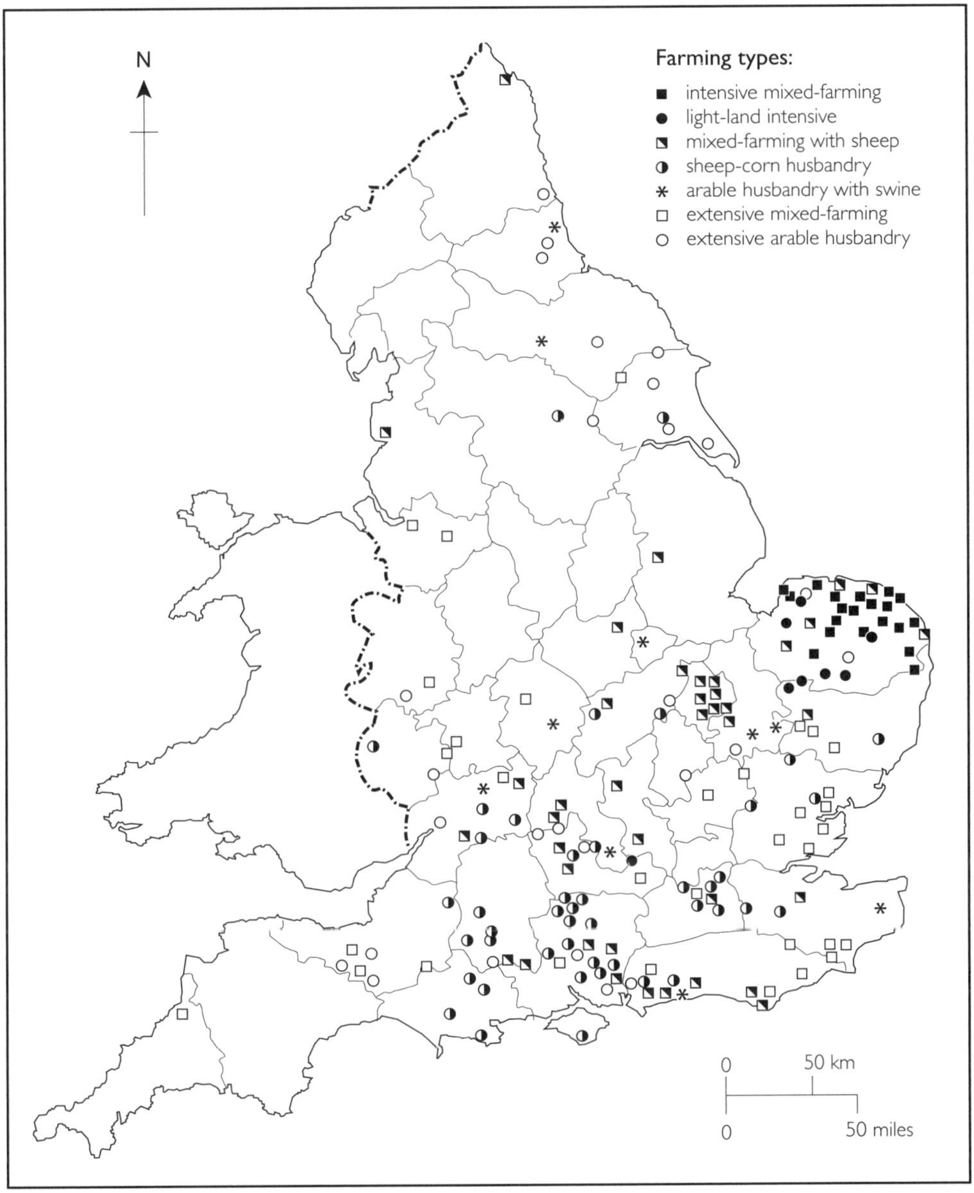

FIGURE 2
National farming types, 1350–1449: core demesnes.

TABLE 5
Farming types in the London region[a] 1375–1400: national classification. Mean characteristics of demesnes

Mean per demesne	*Farming type/cluster group* 1	2	3	4	5	6	7	*All demesnes*
Percentage of sown acreage:								
wheat	17.4	4.9	32.9	30.8	34.1	37.0	31.2	32.3
	7.2	*9.7*	*10.1*	*14.9*	*16.8*	*8.9*	*14.3*	*14.0*
rye	8.9	8.1	0.3	0.0	3.4	2.0	0.0	2.2
	10.6	*16.1*	*1.3*	*0.0*	*6.7*	*5.7*	*0.0*	*6.5*
winter mixtures	8.7	5.2	1.2	0.8	3.0	1.3	0.0	2.1
	10.6	*10.3*	*3.8*	*1.4*	*6.1*	*3.3*	*0.0*	*5.4*
barley	24.7	18.2	28.3	21.5	19.3	8.2	9.2	18.3
	13.6	*16.6*	*11.0*	*17.5*	*15.2*	*7.6*	*15.1*	*14.8*
oats	17.7	46.6	11.1	4.8	22.9	35.3	17.9	22.6
	6.0	*19.5*	*8.1*	*4.9*	*15.4*	*12.9*	*21.1*	*17.1*
spring mixtures	15.5	4.6	6.8	22.7	5.3	4.3	27.5	8.2
	13.7	*5.8*	*11.9*	*27.3*	*6.6*	*9.1*	*29.4*	*14.8*
legumes	7.1	12.5	19.4	19.6	12.1	12.0	14.2	14.4
	4.5	*17.1*	*9.1*	*15.7*	*8.9*	*8.6*	*9.4*	*10.5*
Sown acreage	201.6	78.5	186.0	175.6	154.1	175.1	183.3	171.0
	114.8	*40.3*	*79.1*	*86.7*	*65.8*	*84.1*	*49.1*	*81.5*
Percentage of livestock units[a]								
horses	13.7	6.8	11.4	34.5	14.9	13.0	50.4	16.1
	6.6	*4.3*	*6.1*	*20.5*	*8.4*	*6.9*	*23.5*	*13.9*
oxen	9.6	9.6	18.8	13.1	20.2	12.3	49.4	17.5
	10.5	*7.9*	*12.5*	*15.1*	*15.1*	*9.6*	*24.0*	*15.5*
adult cattle	35.8	37.7	16.5	7.6	14.5	42.1	0.0	23.9
	20.3	*21.7*	*10.5*	*14.5*	*20.2*	*15.4*	*0.0*	*21.0*
immature cattle	10.3	4.7	8.9	0.5	0.1	12.3	0.0	6.7
	4.9	*6.0*	*5.8*	*1.2*	*0.4*	*8.1*	*0.0*	*7.6*
sheep	27.1	40.1	38.3	25.4	47.0	18.4	0.0	31.2
	18.7	*16.0*	*15.6*	*19.0*	*21.4*	*13.6*	*0.0*	*21.2*
swine	3.6	1.0	6.2	19.0	3.3	2.0	0.3	4.6
	2.7	*1.4*	*3.8*	*8.6*	*3.4*	*2.1*	*0.6*	*5.8*
Livestock units[b]	123.6	77.1	103.5	42.5	59.9	88.0	19.4	80.3
	45.2	*33.5*	*62.0*	*27.2*	*28.3*	*43.1*	*4.3*	*50.8*
Stocking density[c]	76.3	118.5	62.0	24.3	56.5	59.9	10.9	57.9
	37.7	*65.2*	*40.8*	*13.0*	*66.9*	*32.9*	*2.6*	*49.0*
Number of demesnes	6	5	33	9	29	37	6	125

Means in roman; standard deviations in italic.
1 = intensive mixed-farming; 2 = light-land intensive; 3 = mixed-farming with sheep; 4 = arable husbandry with swine; 5 = sheep-corn husbandry; 6 = extensive mixed-farming; 7 = extensive arable husbandry.
[a] Beds, Berks, Bucks, Essex, Herts, Kent, Middx, N'hants, Oxon, and Surrey.
[b] [horses × 1.0] + [(oxen + adult cattle) × 1.2] + [immature cattle × 0.8] + [(sheep + swine) × 0.1].
[c] livestock units per 100 sown acres.
Source: Feeding the City II, accounts database.

concurrence in cluster-group membership for this system than either *mixed-farming with sheep* or *extensive mixed-farming*. The three clustering techniques identify a core of twenty demesnes, to which discriminant functions add a further twelve. The latter are less extreme in their characteristics – non-working animals were slightly less

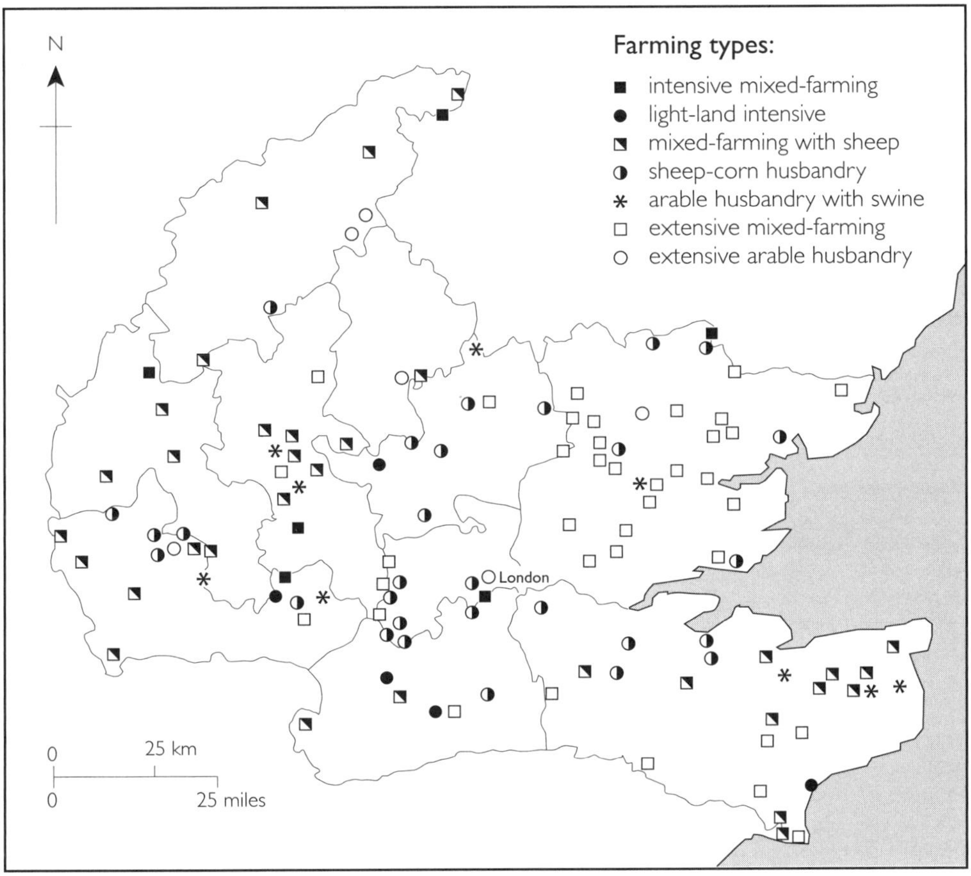

FIGURE 3
Farming types in the London region, 1375–1400: national classification.

important, greater use was made of oxen for draught, and the specialization in barley cultivation was less pronounced (Table 3) – and distributionally mostly peripheral to the main concentration of core demesnes (Figs 2 and 6). By implication, therefore, they were both less specialized and less intensive.

In part this farming system constitutes a modified version of the *intensive mixed-farming* that had evolved during the thirteenth century under the stimulus of rising population and expanding commercial opportunities in areas of fertile and easily cultivated soil, high population density, and good market access (Table 4).[48] Certainly, several of the demesnes that practised *intensive mixed-farming* before 1350 also show up as *intensive mixed-farming* demesnes after 1349 (Figs 1 and 2).[49] Common denominators of both farming

[48] Power and Campbell, 'Cluster analysis', pp 233–6, 240–1; Campbell, 'Agricultural progress'.

[49] For example, Flegg, Halvergate, Hemsby, Martham, and Scratby in east Norfolk, Hindolveston and Hindringham in north-central Norfolk, and Hunstanton in north-west Norfolk.

systems were the priority given to wheat and barley among the winter and spring grains, the disproportionate share of the sown acreage devoted to spring crops, the prominence of horses among working animals, the exceptionally high proportion of non-working animals, and a specific pastoral specialism in cattle-based dairying. These features are all consistent with the relatively intensive use of land, labour, and capital and hence identify this as a comparatively intensive system of husbandry. But whereas before 1350 *intensive mixed-farming* was associated with a fairly equal reliance upon horses and oxen, above average stocking densities, and exceptionally high sowings of legumes (partly to sustain soil fertility and partly to sustain livestock numbers), after 1349 this ceased to be the case. Horses now became the almost exclusive source of draught power; livestock profiles although highly developed were nevertheless associated with stocking densities that were not only below the average for the period 1350–1449 but lower than those associated with pre-1350 versions of *intensive mixed-farming*; and, compared both with other contemporary farming systems and the *intensive mixed-farming* of pre-1350, legumes were no longer cultivated on such an impressive scale.[50] In these respects *intensive mixed-farming* had assumed some of the characteristics of the *light-land intensive* system which occupied a complementary distribution on the light soils of Norfolk in the period 1250–1349 (Table 4 and Fig 1). This light-land system made fuller use of horses for draught power than any other pre-1350 farming system and was also the system in which barley – grown largely for malting – assumed its greatest importance as a crop; both features which became more closely associated with 'intensive mixed-farming' demesnes in the period 1350–1449.

The geographical distribution of demesnes practising *intensive mixed-farming* after 1349 confirms the system's status as a fusion of the *intensive mixed-farming* and *light-land intensive* systems of 1250–1349. Every one of the core demesnes is in Norfolk (Fig 2). Applying discriminant functions to the peripheral demesnes within the national sample adds six more demesnes within the county together with a further six elsewhere, four of them with a strong East Anglian focus (Fig 6).[51] Five others are identified by the application of discriminant functions to the Feeding the City II sample of demesnes (Fig 3).[52] No other post-1349 farming system is spatially so sharply focused, with a marked concentration of demesnes in a single region and only a thin scatter of isolated examples elsewhere (Fig 2). Nor before 1350 was *intensive mixed-farming* so geographically specific in distribution or clearly defined in character (Fig 1). In the earlier period its association with Norfolk is strong but by no means exclusive, with Kent in particular showing up as a second important focus of *intensive mixed-farming*.[53] But after 1349 nowhere went further than Norfolk in the substitution of horses for oxen and reliance upon barley as the principal cash crop. It is these two features which identify the post-1349 version of *intensive mixed-farming* as a more-or-less exclusively Norfolk system (with just a hint of a secondary and far lesser focus in the Soke of Peterborough).

Norfolk is a county particularly well

[50] The advance of horses and retreat of legumes constitute a striking dichotomy. The former represents the continuation of a diffusion process which had begun in the late twelfth century (Campbell, 'Towards an agricultural geography', pp 91–4), the latter, a general throttling back in the intensity of husbandry (Campbell and Overton, 'A new perspective', pp 58–9, 74, 95–6).

[51] Melton (Suffolk), Borley (Essex), Soham (Cambridgeshire), and Longthorpe in the Soke of Peterborough (Northamptonshire).

[52] Adderbury (Oxfordshire) – *a light-land intensive* demesne in the national classification, Boroughbury (Northamptonshire), Culham (Berkshire), Ebury (Middlesex), and West Wycombe (Buckinghamshire) – an *extensive mixed-farming* demesne in the national classification.

[53] Power and Campbell, 'Cluster analysis', pp 233–6, 240–1; *A Medieval Capital*, pp 126, 132–3, 135–8, 140–1; P F Brandon, 'Farming techniques. South-eastern England', in Hallam, *Agrarian History, II*, pp 317–25.

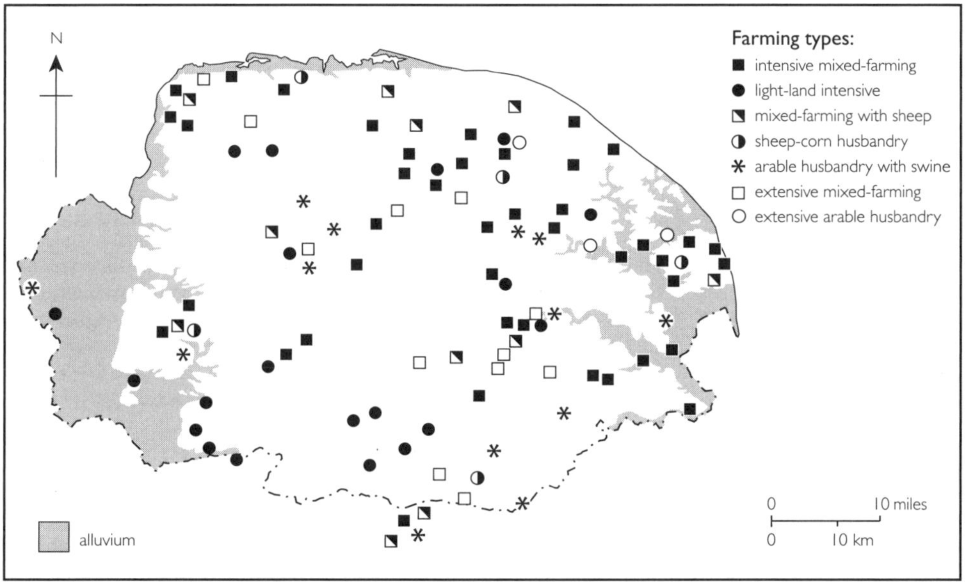

FIGURE 4
Norfolk farming types, 1350–1449: national classification.

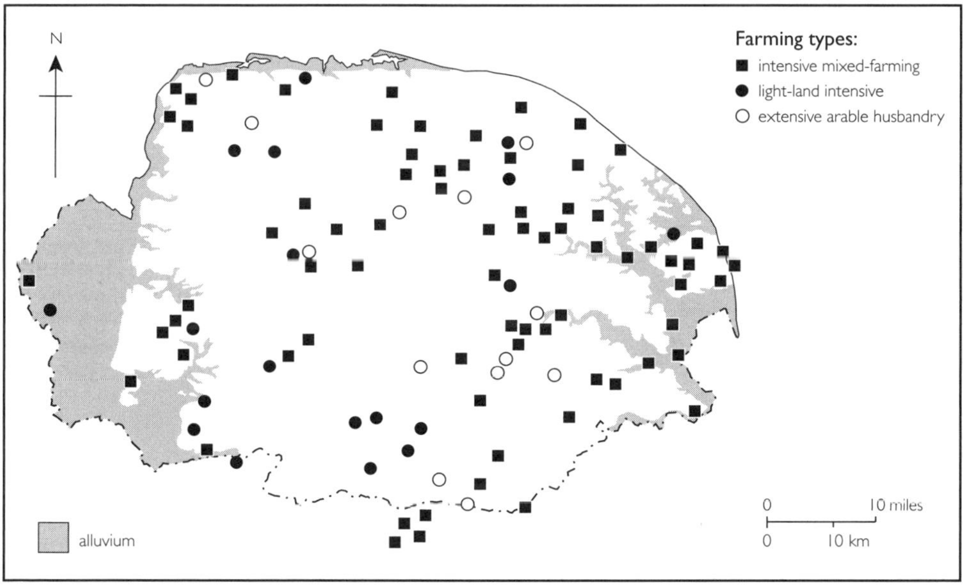

FIGURE 5
Norfolk farming types, 1350–1449: county classification.

served by manorial accounts, with 106 separately documented demesnes in the period 1350–1449 (in fact, five of these – Brandon, Hinderclay, Redgrave, Rickinghall, and Wattisfield – being just across the county boundary into Suffolk).[54] These allow the county's agriculture to be explored in almost greater detail than that of any other part of the country, which is of some advantage given Norfolk's apparent agricultural distinctiveness. Independent classification of these 106 Norfolk demesnes, replicating the method applied at national level, provides an additional perspective on the county's farming systems (Table 6 and Fig 5). At a county scale of analysis the only stable cluster solution is a three-cluster solution which assigns 73 of the 99 core demesnes to a single farming type (to which discriminant functions add one more). So dominant is this single farming system – with three out of four demesnes practising it – that it constitutes a veritable 'Norfolk system'; a medieval precursor of the 'Norfolk system' for which the county was to become so renowned from the late eighteenth century on.[55] Like that more celebrated later system it was a mixed-farming system of above average intensity. The intensity of the system is apparent in its strong arable emphasis (attested by mean stocking densities 25 per cent below the national average), in the predominance of spring-sown crops, and in the primacy accorded barley within the spring course and wheat within the winter. This marked imbalance in the spring and winter courses, coupled with moderate sowings of legumes, is consistent with the flexible and relatively intensive rotations and irregular and infrequent fallowing which may be independently documented from the accounts.[56] That, notwithstanding below average stocking densities, it is rightly regarded as a particularly highly developed form of mixed farming is borne out by the exceptionally developed character of its pastoral sector. This was dominated by non-working rather than working animals, especially dairy cattle, plus some sheep and swine. Cattle-based dairying is one of the most intensive and productive forms of pastoralism and its development was plainly facilitated by the heavy reliance upon horses for draught, for this helped reduce the number of animals it was necessary to keep for ploughing and carting, and horses, by consuming fodder crops, helped free-up scarce meadow and pasture for the cattle.[57] Geographically, examples of this farming system were to be found in virtually all parts of the county, with the conspicuous exception of the sandy Breckland and its margins in the south-west, but it was especially characteristic of an arc of country extending from the richly fertile deep loam soils of Flegg in the east to the shallower and sandier soils in the vicinity of Hunstanton on the north-west coast (Fig 5). This is the self-same area in

[54] The Norfolk accounts database was finalized by Dr Bruce M S Campbell whilst in receipt of a personal research grant from the Social Science Research Council. It contains information on crops and livestock extracted from all known extant grange accounts for the county. Unlike the national and Feeding the City accounts databases, therefore, the mean crop and livestock profiles of individual demesnes are based on all available surviving records. For the period 1350–1449 the database contains information extracted from accounts in the following public and private archives: Bodleian Library, Oxford, British Library, Cambridge University Library, Canterbury Cathedral Library, Chicago University Library, Elveden Hall (Suffolk), Eton College, Holkham Hall (Norfolk), Lambeth Palace Library, Magdalen College Oxford, Norfolk Record Office, Pomeroy & Sons, Wymondham, Public Record Office, Raynham Hall (Norfolk), and the John Ryland's Library, Manchester. A handlist of these accounts is available from Dr Campbell.

[55] William Marshall, *The Rural Economy of Norfolk*, 2 vols, 1787; Arthur Young, *General View of the Agriculture of the County of Norfolk*, 1804; Campbell and Overton, 'A new perspective', pp 92–4, 105.

[56] B M S Campbell, 'The regional uniqueness of English field systems? Some evidence from eastern Norfolk', *AHR*, XXIX, 1981, pp 16–28; *idem*, 'Agricultural progress', pp 28–9; *idem*, 'Arable productivity in medieval English agriculture', unpublished paper presented to the UC-Caltech conference on Pre-industrial Developments in Peasant Economies: the Transition to Economic Growth, Huntington Library, San Marino, May 1987, pp 31–6, 53–7; *Ag Hist III*, pp 199–202.

[57] Campbell, 'Towards an agricultural geography', pp 95–7; *idem*, 'Commercial dairy production on medieval English demesnes: the case of Norfolk', *Anthropozoologica*, 16, 1992, pp 107–118.

TABLE 6

Norfolk farming types 1350–1449: county classification.[a]
Mean characteristics of core demesnes and of core and peripheral demesnes

Mean per demesne	*Core demesnes only*				*All demesnes*			
	Farming type/cluster group			*Overall*	*Farming type/cluster group*			*Overall*
	1	*2*	*3*		*1*	*2*	*3*	
Percentage of sown acreage:								
wheat	17.9	5.5	13.9	15.7	17.8	6.7	13.5	15.3
	9.4	*5.4*	*12.8*	*10.4*	*9.4*	*6.6*	*12.3*	*10.3*
rye	4.6	22.1	2.7	6.9	4.9	18.1	2.5	6.9
	7.4	*10.8*	*5.0*	*10.0*	*7.6*	*12.4*	*4.9*	*9.9*
winter mixtures	0.8	3.2	0.0	1.0	0.7	2.3	0.0	0.9
	2.0	*10.2*	*0.0*	*4.3*	*2.0*	*8.9*	*0.0*	*4.2*
barley	47.0	47.5	49.7	47.4	47.1	45.1	51.2	47.3
	15.2	*16.1*	*9.6*	*15.8*	*15.1*	*18.2*	*10.5*	*15.4*
oats	14.0	19.1	19.8	15.4	14.0	16.8	18.9	15.1
	8.3	*8.9*	*9.7*	*8.9*	*8.2*	*9.0*	*9.8*	*8.8*
spring mixtures	1.2	0.3	0.0	0.9	1.1	1.2	0.0	1.0
	4.0	*0.9*	*0.0*	*3.5*	*4.0*	*4.3*	*0.0*	*3.8*
legumes	14.5	2.3	13.9	12.7	14.5	4.5	14.0	12.6
	6.0	*1.8*	*4.2*	*6.8*	*6.0*	*6.1*	*4.0*	*6.9*
Sown acreage	143.2	139.5	81.7	135.2	143.2	127.2	79.4	132.5
	75.1	*73.9*	*36.5*	*74.1*	*74.5*	*68.2*	*35.9*	*72.8*
Percentage of livestock units[b]:								
horses	15.4	12.0	73.0	21.9	15.3	14.8	73.7	22.4
	8.1	*4.3*	*23.2*	*32.3*	*8.1*	*8.4*	*31.1*	*23.3*
oxen	6.7	1.4	17.3	7.2	6.6	2.8	16.0	7.1
	8.5	*5.2*	*24.0*	*12.3*	*8.5*	*8.5*	*24.5*	*12.3*
adult cattle	41.3	27.9	0.6	34.5	41.3	27.6	0.6	33.8
	17.1	*28.7*	*2.1*	*22.6*	*17.0*	*29.8*	*2.0*	*23.3*
immature cattle	14.0	4.7	0.0	11.0	13.9	4.1	0.0	10.4
	8.0	*6.5*	*0.0*	*9.0*	*8.1*	*6.2*	*0.0*	*9.0*
sheep	19.3	53.0	0.0	21.7	19.6	46.5	0.0	22.0
	22.1	*33.1*	*0.0*	*26.8*	*22.1*	*36.1*	*0.0*	*27.3*
swine	3.4	1.0	0.7	2.7	3.3	4.3	2.1	3.3
	3.9	*1.7*	*1.3*	*3.6*	*3.9*	*13.5*	*5.0*	*6.8*
Livestock units[b]	56.1	55.0	5.8	49.9	56.0	47.8	5.7	48.4
	37.5	*30.0*	*2.6*	*35.7*	*34.3*	*29.1*	*2.5*	*35.2*
Stocking density[c]	42.3	40.7	7.4	37.9	42.2	40.5	7.4	37.6
	20.9	*15.9*	*2.2*	*22.1*	*20.8*	*23.8*	*2.1*	*23.1*
Number of demesnes	73	14	12	99	74	19	13	106

Means in roman; standard deviations in italic.
1 = intensive mixed-farming; 2 = light-land intensive; 3 = extensive arable husbandry.
[a] based upon independent cluster analysis of 106 Norfolk demesnes.
[b] [horses × 1.0] + [(oxen + adult cattle) × 1.2] + [immature cattle × 0.8] + [(sheep + swine) × 0.1].
[c] livestock units per 100 sown acres.
Source: Norfolk accounts database.

which the *intensive mixed-farming* demesnes identified by the national classification of farming systems are most strongly concentrated (Fig 4).

The discriminant functions allow a direct comparison of the national and county classifications of each of these Norfolk demesnes. Significantly, the national classification identifies a wider variety of farming types within this single county, with examples occurring of each of the seven principal farming types (Fig 4). Apart from anything else this exemplifies the superiority of national over regional or local classifications of farming types, at least where the statistical basis of the classification is provided by cluster analysis. This is because minority farming types need to be very strongly differentiated before their distinctive identity is recognized by all three cluster techniques. The wider and more representative the sample, therefore, the greater the prospect of distinguishing between genuine differences in farming system. Numerically, although examples of all seven national farming types may be found in Norfolk it is *intensive mixed-farming* which is by far the best represented. There are 42 examples of this farming system, each of them coinciding with one of the 74 *intensive mixed-farming* demesnes identified by the Norfolk classification. Moreover, the majority of the 42 *intensive mixed-farming* demesnes common to both classifications are concentrated within the same arc of country from Flegg in the east to the north-west coast around Hunstanton, which appears to have been the heartland of this peculiarly Norfolk mixed-farming system (Figs 4 and 5).

Nevertheless, although *intensive mixed-farming* had a stronger affiliation with a single region than any other post-Black Death farming system, it was not so unique that it lacked strong affinities with certain other farming types. Again, the discriminant functions help to identify the systems which these *intensive mixed-farming* demesnes next most closely resembled (Fig 6): in two out of three cases this was *extensive mixed-farming*, and in a further one out of four it was *mixed-farming with sheep*. The same close association between systems at the intensive and extensive ends of the mixed-farming spectrum was evident in the period 1250–1349, and serves as a reminder that the three main mixed-farming systems are best regarded as subsets of a continuum.[58] Both before and after the Black Death intensive and extensive mixed-farming systems were alike in the relatively developed character of their pastoral husbandry (Table 2), the one with strong arable underpinnings, the other with a far greater grassland component. Geographically, the *intensive mixed-farming* demesnes which shared most in common with *extensive mixed-farming* formed a fairly concentrated block, to which demesnes with closer affinities to other systems were mostly peripheral (Fig 6). This was very much the case with the eight demesnes whose second-choice classification was *mixed-farming with sheep*, and even more so with those three demesnes whose second choice was *light-land intensive husbandry*.[59] The one demesne – Aldeby, Norfolk – with a resemblance to *arable husbandry with swine* was a small demesne which served as the home farm to a minor monastic cell with a regular need for fresh pork, but it was also locationally marginal to the main concentration of *intensive mixed-farming* demesnes. Generally, therefore, this analysis of second choices confirms that *intensive mixed-farming* only attained the status of a homogeneous regional farming type in eastern and northern Norfolk and that small but nonetheless real differences existed between the kind of *intensive mixed-farming*

[58] Power and Campbell, 'Cluster analysis', p 239.

[59] Taverham (Norfolk), Melton (Suffolk), and Milton (Oxfordshire) from the national sample and Adderbury (Oxfordshire), Culham (Berkshire), and Ebury (Middlesex) from the Feeding the City II sample.

FIGURE 6
Intensive mixed-farming demesnes, 1350–1449 (core and peripheral demesnes indicating second choice).

practised here and that practised apparently in isolation elsewhere.[60]

Cluster Two: 'light-land intensive farming'

Statistically, *intensive mixed-farming* may have been closer to *extensive mixed-farming* than any other farming system (Table 4), but agriculturally it was *light-land intensive* farming which was the next most proximate system both in terms of intensity and distribution (Table 2, Fig 2).[61] Numerically, *light-land intensive* farming was very much a minority system and is represented in the national sample of demesnes by just a handful of examples. In part this is because the three clustering techniques differ most widely in their definition of this farming type with the result that a core of only seven demesnes is common between them, to which a further seven are added on the basis of their discriminant scores. But the comparative rarity of this farming type was also a function of its specialized character (Table 2). In no other system did working animals comprise such a small share of total livestock units and in none was the reliance upon horses for draught power so complete. Among non-working animals dairy cattle were plainly important, but sheep even more so; in fact, only on sheep-corn demesnes did sheep comprise a greater proportion of livestock units. Swine, in contrast, were mostly conspicuous by their absence. Arable husbandry was equally distinctive. This was the only farming system in which the area sown with rye generally matched or exceeded that devoted to wheat (maslin – a wheat-rye mixture – was also of some importance). Nevertheless, as with *intensive mixed-farming*, it was the spring sown grains that predominated, particularly barley but also significant quantities of oats. Such a pronounced winter-spring imbalance again implies irregular and flexible rotations, although in the absence of significant sowings of legumes (which occupied a smaller share of the sown acreage than in any other system) fallows of varying frequency and duration are likely to have retained a prominent role in the maintenance of soil fertility. This was certainly the case at Brandon (Suffolk) and East Wretham and Eccles (Norfolk), which are known to have employed their sheep flocks to fold arable land left uncultivated for several years in succession, the duration of the 'fallow' depending upon the quality of the land.[62]

Everything about this farming system marks it out as a response to the particular environmental problems presented by light land. Light and easily worked loamy and sandy soils, for instance, help explain the preference for small all-horse plough teams implicit in the modest, horse-dominated draught sector. Sheep also did well on light land, especially where there was much heath and rough pasture, and were an indispensable adjunct of arable husbandry since light sandy soils were quickly exhausted unless regularly dunged (the nightly folding of sheep on the arable performing this function admirably). Provided there was sufficient good grassland, especially in the better watered valley bottoms, cattle fared reasonably well, but in open heathy country with neither pannage nor legumes available in any quantity (peas and beans – widely used for feeding swine – did not do well on light soils) there was little basis for keeping swine. Light acidic soils also explain the predominance of rye over wheat at a time when its cultivation elsewhere was in decline as rising living standards allowed consumers to indulge a strong dietary preference for wheaten

[60] Closer examination confirms that these differences were further underscored by differences in productivity: B M S Campbell, 'Arable productivity in medieval England: some evidence from Norfolk', *J Econ Hist*, XLIII, 1983, pp 379–404.

[61] The reference to light land is entirely descriptive: soil type is not a criterion of classification.

[62] PRO, SC 6/1304/28–36; Chicago University Library, Bacon Roll 643, 645–60; Elveden Hall, Suffolk, Iveagh Collection 148 (Phillipps 26523); Eton College Records, Vol 30, 43–9; Raynham Hall, Norfolk, Townshend Mss.

bread. Similarly, barley and oats grew relatively well on light soils, whose low nitrogen content yielded a barley peculiarly well suited to brewing.[63] Before 1350 so great had been the commercial imperative to produce grain that stocking densities on *light-land intensive* demesnes had been conspicuously below-average, but thereafter they rose to around the national average as demesne managers lengthened fallows and converted to pasture land whose economic rent was now too low to warrant its continued use as arable.

That it is correct to diagnose this farming system as a light-land system is confirmed by the distribution of demesnes that practised it (Figs 2 and 7). Four of the seven core demesnes lie in or on the margins of the Breckland of East Anglia, a locality with some of the lightest and least fertile soils in the country whose distinctive medieval husbandry practices have been the subject of detailed study by Mark Bailey.[64] The other three are all located on soils of similar type.[65] Five of the seven peripheral demesnes (Fig 7) are also associated with light to medium soils.[66] Only Agney-and-Orgarswick in Romney Marsh and Popinho in the Norfolk Fens deviate from the light-land rule, a rule to which the three *light-land intensive* demesnes within the Feeding the City II sample all comply (Fig 3).[67]

The restriction of every one of these *light-land intensive* demesnes to East Anglia or the south-east of England nevertheless implies that environmental factors were by no means the sole influence upon their distribution. Economic factors were at least as important, for it was these which endowed this light-land system with its relatively intensive character. This intensity is apparent in the overwhelming preference for horses rather than oxen, the developed character of pastoral husbandry, the flexibility of arable rotations, and the importance of barley as a crop.[68] East Anglia, in particular, as John Langdon has demonstrated, was in the vanguard of the substitution of horses for oxen.[69] Along with much of the south-east, it lay outside the bounds of the regular commonfield system which allowed the evolution of more irregular and flexible forms of rotation, and was also early involved in an active regional and inter-regional grain trade which provided a stimulus to the commercial production of malting barley.[70] East Anglia was also the preserve of *intensive mixed-farming* and it was this farming system which was the second-choice classification of over half the *light-land intensive* demesnes (Fig 7). These demesnes formed a reasonably cognate group with a strong but by no means exclusive focus upon Norfolk and were complemented in distribution by seven others – six of them in west Norfolk whose sheep farming was on such a scale that their second-choice classification was *sheep-corn husbandry*, an altogether more extensive system of farming. The remaining *light-land intensive* demesnes were also less intensive in character – two of them with a close resemblance to *mixed-farming with sheep* and four bearing a strong affin-

[63] M Bailey, *A Marginal Economy? East Anglian Breckland in the Later Middle Ages*, 1989, p 140.

[64] Brandon and Lakenheath (Suffolk) and East Wretham and Eccles (Norfolk); Bailey, *A Marginal Economy?*

[65] Bircham on the coarse loamy soils of the 'good-sands' region of north-west Norfolk, Costessey, a few miles west of Norwich, on free-draining acidic sandy soils formed from glacial outwash sands and gravels, and Culham in Berkshire on the coarse loamy and sandy soils that are a feature of the gravel terraces of the middle Thames valley.

[66] Feltwell (Norfolk) on the western edge of Breckland; Wrabness on the coarse sandy and loamy soils of north-east Essex, Pyrford (Surrey) on the light soils of the Wey valley, Petworth on the inferior sandy soils of the Sussex Weald, and Adderbury on the light loams of the wold country of north Oxfordshire.

[67] Wargrave, a near neighbour of Culham, Berkshire, on almost identical coarse loamy and sandy soils; Dorking, on broadly equivalent soils in the Surrey Weald; and Great Gaddesden in the Hertfordshire Chilterns, on loamy and clayey soils overlying chalk.

[68] The Feeding the City II accounts database indicates that barley was one of the most 'commercialized' of crops, with 41 per cent of net receipts sold (including sales of malted barley). Only dredge – a barley/oats mixture – was sold in greater proportions (58 per cent of net receipts).

[69] Langdon; *Horses, Oxen and Technological Innovation*, p 43; Campbell, 'Towards an agricultural geography', pp 91–3.

[70] H L Gray, *English Field Systems*, Cambridge, Mass, 1915, pp 305–402; Bailey, *A Marginal Economy?*, pp 140–3, 145–50.

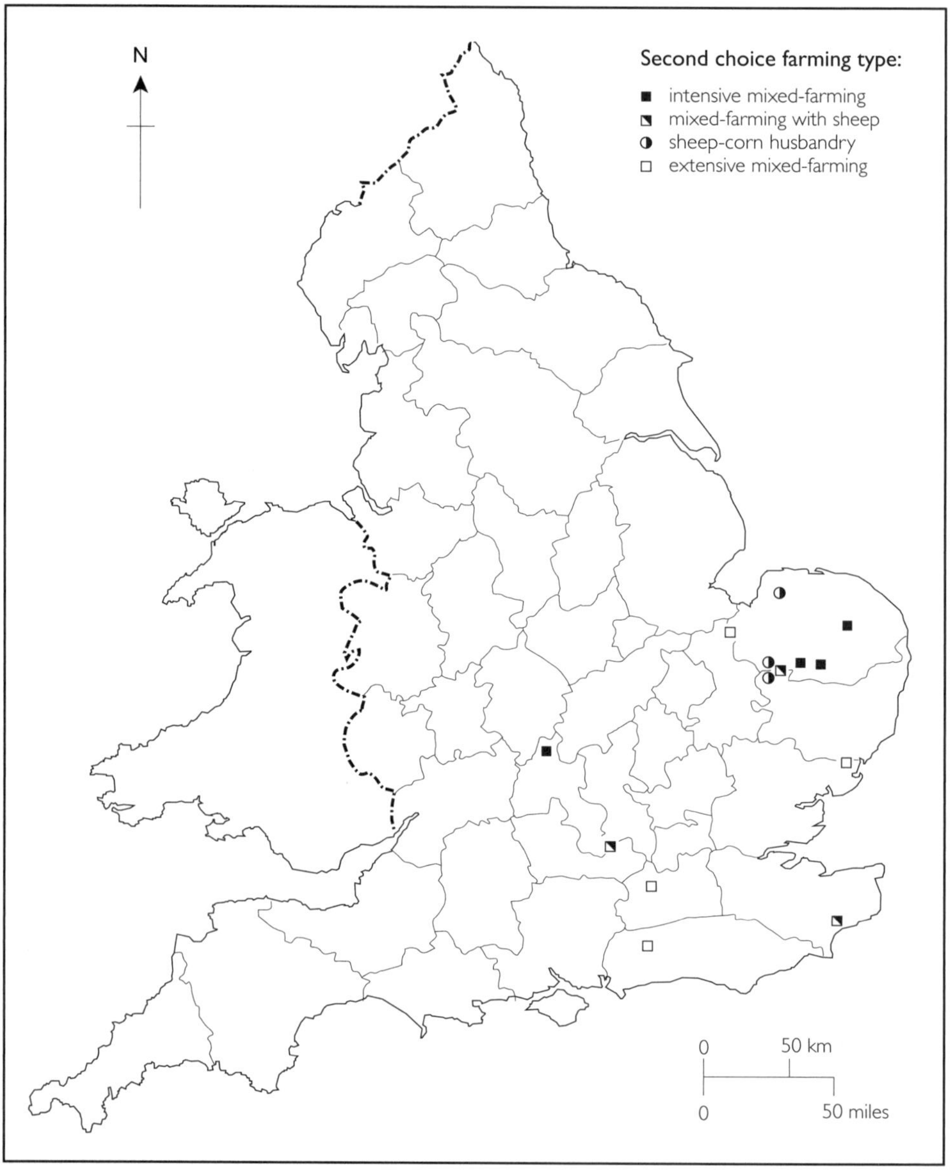

FIGURE 7
Light-land intensive demesnes, 1350–1449 (core and peripheral demesnes indicating second choice).

ity with *extensive mixed-farming* – and were distributionally peripheral to the main concentration of demesnes belonging to this farming type (Fig 7).

The exceptionally detailed Norfolk material brings at least that county's part of this overall picture into sharper focus. Of the 106 documented Norfolk demesnes discriminant functions reveal 20 to have practised some form of *light-land intensive* farming in the period 1350–1449. Half of these demesnes were located in the south-west and west of the county where their dominance was so great as to constitute a sub-regional farming type which clearly extended across the county boundary to embrace the adjoining area of light soils in north-west Suffolk (Fig 4). It is this 'Breckland' system of farming which has been discussed and described in detail by Bailey.[71] Elsewhere examples of *light-land intensive* farming are more scattered, but then so too are patches of light soils in this pedologically most varied of counties. This implies that *light-land intensive* farming mostly occurred as a deviation from prevailing husbandry norms, which points to the paramounce of environmental over economic factors in determining its detailed distribution. Its distinctiveness as a recognizable farming type in its own right is nevertheless confirmed by independent classification of these 106 Norfolk demesnes (Table 6, Fig 5). As already observed, replicating the methodology of the national classification at county level identifies only three basic farming systems: *intensive mixed-farming* is one, *light-land intensive* farming another. The latter system is represented by just 19 demesnes, a number which compares closely with the 20 identified within the county by the national classification. Fifteen demesnes are common to both classifications and consequently represent the most diagnostic examples of this particular farming type. Seven are relatively scattered in distribution and may mostly be explained by localized occurrences of light or poor soil, but eight are located in or near Breckland and thereby confirm the distinctive agricultural character of this small but unique region (Figs 4 and 5).

Cluster Three: 'mixed-farming with sheep'

In contrast to *intensive mixed-farming* and *light-land intensive* farming, which were either regionally or environmentally circumscribed in distribution, *mixed-farming with sheep* was far more widely represented as a husbandry type (Fig 2). Its status as a mixed-farming system is apparent in the number and proportion of non-working animals that were stocked and the above average stocking densities to which these mostly gave rise, notwithstanding sown acreages which themselves were often above average (Table 2). The types and combinations of crops grown imply a relatively intensive agricultural regime, insofar as the acreage of spring-sown crops exceeded that of winter, wheat and barley (respectively, the most valuable and demanding winter and spring grains) were the two most important grains (with oats relegated to a relatively minor role), and roughly a fifth of the cropped acreage was sown with legumes.[72] Nevertheless, such a pattern of cropping is also perfectly compatible with two-course rotations and, hence, relatively extensive systems of crop-

[71] Bailey, *A Marginal Economy?*; *idem*, 'Sand into gold: the evolution of the foldcourse system in west Suffolk, 1200–1600', *AHR*, 38, 1990, pp 40–57.

[72] Core demesnes known to have employed irregular rotations of above average intensity include Holywell (which Raftis believes to have been 'typical of the system on other Huntingdon demesnes') and Wistow, Hunts (J A Raftis, *The Estates of Ramsey Abbey: A Study in Economic Growth and Organization*, Toronto, 1957, pp 184–7; M P Hogan, 'Clays, culturae and the cultivator's wisdom: management efficiency at fourteenth-century Wistow', *AHR*, 36, 1988, pp 117–31); Felbrigg, Langham, and Ormesby, Norfolk (NRO, WKC 2/130/398 × 6, WKC 2/131/398 × 6; Raynham Hall, Townshend Mss; *Ag Hist III*, pp 200–2); and Alciston, Bosham, and Wiston, Sussex (P F Brandon, 'Demesne arable farming in coastal Sussex during the later Middle Ages', *AHR*, 19, 1971, pp 113–34; *Ag Hist III*, pp 273–5).

ping.[73] The latter is certainly more consistent with the intrinsically extensive character of the pastoral side of this farming system. Apart from the relatively high stocking densities, with all that they imply for a higher than average ratio of grassland to arable, there was a strong preference among working animals for oxen rather than horses and among non-working animals for sheep rather than cattle. Oxen and sheep were grassland animals and less labour and capital intensive in their requirements than horses and cattle.[74] Moreover, from the demographic composition of herds it would appear that the emphasis of cattle farming was not primarily upon dairying but upon the less intensive activity of rearing.

Many demesnes possessed these characteristics to some degree with the result that it is one of the least amenable farming systems to precise definition. The three different clustering techniques certainly differ quite considerably in the number of demesnes assigned to this cluster group: Normix assigns 61, Ward's 78, and K-means 128, nevertheless, only 40 demesnes are common between them. To this core group discriminant functions add a further 37 demesnes whose husbandry is less strongly differentiated – their stocking densities were, on average, higher, they kept more cattle and fewer sheep, and sowed more wheat and oats and less barley and legumes – and whose claim to belong to this farming type is correspondingly weaker (Table 3). Together, these core and peripheral demesnes constitute the second most common farming type within the national sample. Coincidentally, *mixed-farming with sheep* is also the second most common farming type within the Feeding the City II sample of demesnes (Table 5).

That this was quintessentially a mixed-farming system and one that lay somewhere between the intensive and extensive extremes of the mixed-farming spectrum is confirmed by an analysis of second choices (Fig 8). Twenty-two of the 77 core and peripheral demesnes more closely resembled *intensive mixed-farming* than any other system, compared with 51 which bore a closer similarity to *extensive mixed-farming*. Of the remaining four, three had a second choice of *light-land intensive* husbandry, and one only a second choice of *sheep-corn husbandry*. The greatest problem of definition is therefore knowing where to draw the line between this and other more or less intensive mixed-farming systems. It is because there is such a continuous gradation from intensive, to semi-intensive/extensive, and ultimately to extensive mixed-farming systems that there is such limited congruence in how the three different clustering techniques define this intermediate category.

Geographically, demesnes practising *mixed-farming with sheep* were found in a wide variety of environmental, economic, and institutional contexts (Fig 2). They occur in the extreme north of the country, as at Holy Island off the Northumberland coast, and in the extreme south, as at Alciston and Bosham on the Sussex coast.[75] They show up in areas of fertile soil, such as Ormesby in the Broadland district of east Norfolk and a group of Ramsey Abbey demesnes in Huntingdonshire, and on thinner and lighter soils of far lower potential productivity, as at Felbrigg and Rougham in north Norfolk and Downton in south Wiltshire. The great majority are in the more developed and commercialized parts of the south and east, but outlying core demesnes also occur on the Fylde coast of

[73] Among core demesnes, those of the bishopric of Winchester at Hambledon, Hants, Witney, Oxon, and Downton, Wilts, appear to have practised three-course rotations, while the demesne at Bishopstone, Wilts, adhered to a two-course system: J Z Titow, 'Land and population on the Bishop of Winchester's estates 1209–1350', unpublished PhD thesis, University of Cambridge, 1962, pp 17–18.

[74] Campbell, 'The livestock of Chaucer's reeve'.

[75] For a case study of Alciston see *Ag Hist III*, pp 272, 273–5, 278, 281, 282.

Lancashire and on the Gloucestershire Cotswolds above Stroud. Peripheral demesnes extend this distribution north-east as far as Coldingham in Berwickshire, west into Worcestershire, and south-west into Somerset and Devon. In distribution they transcend the division between regular and irregular commonfield systems and insofar as there tends to be a common denominator it is accessibility to substantial permanent grazings. Those grazings, however, were of many types – fenland, coastal marsh, heathland, rough pasture, and, above all, the limestone pastures of downland and wold. *Mixed-farming with sheep* is therefore a national rather than regional farming type, although it was more characteristic of some regions and localities than others (Fig 2).

Mixed-farming with sheep assumed its most intensive form in Norfolk and Suffolk and immediately adjacent parts of the east midlands, for it was here that *intensive mixed-farming* comprised the predominant second-choice farming type (Fig 8). Six of the 22 demesnes with this second choice were in Norfolk, plus one each in Suffolk and Lincolnshire, two in Cambridgeshire, and seven in Huntingdonshire. They are thus complementary in distribution to *intensive mixed-farming* demesnes (Fig 6). Like the latter they were characterized by below average stocking densities and hence a stronger bias towards arable husbandry than was otherwise typical of this farming type. That bias was reinforced by a greater reliance upon horses than was normally the case. On the arable side these demesnes grew more barley than wheat and devoted a quarter of their sown acreage on average to legumes.[76] This implies relatively intensive and often quite irregular systems of cropping. Felbrigg in north-east Norfolk, for instance, is known to have operated a convertible-husbandry system whereby land was cropped for three or four years and then left uncultivated for a corresponding period of time.[77] At Ormesby, however, in the same county – on some of the country's most fertile soils – land was cropped for six or more years in succession before being fallowed for one year only.[78] The group of Huntingdonshire demesnes belonging to Ramsey Abbey are also known to have been evolving towards more flexible and intensive systems of rotation during the middle years of the fourteenth century, as was the Crowland Abbey demesne of Oakington, Cambridgeshire.[79] At a further remove, Aylesbury and Water Eaton in Buckinghamshire, Cliffe and Ickham in Kent, and Chaceley in Gloucestershire all shared intensive features (Fig 8). Classification of all 106 documented Norfolk demesnes using the national discriminant functions identifies a total of ten *mixed-farming with sheep* demesnes within the county, plus Redgrave just over the county boundary into Suffolk. Distributionally they form a relatively light scatter and mostly seem to represent less intensive versions of the *intensive mixed-farming* which, as has been shown, was the county's dominant farming system (Fig 4). Corresponding classification of the Feeding the City II demesnes likewise brings the pattern within the ten counties around London into sharper focus (Fig 3). Particularly significant here is the group of four east Kent demesnes – Barton, Bekesbourne, Chartham, and Ickham – all located in an area with a strong tradition of more intensive husbandry.[80] Much the

[76] *Ag Hist III*, pp 214–16.

[77] Campbell, 'Arable productivity in medieval English agriculture'; NRO, WKC 2/130/398 × 6, WKC 2/131/398 × 6.

[78] *Ag Hist III*, p 202; PRO, SC 6/939/1–8.

[79] Raftis, *The Estates of Ramsey Abbey*, pp 184–90; *Ag Hist III*, pp 214–16; F M Page, *The Estates of Crowland Abbey: A Study in Manorial Organisation*, 1934, p 119; J R Ravensdale, *Liable to Floods: Village Landscape on the Edge of the Fens AD 450–1850*, 1974, pp 115–20. For developments at Wistow, Hunts, see Hogan, 'Clays, culturae and the cultivator's wisdom'.

[80] *A Medieval Capital*, pp 131, 140–1; *Ag Hist III*, p 271, 272, 275–8, 282, 283.

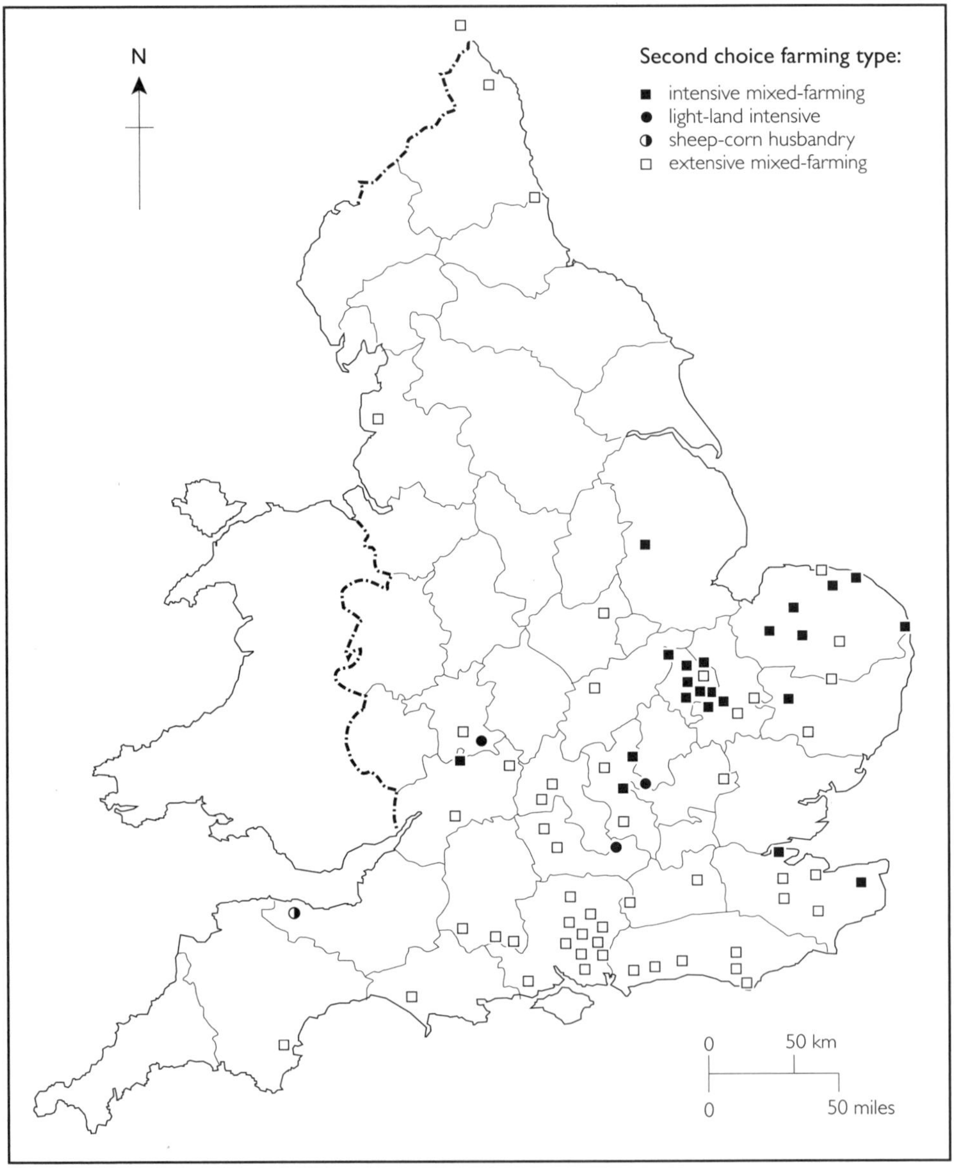

FIGURE 8
Mixed-farming with sheep demesnes, 1350–1449 (core and peripheral demesnes indicating second choice).

same applies to Eye and Maidwell in Northamptonshire.[81]

The majority of *mixed-farming with sheep* demesnes were, however, more extensive in their traits and closer in character to *extensive mixed-farming* (Fig 8). These demesnes stocked more animals and cropped fewer acres and hence supported stocking densities which were well above the national average. Oxen remained the fundamental source of most draught power and cattle herds appear to have been geared primarily to the breeding of replacement work animals. On the arable side these demesnes grew more wheat and oats and less barley and legumes and hence probably employed less intensive types of rotation.[82] Demesnes of this type are particularly well represented in the south midlands and in southern England, with a string of examples extending along the chalkland belt from East Sussex through Hampshire into southern Wiltshire and beyond into Dorset and Devon. A scatter of examples even appears in East Anglia.[83] The Norfolk sample adds two further examples within this most intensively cultivated of counties (Fig 4) while the Feeding the City II sample confirms that a loose string of demesnes of this type followed the North Downs in Surrey and Kent (Fig 3). The same sample also yields further examples of *mixed-farming with sheep* in the wold and vale country north and west of the Chilterns.

Cluster Four: 'arable husbandry with swine'

This is another minority farming type, represented in the national sample by a comparatively small number of demesnes (Table 2). Normix identifies only ten demesnes of this type, K-means 21, and Ward's 23. A core of only 10 demesnes is therefore common to all three cluster solutions to which only two further peripheral demesnes are added on the basis of the discriminant functions (Table 3). What sets these demesnes apart and justifies a separate farming classification are their absolutely low stocking densities, borne of the small numbers of animals kept, the predominance of working animals within the pastoral sector (with horses and oxen of almost equal importance), the dominance of the limited number of non-working animals by swine, and the devotion of a substantial proportion of the sown acreage to legumes (an important source of fodder for both horses and hogs).[84] This is also the farming system in which, on average, oats occupied their smallest, and dredge its largest, share of the sown acreage (the latter being very much a direct substitute for the former).

In the smallness of their livestock sector and consequent lowness of their stocking densities demesnes practising *arable husbandry with swine* bore their closest resemblance to the *extensive arable husbandry* of Cluster Seven; indeed, for six of the twelve the latter is their second-choice farming classification (Fig 9). But four demesnes displayed a stronger affinity with the *sheep-corn husbandry* of Cluster Five (the four which carried most sheep), and two had as their second choice the *mixed-farming with sheep* of Cluster Three. Significantly, these last six demesnes were those which grew legumes on the largest scale. Hardwicke (Gloucestershire) and Angmering (Sussex), for instance, devoted over a third of their sown acreage to this crop and it can be no coincidence that on both demesnes swine accounted for a greater proportion of livestock units than on any others in this cluster group.[85]

Distributionally, this is the most dis-

[81] The other isolated east midland demesnes of this type are Shillington (Bedfordshire) and Horsenden (Buckinghamshire).

[82] *Ag Hist III*, pp 286–7.

[83] Langham and Wymondham (Norfolk), Acton and Redgrave (Suffolk), Burwell and Ditton Valence (Cambridgeshire), and Much Wymondley (Hertfordshire).

[84] On the use of legumes as fodder for horses and/or hogs see note 17 above; Biddick, *The Other Economy*, pp 122–3, 132, 202; *Ag Hist III*, pp 191, 270.

[85] See *Ag Hist III*, p 387, for the droving of swine from Hardwicke to Islip, Denham, and Westminster Abbey's kitchens.

FIGURE 9
Arable husbandry with swine demesnes, 1350–1449 (core and peripheral demesnes indicating second choice).

persed of the seven farming types. The ten core and two peripheral demesnes are widely separate in space with the result that in no single locality was this the predominant farming type (Figs 2 and 9). The much fuller sample of demesnes for Norfolk and the ten-county area around London corroborate this conclusion (Figs 4 and 3), the former furnishing two and the latter eight examples.[86] It is possible that in some of these instances *arable husbandry with swine* represented a residual, and therefore somewhat artificial, farming system on demesnes which had opted to let their dairy herds at farm. But there were others – Potter Heigham and Hoveton in Norfolk are good examples – where it undoubtedly constituted a genuine specialism.

Swine were the only animal reared exclusively for their meat.[87] Permanent households with large numbers of mouths to feed consumed pork and bacon in significant quantities.[88] At Potter Heigham and Hoveton, both home farms of the Benedictine abbey of St Benet at Holme some three miles away, swine were stocked in large numbers and sty fed on the legumes grown on a correspondingly large scale.[89] Monkwearmouth (Durham), Oakham (Rutland), and Exning (Suffolk) probably similarly functioned as home farms to resident households, rearing pork for the table.[90] Sty-feeding swine in this way was a comparatively intensive activity, but swine could also be reared extensively on woodland pannage.[91] The latter is probably the method by which swine were fed at Monks Risborough, on the edge of Chiltern woodlands, and at West Tanfield in the North Riding of Yorkshire. More than any other farming type *arable husbandry with swine* is therefore to be explained by a combination of local and institutional factors.

Cluster Five: 'sheep-corn husbandry'

This is one of the most clearly differentiated of all the farming systems with the result that there is a remarkably close congruence in the result obtained from each of the three clustering techniques. The 45 core demesnes make this the largest of any of the core groups, but to these only a further 11 peripheral demesnes are added with the result that *mixed-farming with sheep* and *extensive mixed-farming* eventually emerge as more common farming types (Tables 2 and 3). The single most striking feature of this farming system is the exceptionally high proportion of livestock units – on average well over half – accounted for by sheep. Sheep, in fact, were kept to the virtual exclusion of other nonworking animals with the result that demesnes must have relied upon market purchase or intermanorial transfer to obtain replacement draught animals, which in the majority of cases were predominantly oxen.[92] The combination of sheep and oxen immediately identifies the pastoral component of this farming system as grass-based and essentially extensive and implies a heavy reliance upon permanent pasture. Since demesnes were themselves very unequally endowed with several and communal pastures, stocking densities, although on average well above the national mean, were highly variable. The sheep-farming component of this husbandry system therefore tended to attain its fullest development in localities with, for whatever reason, a high

[86] Bray (Berkshire), Monks Risborough (Buckinghamshire), Writtle Rectory (Essex), Ashwell (Hertfordshire), Adisham, Copton, and Eastry (Kent), and South Stoke (Oxfordshire).

[87] Campbell, 'Measuring the commercialisation of seigneurial agriculture', pp 164, 168.

[88] Biddick, *The Other Economy*, pp 37–40, 121; Dyer, *Standards of Living*, pp 59–60.

[89] NRO, Diocesan Est/11; Church Commissioners 101426 7/13, 2/13, 11/13.

[90] *Ag Hist III*, p 213.

[91] Biddick, *The Other Economy*, p 45; *Ag Hist III*, pp 190, 243, 267, 417–18; D L Farmer, 'Woodland and pasture sales on the Winchester manors in the thirteenth century: disposing of a surplus or producing for the market?', in Britnell and Campbell, *A Commercialising Economy*, pp 105–7, 112–13.

[92] *Ag Hist III*, p 298.

ratio of grassland to arable. As was consistent with a forage- rather than fodder-based pastoral sector, legumes – a key fodder crop – were grown on only a modest scale. True, oats, an alternative fodder crop, occupied a significant share of the sown acreage, but since these were only occasionally fed to sheep, and horses (which consumed them in greater quantity) were stocked in only modest numbers, it is probable that most of the oats harvest was destined for human consumption. In other respects there was little that was particularly distinctive about the cropping regime of these demesnes, which approximated remarkably closely to the national average (Table 2).

Not surprisingly, *sheep-corn husbandry* bore much in common with *mixed-farming with sheep* and in 22 cases this was the second-choice farming type (Fig 10). These were the demesnes with the smallest proportions of non-working animals, greatest proportions of sheep, and highest stocking densities. Eleven further demesnes had *extensive mixed-farming* as their second choice. These supported far lower stocking densities, carried smaller flocks, kept modest cattle herds, and made the greatest relative use of horses. They also grew the least barley and legumes and most oats. Fourteen others bore their closest resemblance to *arable husbandry with swine*. These were the only demesnes on which swine were of any significance and were also those which grew legumes on the largest and oats on the smallest scale. Other second choices were *extensive arable husbandry* (six demesnes), *light-land intensive* farming (two demesnes), and *intensive mixed-farming* (one demesne). This wide diversity of second choices, wider than that of any other farming system, testifies both to the variety of cropping regimes with which *sheep-corn husbandry* was associated (it is this that the regional distribution of second choices largely reflects – Fig 10) and the range of stocking densities by which *sheep-corn* demesnes were characterized. What endowed this farming system with the unity that enabled it to transcend this diversity were two overwhelming common denominators, namely retention of the ox as the predominant working animal and the domination of non-working animals by sheep.

Sheep-corn husbandry was already well established as a farming system before 1350, at which time stocking densities on sheep-corn demesnes were mostly lower and the emphasis upon sheep less overwhelming. After 1349 there was a real increase in both the sheep-farming component of this husbandry system and the number of sheep-corn demesnes (Figs 1 and 2). Geographically, it remained one of the more far-flung farming types, with core demesnes occurring in such widely separate locations as the East and West Ridings of Yorkshire, the Welsh border, the Sandlings of East Suffolk, east Kent, the Isle of Wight, and Somerset (Fig 2). But as *sheep-corn husbandry* became more common so it began to become the characteristic demesne-farming system of certain localities. This was most conspicuously the case in the downland and woodland country of southern England, especially in Hampshire and Wiltshire and adjoining portions of neighbouring counties.[93] Twelve of the 27 sampled Hampshire demesnes (many of them properties of the bishops of Winchester on which, as Martin Stephenson has shown, sheep numbers more than doubled between the beginning and end of the fourteenth century) practised some form of *sheep-corn husbandry*, as did five of the ten sampled Wiltshire demesnes, and four of the five sampled Dorset demesnes.[94] Moreover, the majority of the

[93] R Scott, 'Medieval agriculture', in R B Pugh, ed, *Victoria History of the County of Wiltshire, IV*, 1959, pp 19–21; J N Hare, 'Change and continuity in Wiltshire agriculture in the later Middle Ages', in W Minchinton, ed, *Agricultural Improvement: Medieval and Modern*, Exeter Papers in Economic History, 14, 1981, pp 4–9; *Ag Hist III*, pp 144, 292–8.

[94] M J Stephenson, 'Wool yields in the medieval economy', *Econ Hist Rev*, 2nd series, XLI, 1988, pp 368–91.

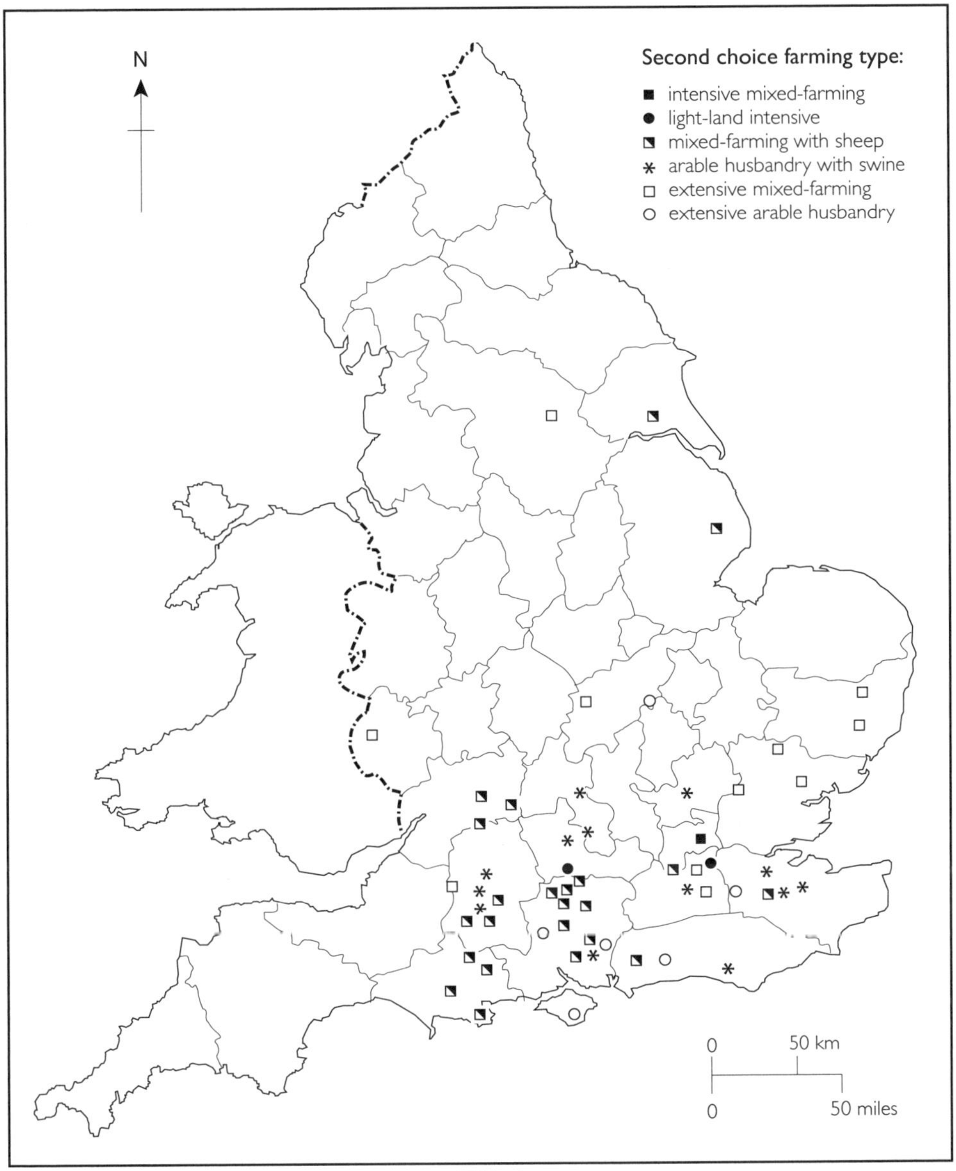

FIGURE 10

Sheep-corn husbandry demesnes, 1350–1449 (core and peripheral demesnes indicating second choice).

remaining sampled demesnes in these three counties practised *mixed-farming with sheep* (Fig 2). Other localities where *sheep-corn husbandry* was firmly established are the Gloucestershire Cotswolds and Sussex Downs.[95]

Sheep-corn husbandry is also well represented among the Feeding the City II sample of demesnes, with no less than 29 examples (Table 5 and Fig 3). Thus, several demesnes in the Hertfordshire Chilterns appear to have developed this specialism, as did a group of demesnes in the Vale of the White Horse in north-west Berkshire and a scatter of demesnes on or accessible to the North Downs of Kent and Surrey.[96] In contrast, at Langenhoe and Southchurch (Essex) and Sharpness and Ham (Kent) *sheep-corn husbandry* appears to have been a response to the pastoral potential provided by extensive areas of coastal marsh. Most notable of all, however, is the concentration of sheep-corn demesnes near London; at Eltham a few miles to the south-west in Kent, at Wandsworth just across the Thames in northern Surrey, at Hyde immediately west of the city in Middlesex, and at a string of demesnes in the Colne and Thames valleys. There are intimations of this metropolitan grouping at the beginning of the fourteenth century but by the end of that century it had evidently become much more pronounced. Whether *sheep-corn husbandry* developed here as a response to the capital's demand for wool, for mutton, or for the grain produced from fields dunged by the flocks nevertheless awaits investigation.[97]

Although an intrinsically extensive system of farming most typical of areas of comparatively low economic rent, the presence of so many sheep-corn demesnes so close to London does demonstrate that it could and did develop in areas of strong commercial demand. Yet with certain exceptions the immediate hinterland of the capital had never been noted for the intensity and productivity of its demesne agriculture, mainly because of the unrewarding character of the soils with which, for the most part, cultivators had to contend.[98] Instead, it was at some distance from the city that medieval agriculture attained its greatest intensity, in north-eastern Kent, and above all in Norfolk and adjacent portions of the east midlands. At the end of the fourteenth century it was here that the most intensive farming systems were still to be found with the result that *sheep-corn husbandry* was conspicuous by its virtual absence. There is not a single core or peripheral sheep-corn demesne in Norfolk, Cambridgeshire, Huntingdonshire, or the Soke of Peterborough, and only two in Suffolk (Figs 2 and 10). Putting Norfolk under the microscope using the entire available sample of 106 documented demesnes identifies just five that practised *sheep-corn husbandry* (Fig 4).[99] Each was widely separate from the others and appears to have developed its sheep-corn regime in response to essentially local circumstances that were either or both environmental (the availability of fenland, marshland, or heathland pastures) and institutional (seignorial rights of foldcourse over the fallow arable and associated permanent pastures).[100] Rather than adopt *sheep-corn husbandry per se*, Norfolk demesnes tended to develop their sheep farming within the context provided by either *mixed-farming with sheep* (12 demesnes) or *light-land intensive* farming (20 demesnes). Apart from

95 *Ag Hist III*, pp 264–5.

96 *Ibid.*

97 In the Dutch province of Drenthe, for instance, an expansion in arable production after 1650 was associated with declining numbers of cattle and rising numbers of sheep kept primarily for manuring: J Bieleman, 'Changing manuring techniques in open field farming in the Dutch province of Drenthe 1650–1850', in *I Jornadas Internacionales sobre Technologie Agraria Tradicional*, Museo Nacional del Pueblo Español, Madrid, 1992, pp 251–6.

98 *A Medieval Capital*, pp 138–40.

99 Barton Bendish, Blickling, Burgh in Flegg, Burston, and Holkham.

100 Campbell, 'Regional uniqueness'; Bailey, 'Sand into gold'; *Ag Hist III*, p 200; Campbell and Overton, 'A new perspective', pp 77–80; W Hassall and J Beauroy, eds, *Lordship and Landscape in Norfolk 1250–1350: The Early Records of Holkham*, 1993, pp 35–7, 52.

anything else the early replacement of the ox by the horse on the light soils of the county eliminated one of the most fundamental components of classic *sheep-corn husbandry*, which shows up better as the second-choice rather than first-choice farming type of a string of sheep-dominated mixed-farming demesnes on the light soils of the west of the county.

Cluster Six: 'extensive mixed-farming'

This is at once effectively the least well defined and most widely distributed farming type. Each of the three cluster techniques defines it differently, with the result that although K-means allocates 48 demesnes to this group, Ward's 71, and Normix 92, only 33 demesnes are common between them (Table 2). This core group is more than doubled in size by the addition of no less than 45 peripheral demesnes, thereby making *extensive mixed-farming* the single most common farming type (Table 3). The problem of definition derives in the main from the difficulty of distinguishing *extensive mixed-farming* from the other mixed-farming systems, particularly the *intensive mixed-farming* of Cluster One and *mixed-farming with sheep* of Cluster Three. This is highlighted by the fact that one or other of these alternative mixed-farming systems is the second-choice classification of all but three of these 78 demesnes (Fig 11). Moreover, the three exceptions are all peripheral demesnes (*sheep-corn husbandry* is the second choice of two, *extensive arable husbandry* the second choice of one). Numerically the pattern of second choices implies that *extensive mixed-farming* demesnes had most in common with *mixed-farming with sheep* but also a good deal in common with *intensive mixed-farming*. For this reason they show up in areas of both relatively intensive and extensive agriculture, in some of the economically most developed and economically most remote parts of the country, with the result that *extensive mixed-farming* is the most truly 'national' of all the farming systems (Figs 2 and 11). Nevertheless, as the pattern of second choices makes plain, it was a national system with important regional variations.

Overall, the core demesnes most diagnostic of this farming type were characterized by higher than average stocking densities, significantly above those of *intensive mixed-farming* demesnes but below those of *mixed-farming with sheep* demesnes (Table 2). Working animals tended to comprise a higher proportion of livestock units than in either of these other two systems, with oxen mostly of far greater importance than horses (although this was subject to quite wide variation). Among non-working animals it was cattle that predominated, and demesnes of this type generally carried quite well developed cattle herds which served the dual functions of dairying and rearing replacement oxen. Sheep were often kept but very much took second place to cattle. On the arable front these demesnes were characterized by a more equal balance between the winter- and spring-sown grains than in any other farming system. Moreover, this was the system in which wheat and oats both respectively assumed their greatest relative importance. By comparison, rye, barley, and the various winter and spring mixtures were of little significance. Legumes, too, occupied a smaller than average share of the sown acreage. Such a combination of crops is strongly suggestive of a three-course system of rotation of a fairly unimproved and unintensive nature.[101] It is also consistent with the cultivation of medium to heavy

[101] In Essex and Suffolk, for instance (the location of ten of the core demesnes), 'arable was usually fallowed one year in three in preparation for a winter-sown crop, to be followed the next year by a spring-sown crop': *Ag Hist III*, p 195. Three-course cropping also prevailed on the core demesne at Rimpton, Somerset: C C Thornton, 'The determinants of land productivity on the Bishop of Winchester's demesne of Rimpton, 1208 to 1403', in Campbell and Overton, *Land, Labour and Livestock*, pp 183–210. At Broadway, Worcs, however, two-course cropping appears to have been the norm: Gray, *English Field Systems*, p 503.

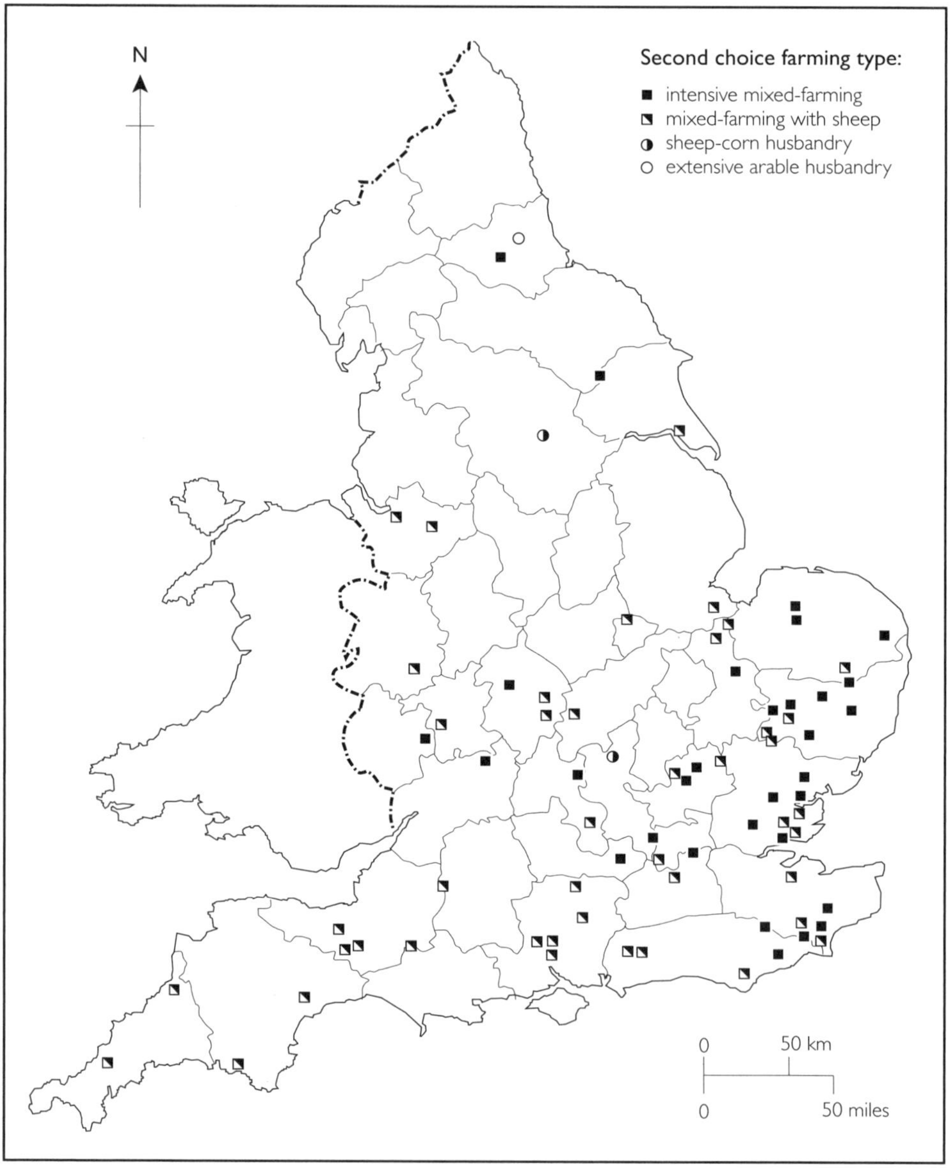

FIGURE 11
Extensive mixed-farming demesnes, 1350–1449 (core and peripheral demesnes indicating second choice).

soils, an impression reinforced by the strong pastoral emphasis upon cattle.[102] In many of these respects there is a strong affinity between this farming system and the *mixed-farming with cattle* of the pre-Black Death period whose association with the heavier lowland soils of counties such as Essex and Warwickshire was especially marked (Fig 1).[103]

The arable component of *extensive mixed-farming* was in a very real sense its most consistent feature. Both the core and the peripheral demesnes, those whose second choice was *intensive mixed-farming* and those for which it was *mixed-farming with sheep*, sowed the same basic combination of crops. On all of them wheat and oats were the two leading crops, all maintained a fairly equal balance between the winter and spring grains, and none differed greatly in the share of the cropped acreage devoted to legumes. Those demesnes which departed furthest from this general pattern of cropping were those peripheral demesnes which most resembled *intensive mixed-farming*. The latter are the demesnes with the strongest bias towards spring grains and which grew legumes on the largest scale. Nevertheless, the winter grains still occupied at least a third of the sown acreage and legumes only an eighth. Differences of cropping regime between *extensive mixed-farming* demesnes were therefore remarkably slight. Of far greater moment were the differences which existed in the pastoral component of husbandry.

Extensive mixed-farming demesnes varied first and foremost in the relative size of their arable and pastoral sectors as measured by the ratio of livestock units to cropped acres. Stocking densities were lowest of all, and the bias towards arable husbandry consequently greatest, on those peripheral demesnes with the closest resemblance to *intensive mixed-farming*. Conversely, they were highest on those peripheral demesnes whose closest affinity was with *mixed-farming with sheep*. On the former group of demesnes stocking densities were generally well below the national average, on the latter they were well above. Somewhat paradoxically, working animals – horses and oxen – accounted for a smaller share of total livestock units on demesnes with a second choice of *intensive mixed-farming* than on those with a second choice of *mixed-farming with sheep*, the reason being that horses made their greatest relative contribution to draught power on the former. The more intensive demesnes were also those with the most developed cattle herds, as manifest in a demographic structure strongly biased towards mature females and a far greater relative importance attached to cattle than other categories of animal. Where pastoral husbandry was more extensive, for instance, the specialist interest in cattle-based dairying was less developed and sheep comprised a larger share of livestock units. Even so, it was these more extensive forms of pastoral regime which were associated with the greatest numbers of livestock units.

The core demesnes whose traits most embody this husbandry system are geographically to be found in some of both the most and least economically developed parts of the country (Fig 2). Two-thirds were located in the counties of the south and east of England, with particularly notable concentrations in Essex (seven examples) and southern Kent/East Sussex (six examples). This south-eastern distribution of *extensive mixed-farming* demesnes is consistent with the distribution of demesnes practising *mixed-farming with cattle* in the period 1250–1349 and implies considerable continuity of farming system on

[102] At Marley, Sussex, on heavy clay, cattle farming was Battle Abbey's prime concern, underpinned, apparently, by a system of convertible husbandry: E Searle, *Lordship and Community: Battle Abbey and its Banlieu, 1066–1538*, Toronto, 1974, pp 272–303. Pastoral farming also loomed large on the Romney Marsh manors of Appledore, Ebony, and Ruckinge, Kent, where oats were typically sown on reclaimed marshland: *Ag Hist III*, p 131; R A L Smith, *Canterbury Cathedral Priory: A Study in Monastic Administration*, 1943, pp 181–3, 188.

[103] Power and Campbell, 'Cluster analysis', pp 236–7, 240–1.

many demesnes in this area (Fig 1).[104] Similar continuity appears to have prevailed in parts of the west midlands, for here too a loose concentration of *extensive mixed-farming* demesnes is present in an area where *mixed-farming with cattle* had been well represented in the earlier period. The spatial coverage of the sample is unfortunately thinner in the period 1350–1449, but five *extensive mixed-farming* demesnes do nevertheless show up at Mathon (Herefordshire), Knowle (Warwickshire), Broadway and Leigh (Worcestershire), and Cleobury Barnes (Shropshire), the first three more intensive than extensive in character. Elsewhere, *extensive mixed-farming* shows up on a loose scatter of demesnes in southern England – all of them more extensive than intensive in character – including three in south Somerset. The only two recorded Cheshire demesnes were also both of this type, as was Howsham in the East Riding of Yorkshire. Core demesnes of this type are, however, absent from Norfolk, the east midlands, and much of south central England.

Adding the peripheral demesnes greatly extends and amplifies this picture (Fig 11). Not only are the existing concentrations in the south-east and the west midlands, reinforced, but several further examples are added both in Somerset and in the north-east.[105] *Extensive mixed-farming* also shows up in areas where it is otherwise unrepresented by core demesnes, most notably in Devon and Cornwall (where there are four peripheral demesnes) and Norfolk (where there are six). The combined distribution of core and peripheral demesnes reveals a spatially significant pattern of second choices (Fig 11). In southern and south-western England, for example, *extensive mixed-farming* was most closely related to *mixed-farming with sheep*. East Anglian examples of *extensive mixed-farming* demesnes, in contrast, mostly bore a closer resemblance to *intensive mixed-farming*. Exceptions to the latter rule are mostly explicable in terms of local circumstances. Wisbech, Gedney and West Walton in the East Anglian Fens, for instance, all have the more extensive *mixed-farming with sheep* as their second choice, a reflection, no doubt, of the opportunities for extensive pastoral husbandry offered by the abundance of marshland grazing which each enjoyed. A superabundance of pasture may explain why similarly extensive versions of *extensive mixed-farming* also prevailed at Burstwick in Holderness, adjacent to the Humberside marshes, at Eastwood, Lawling, and Milton in Essex, with access to coastal marshes, and at Appledore and Barksore in Kent, both strategically placed to take advantage of the lush alluvial grazings available in Romney Marsh.

Analysis of the Feeding the City II sample of demesnes brings the distribution of *extensive mixed-farming* demesnes within the south-east into sharper focus (Fig 3). Although represented in greater numbers than demesnes of any other farming type, the 37 examples of *extensive mixed-farming* – 30 per cent of the total – are geographically highly specific in their distribution. No less than 22 are concentrated in Essex (plus Sawbridgeworth just across the county boundary into Hertfordshire) and six in Kent. In Essex *extensive mixed-farming*, generally in its more intensive form, was plainly the dominant farming type, practised by three out of four documented demesnes. Examples occur throughout most parts of the county and are particularly well represented on the heavy boulder clay soils which dominate so much of its centre.[106] Here, if further proof were

[104] Power and Campbell, 'Cluster analysis', pp 236–7, 240–1; *Ag Hist III*, pp 195–6, 206–7, 282–3.

[105] Elvethall and Witton (Durham), Burstwick and Methley (East and West Ridings of Yorkshire).

[106] R H Britnell, 'Agriculture in a region of ancient enclosure, 1185–1500', *Nottingham Medieval Studies*, XXVII, 1983, pp 37–55. For a case study of one Essex manor see, R H Britnell, 'Agricultural technology and the margin of cultivation in the fourteenth century', *Econ Hist Rev*, 2nd series, XXX, 1977, pp 53–66.

wanted, is clear evidence of a direct association between *extensive mixed-farming* and heavy clay soils. It is this same association which accounts for the distribution of this farming type in Kent, where it is exclusively confined to the wealden clays of the southern half of the county and the immediate environs of Romney Marsh. Elsewhere, apart from three examples in the Colne valley, *extensive mixed-farming* is not a particularly common farming type.[107] Nor is it well represented in Norfolk, where the application of discriminant functions identifies only 13 of the 106 documented demesnes as belonging to this farming type (Fig 4). These demesnes are more scattered than concentrated in distribution but do display a general bias towards localities of heavier soil or extensive alluvial grazings which coincidentally also tended to be those localities where oxen remained of significance longest.[108]

Cluster Seven: 'extensive arable farming'

By the end of the fourteenth century pastoral farming had grown in importance on most demesnes with the result that there were relatively few on which arable husbandry remained the almost exclusive focus of attention. Whereas before 1350 one in five of all sampled demesnes carried virtually no livestock other than essential working animals, after 1349 the equivalent proportion is only one in ten.[109] These are the most 'arable' of all demesnes and, as a farming type, are immediately recognizable. No other group of demesnes is more consistently identified by all three clustering methods. Ward's, K-means, and Normix all identify a common core of 27 demesnes, which only in the case of Normix is enlarged by the inclusion of two others. The net result is a group of 28 demesnes with a core membership of 27 and a peripheral membership of one (Tables 2 and 3). This group is defined less by the combination of crops which it grew – which differed in no significant respect from the mean pattern prevailing in the country as a whole – than by the scale and composition of its pastoral sector. The latter was small and composed almost exclusively of working animals – horses and oxen (with, on average, twice as many of the latter as the former). Stocking densities, consequently, were the lowest of any farming type. Arable production may have been everything but with such a deficiency of livestock high levels of productivity were unattainable. Production was perforce extensive.

One reason why patterns of cropping on *extensive arable husbandry* demesnes approximated so closely to the national average is that like *sheep-corn* demesnes (the other farming type of which this is true) they occurred in a wide variety of farming contexts. The 28 demesnes in question were scattered through 17 different counties ranging from Durham in the north to Hampshire in the south, from Norfolk in the east to Somerset in the west (Figs 2 and 12). As with *arable husbandry with swine* the explanation for this wide dispersal lies partly with institutional factors.

Rather than manage their flocks and herds directly some landlords chose to lease them out and keep only their arable land in hand. Others organized their sheep flocks on an intermanorial basis and accounted for them separately. The result in both cases is that some or all of the non-working animals disappear from the manorial accounts. Twelve of Norfolk's 106 documented demesnes fall into the *extensive arable husbandry* category and most probably do so for this reason (Fig 4). So conspicuously different from other demesnes are they in this otherwise intensively farmed

[107] Denham and Iver (Buckinghamshire) and Yeoveney (Middlesex).

[108] Especially Thorpe Abbotts, Tivetshall, Topcroft, and Rickinghall (Suffolk) on the heavy soils of south-east Norfolk and north Suffolk, Kempstone and Mileham on the boulder-clay soils of mid-Norfolk, West Walton in the Fens and Boughton on the Fen edge, and Halvergate in the Broadland marshes.

[109] Power and Campbell, 'Cluster analysis', p 238.

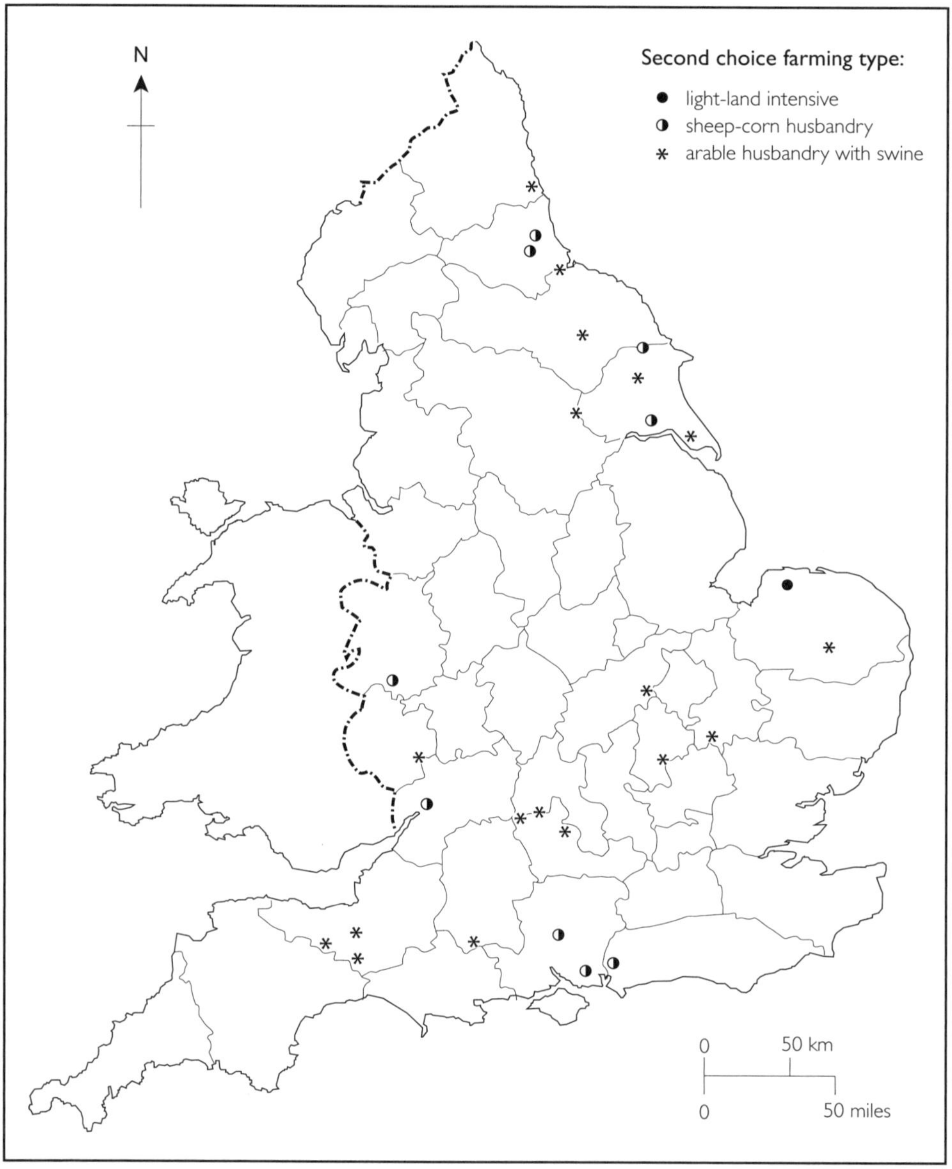

FIGURE 12
Extensive arable husbandry demesnes, 1350–1449 (core and peripheral demesnes indicating second choice).

county that they retain their distinctive identity even at a county level of analysis as the third of Norfolk's three basic farming types, the others being versions of *intensive mixed-farming* and *light-land intensive* farming (Table 6, Fig 5). The same explanation probably applies to the six examples of this farming type within the Feeding the City

II sample of demesnes (Table 5, Fig 3). Nevertheless, in parts of the north-east of England *extensive arable husbandry* more probably represents a genuine economic preference.

Ten of the 28 *extensive arable-husbandry* demesnes are located in Durham and Yorkshire where they comprise just over half of all sampled demesnes (Fig 12). This is a significant concentration, the more so since an equivalent concentration of *extensive arable-husbandry* demesnes is apparent in this self-same part of the country in the period 1250–1349 (Fig 1). As before, a spatial division of production appears to have prevailed with pastoral production concentrated in the main on the uplands and upland margins, often on specialist vaccaries and bercaries, and lowland demesnes, particularly on the larger estates, only carrying enough animals to satisfy the draught requirements of the arable in which branch of production they in turn specialized.[110] This was especially the case on estates which combined properties in both locations since replacement draught animals could simply be transferred in from the reservoir of animals maintained on the stock farms. Whereas demesnes elsewhere in England were becoming more mixed, many of those in the north-east maintained their traditional specialism in one or other branch of production with the result that the swing towards pastoralism is less apparent than might otherwise have been expected of an area naturally so well endowed with pastures. *Extensive arable-husbandry* demesnes at Stanton Lacy, Shropshire and Durneford, Herefordshire may possibly reflect a similar upland/lowland separation of pastoral and arable production in the counties of the Welsh border.[111]

III

Of the seven basic demesne-farming systems that have been identified in the period 1350–1449 some were more distinctive and some more homogeneous than others.[112] Cluster analysis has particular utility here. Not only does it provide a statistical method of classifying farming systems based upon a consistent set of variables and contain a procedure for establishing the number of farming types present (thereby ensuring that classification reflects real similarities and dissimilarities within the data), but it can also be used to establish which farming systems are most alike or unlike each other.

Extensive arable husbandry and *sheep-corn husbandry* are the two farming systems most consistently defined by all three clustering methods. Significantly, the same two farming types are equally clearly defined in the period 1250–1349. After 1349 the K-means, Normix, and Ward's methods approach virtual unanimity in their identification of *extensive arable-husbandry* demesnes. So conspicuously different is this farming type that, as independent classification of the 106 Norfolk demesnes demonstrates, it shows up at a county as well as a national level of analysis. Whereas *extensive arable husbandry* is represented within the national sample by 27 core demesnes, *sheep-corn husbandry* is represented by 45. The larger number obviously provides greater scope for differences between the outcomes obtained from the three chosen clustering methods. Even so, there is a close correspondence between all three solutions. Nor are significant numbers of peripheral demesnes added to either farming category on the basis of their discriminant scores.

Each of these two farming systems is immediately and consistently defined because the demesnes that practised them

[110] I S W Blanchard, 'Economic change in Derbyshire in the late Middle Ages, 1272–1540', unpublished PhD thesis, University of London, 1967, pp 168–74; *Ag Hist III*, pp 189–90.

[111] Davies, *Lordship and Society*, pp 112–17; *Ag Hist III*, pp 241–4.

[112] This analysis does not, of course, preclude the existence of other demesne-farming types not adequately represented in the available sample of accounts.

display certain features that are markedly atypical. With *extensive arable husbandry* this is its stunted pastoral sector composed almost exclusively of working animals; in the case of *sheep-corn husbandry* it is the predominance of oxen among working, and sheep among non-working animals. By comparison, the arable profiles of these two systems are far less distinctive and actually transcend differences in cropping regime, for geographically both systems occur in a variety of arable contexts; hence the wide range of alternative farming systems with which individual sheep-corn demesnes shared strong affinities. In contrast, *light-land intensive* husbandry was characterized by one of the most distinctive and environmentally specific of all arable regimes. No other system devoted so small an acreage to wheat and legumes and so large an acreage to rye. Yet (and contrary to the situation before 1350), this is the least consistently defined of all farming systems with only a relatively small core of demesnes common to all three cluster solutions. This is partly because of a high degree of variation in the precise form of husbandry practised by each constituent demesne but also because this farming system shared a number of features in common with several other farming systems (as witnessed by the fact that the 14 *light-land intensive* demesnes display three different second choices). Nevertheless, it is sufficiently distinctive to show up at a county as well as a national level of analysis and geographically displays a general association with a particular region and a specific association with a particular type of environment. There is therefore considerable merit distinguishing it as a separate system.

The remaining farming systems, embracing 70 per cent of all sampled demesnes, fall between these two extremes of definition. Apart from *arable husbandry with swine*, whose pastoral component was comparatively small, all these systems are mixed-farming systems of one sort or another and all display strong affinities with each other. Distinguishing between the almost seamless gradation of mixed-farming types is the main challenge to the derivation of an effective classification of farming systems in this period. In fact, a stable cluster solution based on the first six principal component scores lumps all these demesnes into a single mixed-farming category.[113] This is unhelpful. Genuine, if often subtle, differences did exist in the character, intensity, and distribution of mixed-farming systems and it is the task of any effective classification system to draw meaningful distinctions between them. That is why it was found more effective in this instance to base classification upon five rather than six principal component scores. Use of five rather than six component scores focuses upon broader differences which blur when the additional detail of the sixth component is included.

A striking feature of the three main mixed-farming systems identified here – *intensive mixed-farming*, *mixed-farming with sheep*, and *extensive mixed-farming* – is that although each shared certain features in common with the others, commonalties with other farming systems are far less marked. In fact, taking all three mixed-farming systems together, only 3 per cent of the 93 core demesnes and 6 per cent of the 187 core plus peripheral demesnes have a second choice other than one of these mixed-farming systems. *Intensive mixed-farming* stands out as the most clearly and consistently defined of the three systems, with the greatest correspondence in cluster membership between the three cluster solutions. Its intensive character endowed it with several extreme features – the scale of barley cultivation, specialist dairy function of its cattle herds, and majority reliance upon horse power – which help to differentiate it from the other mixed-farming

[113] Above, p 139.

systems. *Mixed-farming with sheep*, in contrast, exhibits fewer exceptional features, constitutes a system of middling intensity, and consequently is far less easily defined. All three clustering techniques identify substantial numbers of demesnes of this type, but the core of demesnes that is common between them is smaller than for any other mixed-farming type. Nor is *extensive mixed-farming* much better defined, for all the distinctiveness of its cropping regime and the large number of demesnes that practised some version of it. Difficult though it is to draw a line between these three mixed-farming systems, the result is nonetheless well worth the effort for these are among the most coherent of farming types in geographical distribution. Of no farming system is this truer than *intensive mixed-farming*, which is more closely associated with a single specific region than any other post-1349 farming type.

Those farming systems which, like *intensive mixed-farming*, were spatially most sharply focused are those which owed most to specific environmental and economic circumstances and which embodied regionally specific forms of technology (most notably a reliance upon horses rather than oxen for draught power, which, as Langdon has observed, was a major source of regional differentiation within medieval agriculture).[114] But farm enterprise was a function of more than environmental, economic, and technological factors. What type of landlord, what size of estate, and the position of the individual demesne within the overall estate production system were also important considerations. Institutional factors such as these all exercised a far greater influence upon the pastoral than the arable sector. As analysis of the Feeding the City II sample of accounts demonstrates, conventual, collegiate, episcopal, lay, and royal estates differed in their capacities and willingness to invest in and maintain flocks and herds.[115] Perpetual institutions, such as convents and colleges, clearly enjoyed an edge over episcopal and lay estates which were vulnerable to asset stripping whenever the current incumbent died. Individual estate policy also exercised a bearing upon the choice of animals stocked and whether or not these were managed directly at demesne level, centralized and accounted for separately, or leased out. Those farming systems which exhibit the strongest imprint of the operation of these kinds of institutional factors – *arable husbandry with swine* and *extensive arable husbandry* – are also those which are geographically for the most part the least specific in distribution. As pastoral farming became more important so institutional factors became of correspondingly greater significance in determining the geographical distribution of farming types.

Under these circumstances it is hardly surprising that when farming systems are viewed collectively rather than individually the picture that emerges is more heterogeneous than homogeneous (Figs 2 and 3). Few types of farm enterprise were unique to a locality or region with the result that many demesnes remote from one another had more in common than those that were neighbouring. There is a lesson here for those who would seek to define and describe farming types in exclusively regional terms. Unfortunately, the national sample of demesnes is mostly too sparse in coverage to reveal local and regional configurations of farming types in sufficiently sharp focus (Fig 2). Only the Norfolk and portions of the Feeding the City II samples of demesnes are adequate for this purpose (Figs 3 and 4). From these it is clear that individual localities were sometimes dominated by particular farming types. On the boulder-clay plateau of central Essex, for instance, few demesnes deviated from some

[114] Langdon, *Horses, Oxen and Technological Innovation*, pp 273–6.

[115] Campbell, 'Measuring the commercialisation of seigneurial agriculture', pp 142–7.

version of *extensive mixed-farming*, and in eastern and northern Norfolk *intensive mixed-farming* was the preferred system of most demesnes. The latter example is particularly instructive for the Norfolk evidence is exceptionally full, allowing the agricultural geography of this one county to be seen more clearly than that of any other part of the country. During the period 1350–1449 a system of intensive mixed farming crystallized here on light to medium soils which was almost unique to this county. Forty-two of the 106 documented demesnes employed this system and 20 of the remainder, mostly on the county's lightest and poorest soils, followed a light-land alternative.[116] Reclassifying Norfolk demesne-farming systems at county level elevates these two farming types to almost complete hegemony, with three out of four demesnes practising a version of *intensive mixed-farming* and one out of five a version of *light-land intensive* farming. Nevertheless, it is the national-level classification which is the more detailed and subtle and captures most fully the range of farm enterprise within the county. It serves to bring home the intrinsically heterogeneous character of late medieval farming systems, for examples of all seven national farming types occur within this one county, often as many as four or five of them within a few miles of each other. As Mark Overton found in the same county at the end of the sixteenth century '[farming] types do not form discrete regions so there is little point in drawing boundaries'.[117] Paradoxically, therefore, although the form of mixed farming practised on most late fourteenth-century Norfolk demesnes stands out as the only truly regional farming type, its distribution is more accurately represented by points than lines on the map. This bears out E L Jones' observation that 'lines drawn on the topographical map ... must arbitrarily bisect interesting zones of overlap and may distract attention from them'.[118] Certain localities and regions, in fact, were distinguished more by the variety than the uniformity of farming systems that existed within them. This was conspicuously true of the Thames valley – one of the country's key economic arteries – and of Kent, a county which presented a range of environmental and economic opportunities (Fig 3).

Norfolk and the counties surrounding London together demonstrate that whereas the more extensive systems of husbandry might be found in almost any part of the country, even, on occasion, in areas of apparently high economic rent, the more intensive systems were much more circumscribed in distribution. Norfolk exhibited the full range of national farming types because it remained relatively populous and there were strong (if albeit waning) economic incentives to maintain more specialized and intensive forms of agriculture. The same, to a degree, was true of the immediate hinterland of London (whose reduced post-plague population nevertheless probably represented a greater share of the national total and continued to constitute a major focus of demand), where a correspondingly wide range of farming types may be identified. In economically less developed and commercialized parts of the country, however, a narrower range of farming types appears to have existed since the forces promoting economic differentiation were much weaker. In particular, the most intensive systems were either rare or absent.

[116] Most of these demesnes also employed wheeled ploughs and sowed their seed relatively thickly.

[117] M Overton, *Agricultural Regions in Early Modern England: An Example from East Anglia*, University of Newcastle upon Tyne, Department of Geography seminar paper, XLII, 1983, p 18.

[118] E L Jones, 'The condition of English agriculture', p 616.

Compared with the situation before 1350 there was certainly a shift in the relative importance of these seven main farming systems. Demesnes practising *light-land intensive* and *extensive arable husbandry* both became relatively less common. *Intensive mixed-farming* lost ground and *arable husbandry with swine* merely maintained its minority status. The more extensive mixed-farming systems, however, became both more numerous and more mixed (insofar as their stocking densities tended to register the greatest increases). *Sheep-corn husbandry* – a particularly extensive system – gained most of all. Nowhere was the growth of mixed farming more pronounced than in central and southern England. It was here, therefore, in the half century or so immediately following the Black Death that farming systems changed most and the greatest gains in demesne stocking densities accrued.[119] These areas stood in the vanguard of the shift from corn to horn – as they were to do again in comparable periods of agrarian change – for the simple reason that they were best placed environmentally and economically to switch resources from arable to pastoral production.[120] Above all this meant converting arable land to pasture. Further east, and especially on the lighter soils of East Anglia, the mixed-farming systems that had developed were traditionally both more intensive and more arable based, insofar as they were underpinned by fodder cropping and temporary grazing/folding on the arable.[121] Demesne producers here lacked an equivalent comparative advantage when it came to substituting more extensive and, consequently, more grass-based mixed-farming systems. They had prospered when the demand for grain was strong and now, as in all subsequent agricultural recessions, suffered as the terms of trade shifted in favour of low-cost pastoral producers.[122] Demesne-farming systems in the west and north were similarly slow to change.[123] Here methods of pastoral farming had always been relatively extensive and an inherent land-use bias towards grass meant that good arable land remained at something of a premium. There was therefore no great incentive to alter the established balance of production.

IV

By the close of the fourteenth century, therefore, the country's agricultural geography had been subtly but profoundly transformed. Underpinning this transformation was a reconfiguration of economic rent, for it was upon this that the pattern and intensity of agricultural land use depended.[124] Economic rent is a function of land quality, of population density (and therefore the demand for land), and of distance from the market, and in all three respects the later fourteenth century witnessed important changes. Changing factor and commodity prices played to the comparative advantage of some regions more than others. Population densities fell but the rate of fall was spatially uneven due undoubtedly to local and regional differen-

[119] Campbell, 'Land, labour, livestock and productivity trends', pp 158–9.

[120] For the expansion of grass at the expense of arable after 1349 in the midlands see, *Ag Hist III*, pp 78–80; C Dyer, *Warwickshire Farming 1349–c 1520: Preparations for Agricultural Revolution*, Dugdale Society Occasional Papers, 27, 1981, pp 9–12. For the shifting frontier between corn and horn in the seventeenth century, see Kussmaul, 'Agrarian change in seventeenth-century England'.

[121] The story on the heavy clay soils of East Anglia was rather different, although it was in the fifteenth century rather than the fourteenth that their conversion to grass made greatest progress. For a case study of land-use trends in Essex over this period see L R Poos, *A Rural Society after the Black Death: Essex 1350–1525*, 1991, pp 46–51.

[122] Cf J D Chambers and G E Mingay, *The Agricultural Revolution 1750–1880*, 1966, p 181; P J Perry, 'Where was the "Great Agricultural Depression"? A geography of agricultural bankruptcy in late Victorian England and Wales', *AHR*, 20, 1972, pp 30–45.

[123] *Ag Hist III*, pp 45–8, 152–63.

[124] P Hall, ed, *Von Thünen's Isolated State: An English Edition of Der Isolierte Staat by Johann Heinrich von Thünen*, trans C M Wartenberg, 1966; M Chisholm, *Rural Settlement and Land-use: An Essay on Location*, 1962, pp 20–32; D Grigg, *The Dynamics of Agricultural Change: The Historical Experience*, 1982, pp 135–40; *A Medieval Capital*, pp 4–7.

tials in fertility and mortality and the redistribution of population by migration.[125] Towns for the most part contracted in size, the composition of urban demand for rural provisions altered, and the extent and orientation of their provisioning hinterlands were redrawn.[126]

If the pattern of farming systems mapped here represents the response to these changes how should it be interpreted? It certainly represents a reversion to more land extensive and especially more pastoral forms of agriculture, and as such was consistent with prevailing economic trends. But as demesne managers increasingly counterbalanced grain production with animal husbandry were they also reverting to a form of agriculture in which estates on the one hand and localities and regions on the other were increasingly self-sufficient in what they produced? Certainly, as pressure eased off the land more localities and regions would have been able to meet their consumption requirements internally from within their own resources. Whereas *c* 1300 it is possible to recognize the clear impact of growing centres of concentrated urban demand, both at home and overseas, upon the pattern of farming systems, by the end of the fourteenth century such centripetal influences had diminished in scale and their impact is far less self-evident.[127] The very fact that demesne agriculture became less differentiated suggests that incentives to specialize and intensify had weakened. The new agricultural landscape that emerged therefore appears more Ricardian then Thünenesque, its configuration influenced more by land quality and local demand for the land and its products than by distance from major markets.[128] The latent patterns of specialization and intensification that may be detected at the climax of medieval economic expansion *c* 1300 had mostly fallen into abeyance in an agrarian world seemingly reoriented along more local and regional lines. This, however, is almost certainly to underestimate the commercial potential of animals and animal products, both of which were capable of being marketed at a far greater range than grain.[129]

If the orbit of grain markets became more circumscribed it does not necessarily follow that the same applied to the markets for live animals, for dairy products, hides, skins, and wool.[130] In fact, an active trade in live animals undoubtedly helped sustain the expansion of flocks and herds that was such a feature of this period. Most demesnes relied upon the market for replacement work horses, for the plough and especially for the cart, and the same often applied to oxen. The conduct of pastoral husbandry also regularly generated surplus animals for sale: redundant, decrepit, and sickly animals requiring replacement, surplus calves and lambs from specialist herds and flocks, animals purpose-bred for sale and others fattened for meat. On what scale that trade was conducted and over what distances remains to be established, but neither is likely to have

[125] The changing distribution of population is implicit in the changing distribution of taxable wealth: R S Schofield, 'The geographical distribution of wealth in England, 1334–1649', *Econ Hist Rev*, 2nd series, XVIII, 1965, pp 483–510; H C Darby, R E Glasscock, J Sheail, and G R Versey, 'The changing geographical distribution of wealth in England 1086–1334–1525', *J Hist Geog*, 5, 1979, pp 257–61; A Dyer, *Decline and Growth in English Towns, 1400–1640*, 1991, pp 40–2.

[126] Dyer, *Decline and Growth in English Towns*, pp 20–4; D Keene, *Cheapside before the Great Fire*, 1985, pp 19–20; Dyer, *Standards of Living*, pp 199–202; Galloway *et al*, 'Changes in grain production and distribution'; *Ag Hist III*, pp 372–3.

[127] Power and Campbell, 'Cluster analysis', pp 242; *A Medieval Capital*, pp 172–83.

[128] For the difference between Ricardian and von Thünen economic rent see, Grigg, *The Dynamics of Agricultural Change*, pp 50–1, 135–40.

[129] T H Lloyd, *The English Wool Trade in the Middle Ages*, 1977; D L Farmer, 'Marketing the produce of the countryside, 1200–1500', in *Ag Hist III*, pp 377–408; Overton and Campbell, 'Norfolk livestock farming', pp 377–8, 393–4.

[130] Lloyd, *The English Wool Trade*; M Kowaleski, 'Town and country in late medieval England: the hide and leather trade', in P J Corfield and D Keene, eds, *Work in Towns 850–1850*, Leicester, 1990, pp 57–73.

been inconsiderable.[131] Whether it was sufficient to maintain established levels of commercialization within the demesne sector is another matter.[132] Nevertheless, in the ongoing debate about the character and trend of the late medieval economy the classification of demesne-farming systems outlined here should provide an appropriate framework for a more critical evaluation of these and other issues.[133]

[131] Campbell, 'Measuring the commercialisation of seigneurial agriculture', pp 142–3, 148–9, 152–3, 163–74; I Blanchard, 'The Continental European cattle trade, 1400–1600', *Econ Hist Rev*, 2nd series, 39, 1986, pp 428–31; M K McIntosh, *Autonomy and Community: The Royal Manor of Havering, 1200–1500*, 1986, pp 141–3.

[132] Campbell, 'A fair field', pp 68–9.

[133] For recent contrasting views of the period compare R H Britnell, *The Commercialisation of English Society 1000–1500*, 1993, pp 179–203; S R Epstein, 'Regional fairs, institutional innovation, and economic growth in late medieval Europe', *Econ Hist Rev*, 2nd series, XLVII, 1994, pp 441–58.

XIII

Inquisitiones Post Mortem, GIS, and the creation of a land-use map of medieval England

Ken Bartley and Bruce M.S. Campbell

1. Introduction

Today, maps are an accepted means of recording and representing data. Not so in the Middle Ages, when the art of cartography was in its infancy. The Domesday Survey, an inherently spatial exercise, relies upon words rather than maps to record its vast compendium of information. Operating with hand-tools, it took the late H. C. Darby and his team of collaborators 25 years of painstaking effort to reconstruct the social and economic geography of post-Conquest England from this exclusively verbal compilation (Darby 1977: 375–84). The Domesday Book may be England's most celebrated public record, but the climax of medieval record creation actually came 250 years later, during the successive reigns of the first three Edwards (1272–1377) (Clanchy 1979). No other country can match the mass of public and private records which survive from the 80 years or so before the Black Death of 1348, virtually all of them written on parchment in abbreviated Latin using Roman numerals. Collectively these sources make feasible the reconstruction of the geography of England at a key time, following two centuries of more-or-less sustained expansion and growth and on the eve of the greatest demographic catastrophe in recorded English history. Modern computerized methods of data collection and analysis mean that the task can be undertaken in a fraction of the time required by the Darby team (Darby 1977), while the application of a GIS means that it can be taken much further. In particular, the development of a multivariable land-use classification is a real possibility. Such a classification is not only of interest in itself but can supply a much-needed context for in-depth case studies of individual manors, estates, localities, and farming systems.

2. The *Inquisitiones Post Mortem*: problems and potential

Among pre-Black Death records of national scope it is the *Inquisitiones Post Mortem* (*IPM*s) which are most comparable to the Domesday Survey in the range and nature of the information they contain (Campbell et al 1992, 1993: 17–18). No other source contains as much explicit land-use information until the tithe files of the early-nineteenth century. The *IPM*s, as their name suggests, were the product of official enquiries conducted by royal escheators following the death of a tenant-in-chief of the Crown (i.e. lay lords who held their estates directly from the king) (Stevenson 1947). Typically, this involved making a detailed extent (i.e. survey) of each of the deceased tenant's manors or properties. A local jury supplied the information on oath. Particular attention was paid to the physical extent and value of the tenant-in-chief's own demesne lands, since the Crown stood to gain directly from their management if a tenant died without heir or the heir was either female or under-age. The quantity of information required and supplied was often considerable, although individual escheators and local juries naturally varied a good deal in how they interpreted their remit. Escheators were busy men with wide territorial responsibilities and relied upon a team of county-based sub-escheators backed up an appropriate number of clerks to get their job done. On average, an *IPM* return was being compiled somewhere in the realm each working weekday throughout the first half of the fourteenth century. The documentary legacy which this has bequeathed is formidable.

Those unfamiliar with the period are often surprised by the efficiency of medieval government administration, whose arm was capable of reaching to the furthest corners of the realm. For instance, when Gilbert de Clare, earl of Hertford and Gloucester was killed on 24 June, 1314, in an ill-fated charge against the main Scots army at the Battle of Bannockburn; a well-practised administrative process was activated (Prestwich 1980: 54). Within a little over two weeks writs had been issued to the royal escheators in England south of the Trent, Wales, and the Lordship of Ireland to inquire of what lands and estates the earl was possessed (Public Record Office, London: C/134/41, 42, 43). The first jury to report was on 1 August 1314 at Chipping Campden in Gloucestershire. The administrative process continued over the next three years until the estate was finally settled in November 1317 (Holmes 1957). De Clare was a great earl with far flung estates – few subjects in the land were mightier – hence no less than 161 *IPM* extents resulted from his untimely death. Yet these pale in number against the 9,291 extents (plus a further 322 for the Anglo-Norman lands in Wales and 324 for lands in the Lordship of Ireland) created over the half century 1300–1349 and still preserved at the Public Record Office, London. Collectively they provide at least some land-use information for approximately one in three of all townships in the kingdom.

Analysing the IPMs presents a number of problems. First, in order to obtain a geographical coverage capable of producing meaningful results it is necessary to take a 50–year time span. This inevitably introduces a degree of temporal bias to any result, not least because of administrative changes in the nature of the IPM operation. As with any revenue-raising exercise, the escheators varied in the efficiency and rigour with which they did their work on the ground. The requirements imposed upon them by the officials at the administrative centre were also periodically revised. Driving the whole operation were the financial exigencies of the Crown. Thus, surviving numbers of IPMs peak in the opening decade of the fourteenth century when the financial pressures of Edward I's various military enterprises were at a maximum. It was at that

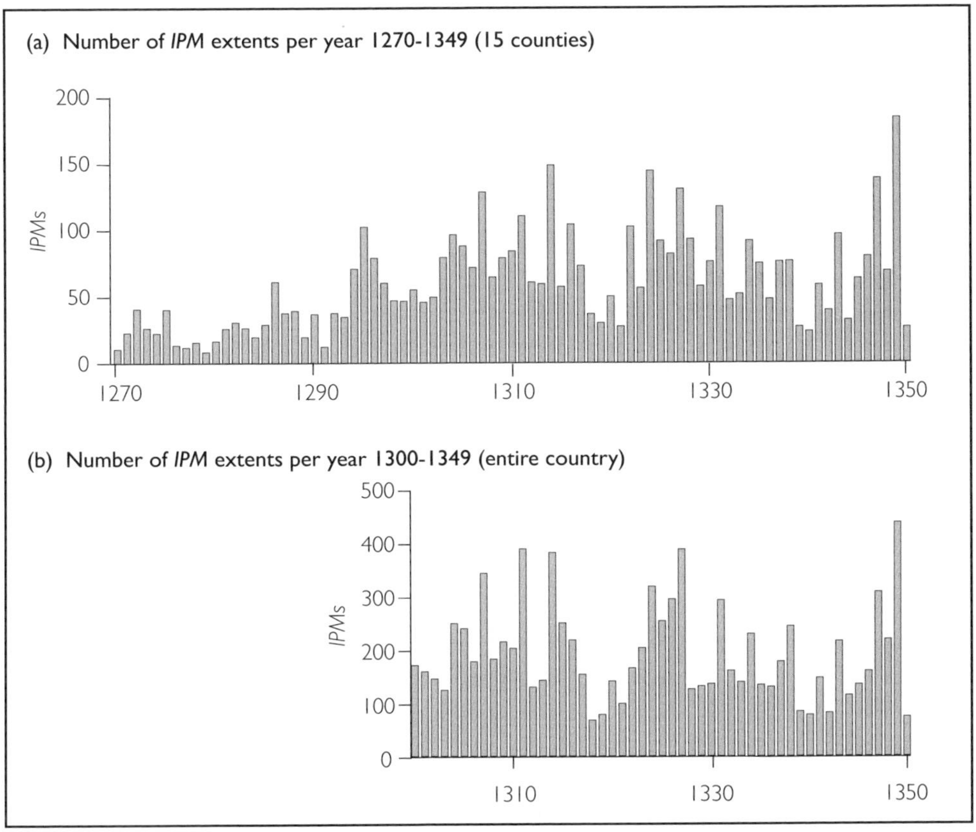

Figure 1. Annual totals on *Inquisitiones Post Mortem.*

time that the estates of many minor tenants-in-chief first came under the purview of the escheators (Figure 1). Edward II's administration was less energetic in its pursuit of these smaller tenants-in-chief and the efficiency of the whole exercise tended to lapse until 1324, when a major overhaul served to reinstate it on an even more thorough basis. Thenceforth, until 1342, the escheators were expected to supply more detailed and explicit information than ever before. Only in the immediate aftermath of Edward III's financial crisis of 1341 was there a reversion to a more generalized and simplified format, with the result that thereafter the IPMs become progressively less informative as a source.

Temporal change was matched by spatial variation. For much of the period, England south of the Trent and east of the Tamar was under the jurisdiction of a single escheator. In contrast, Cornwall, England north of the Trent, Wales, and the Lordship of Ireland were each mostly the responsibilities of separate escheators who appear from the surviving records to have gone about their business in rather different ways. IPMs from these relatively remote escheatries are usually thinner on the ground, often less detailed, and appear to have surveyed and valued properties on a somewhat different basis. As result, direct comparison between IPMs from these regions and those produced south of the Trent can pose problems. Regional variations in terminology and the size of customary measures further compound the difficulty: the customary acres which prevailed in some parts of the north were twice the size of a statute acre. Hence the reputation which the IPMs have acquired as 'an extremely unreliable source' (Kosminsky 1956: 63), a reputation which has hitherto deterred much systematic use of them.

Nor do the analytical and interpretative problems end here. IPMs are representative solely of property in lay hands. Ecclesiastical estates – territorially, tenurially, and jurisdictionally different in many ways from their lay counterparts – are excluded (Kosminsky 1956: 103–13). In localities dominated by ecclesiastical land-holdings this can have a considerable impact on the picture that emerges since IPM coverage is inevitably thinner and the lay properties which are recorded often bear little resemblance to the neighbouring holdings of the Church. Most of Hampshire, western Middlesex, and the Soke of Peterborough in Northamptonshire are cases in point (Figure 2). Moreover, only resources in which lords had an absolute property right were generally recorded. Communal resources, including fallow arable subject to common grazing rights and common pastures which locally or regionally might be of considerable physical and agricultural significance, generally passed without mention. So, too, did resources which generated little or no revenue; hence the paucity of references to woodland in much of the north of England. The picture that emerges can therefore be no more than partial. The single greatest omission is the non-demesne land, which amounted to perhaps two-thirds of the total (Kosminsky 1956: 93). Nevertheless, even a map of recorded demesne land-use has great historical utility if the methodological problems involved in its construction can satisfactorily be overcome.

An essential first step is the creation of a database from the mass of verbal information contained in the *IPM*s. The database has to be sufficiently flexible to cope with the many different ways in which information is given. Arable, for instance, is the most commonly extended of all resources. It may be recorded with a total value, a total acreage, a value per acre, or some combination of these. Sometimes a distinction is made between the sown and unsown arable and occasionally grazing rights and meadow appurtenant to the arable are subsumed in the overall valuation. Quite often supplementary information is provided of a descriptive or explanatory nature which itself can be quite revealing when subject to spatial and temporal investigation. Once extracted from the original documents and input to the database – a task for skilled historians – mechanical manipulation of this information is contingent upon the design and application of an effective coding scheme to allow data selection by category, item and/or subitem. All the lands and assets recorded in the database are coded according to a system of overlapping hierarchies. For example, land is subdivided into major land types (arable, grassland, woodland, etc.) and these are further subdivided into lesser divisions (fallow, mowable, common, etc.). Certain types of subdivision may occur across groupings; for instance, most land types may be recorded as being held in common. These subgroups can be identified by secondary codes which allow

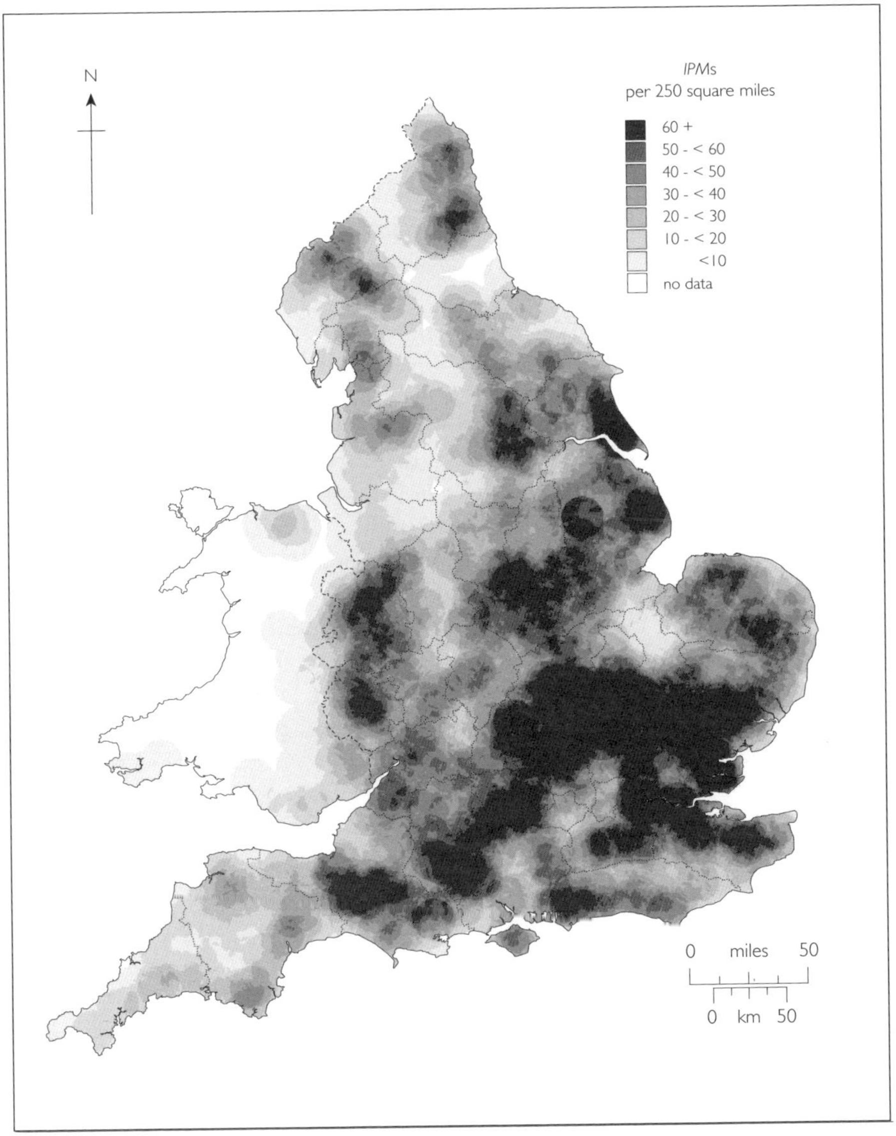

Figure 2. Number of *IPM* extents per 250 square miles, 1300–49.

for data selection across the land hierarchies. By coding the database, the entries transcribed from the original documents remain intact and are thereby available for other types of historical analysis, such as in the study of linguistics.

Spatial analysis is dependent upon accurate identification and georeferencing of each of the unique places with *IPM* extents, over 6,000 of them. This arduous task is rendered all the more onerous by the vagaries of medieval place-name spellings. Only with a full developed, coded, and georeferenced database can systematic spatial analysis of demesne land-use commence.

3. The individual land-use variables

The scope and content of any classification of historical land-use are constrained by the nature of the available data. In the case of the immediate pre-Black Death period, the variables selected and manner of their specification must necessarily respect the limitations of the *IPM*s as a source. Thus, since the pound had the same financial value throughout the kingdom whereas the acre was prone to customary variation, the respective quantities of most land types are more reliably measured by value than of area. For similar reasons, relative measures are to be preferred to absolute measures. Useful relative measures are the proportion of *IPM*s recording a particular land-use, the ratio of one land-use to another either in quantity or unit value, and the share of a given land-use or combination of land-uses in a larger portfolio of resources. It is also prudent to ensure that proportions and percentages, when used, are each based upon unique denominators. In this way, correlations created by variables sharing the same base – the closed-numbers problem – are avoided.

On the basis of these guiding principles, 12 components of land-use can be identified capable of independent analysis at a national level (Figure 3): arable as percentage of [arable + grassland], pasturage as percentage of [arable + pasturage], meadow as percentage of all pasturage, [pasture + herbage] as percentage of [all pasturage – meadow], percentage of IPMs recording wood, forest, chase, and park (WFCP), percentage of WFCP IPMs recording wood, percentage of WFCP IPMs recording forest, chase, park (FCP), percentage of IPMs recording 'other' land uses, the relative unit values of grassland to arable, meadow to arable, meadow to pasture, and value per acre of meadow.

Four of them measure the relative composition of land use in terms of some specified combination of either area or value (arable as percentage of [arable + grassland], pasturage as percentage of [arable + pasturage], meadow as percentage of all pasturage, and [pasture + herbage] as percentage of [all pasturage – meadow]). Since these require comparable information for more than one major category of land use, they are based upon a subset of the most complete *IPM*s relating to properties with the territorial and jurisdictional attributes of full manors. They focus in the main upon the relative importance and composition of the arable and pastoral sectors (the latter defined as comprising meadow, pasture, 'land', foreland, verges, herbage, waste, heath, broom, moor and marsh either as principal or subsidiary land-uses).

A further four variables (percentage of *IPM*s recording WFCP, percentage of WFCP *IPM*s recording wood, percentage of WFCP *IPM*s recording FCP, percentage of *IPM*s recording 'other' land uses) relate to the other major and minor categories of land-use: woods, forests, chases, parks, bog, turbary, bracken, rushes, reeds, heronries, and warrens. Since these are rarely recorded in such consistent detail as the arable and grassland (meadow and pasture) their respective contributions to the overall land-use

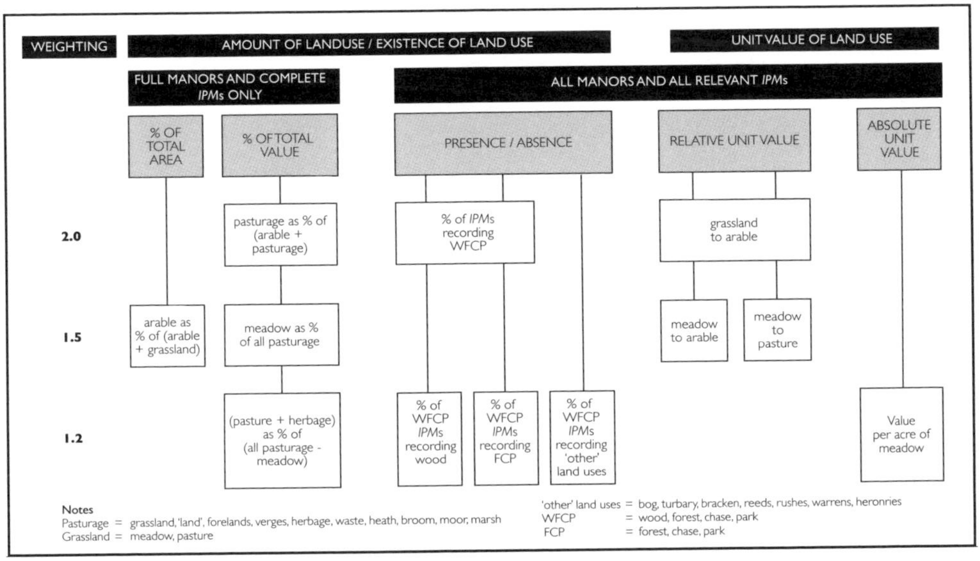

Figure 3. Variables used for classification of demesne land use.

portfolio are measured in terms of presence/absence. To give this as broad and firm a basis as possible, analysis is extended to all manors and all relevant *IPMs*.

The same approach applies to the four last variables which encapsulate different aspects of the relative and absolute unit values of resources and thus their respective availabilities, qualities and profitabilities. Since acres varied in their size and the methods by which they were assessed, three variables (grassland to arable, meadow to arable, and meadow to pasture) concentrate upon the ratio between the unit value of one land use and that of another. To prevent ratios with numerators smaller than denominators being limited to the range 0 to 1, each ratio is converted to logs to the base ten. Thus a ratio of 1:10 will be entered as -1, while the opposite situation of 10:1 will mirror this by being recorded as + 1. Ceiling values of the mean plus four standard deviations are also applied to prevent extreme ratios from having a disproportionate effect upon the overall result. Without a ceiling value, cluster analysis would be prone to isolating small groups of extremes while leaving the main body of data as a seemingly homogenous group. Only the fourth variable — the value per acre of meadow (Figure 4) – measures the actual unit value of a specific land-use. This acts as a control on the three ratios, meadow being chosen for this purpose because its general scarcity and high value rendered it the land-use whose unit value was most carefully and consistently recorded by the escheators.

Obviously, some of these variables are of more importance in designating composite land use types than others. A system of weightings is therefore employed to ensure that the most diagnostic variables exercise the greatest influence upon the final classification, and vice versa. The rationale of these weightings is that the three most important variables carry the same aggregate weight in the final classification as both

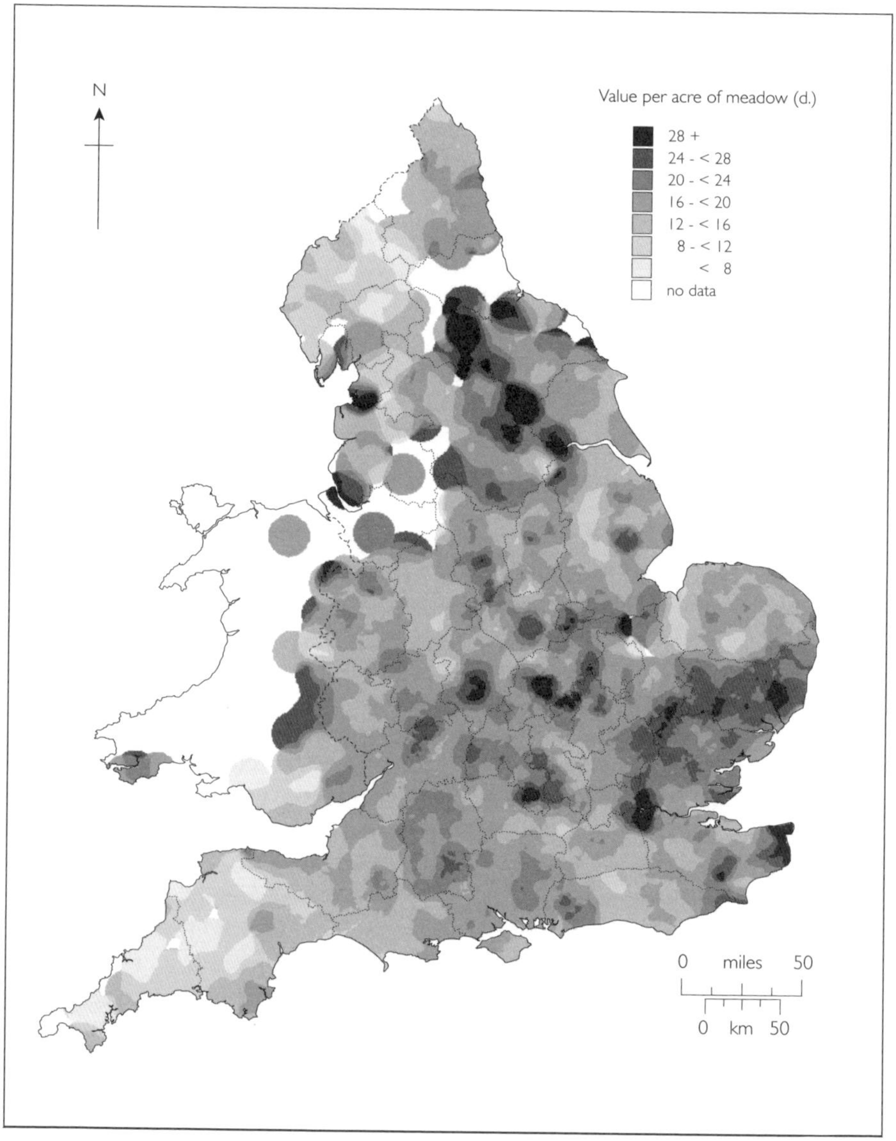

Figure 4. Value per acre of demesne meadow, 1300–49.

the four next-most-important variables and the five least-important variables. In other words, each of the three tiers of variables makes the same relative contribution to the final result.

Translating this hypothetical scheme of variables into cartographic reality depends upon the potential-mapping facility available within SPANS GIS (Milton Keynes, PCI Geomatics Group). With such a large data set, this represents a significant advance upon simple point mapping. Potential mapping also represents a great improvement upon choropleth mapping based upon administrative units (which are hardly appropriate to an analysis of land-use types), hitherto the most common solution to the widespread historical problem of generating surface maps from spatially discontinuous data. Potential mapping has the further merit of providing a systematic method of resolving gaps in data coverage by allowing known true values to be extrapolated across areas lacking information. In this way a surface map of probable, or 'potential', values is produced (Bonham-Carter 1994; Wattel and Van Reenen 1995). Since only a minority of IPMs provide the requisite information for the full range of variables, each variable is mapped independently, drawing upon all IPMs which contain relevant information (where there are multiple IPMs for a single location, the separate returns for each manor have been averaged). These potential maps can then, in turn, provide the basis for calculating ratios between unit land values using map overlays. For instance, by overlaying a potential map of the value per acre of arable with a corresponding map of the value per acre of meadow, each generated from all available valuations, it is possible to construct a third map of the ratio between the unit values represented in these two distributions. In this way, maximum use is made of all available historical information.

Each individual potential map represents the average of a given variable derived from all eligible *IPMs* falling within the surrounding 250 square miles of a given location, as exemplified by Figure 4 (the mean value per acre of meadow). The potential mapping algorithm employed made use of distance weighting so that the effect of individual *IPMs* on the final result is greater for those *IPMs* at the centre of the kernel than for those at the edges. The kernel size is expressed in miles because these are the unit most appropriate to the era under investigation. Empirical testing was employed to determine the most appropriate kernel size given the spatially incomplete and uneven coverage of the data. A smaller kernel size would not have yielded sufficiently continuous surfaces across the full range of 300 maps which it was intended to produce from the *IPMs* (Campbell and Bartley, forthcoming 2006); a larger kernel size would have resulted in surfaces that were too over-generalized for the purpose of this classification. As it is, the surfaces fragment and lose focus in those parts of the north and south-west of England where *IPM* coverage is particularly sparse (Figures 2 and 4). Although a variable kernel size could have been achieved by combining GIS with an external statistical package, for example Brunsdon's use of SAS (Cary, NC, SAS Institute) to model Californian Redwood (Brunsdon 1995), a fixed kernel size was preferred. First, it better fulfils the objective of creating a series of maps which are immediately comprehensible to historians. Second, it enables the methods to be replicated within currently available commercial GIS.

4. The method of classification

Once each of the twelve variables has been analysed and mapped it is possible to proceed with the more complex task of integrating them into a single, composite classification. Here the aim has been to allow the final classification to emerge from genuine differences inherent within the data independently of such potential explanatory factors as soils, terrain, and climate. This is in contrast to some previous, more qualitative classification schemes produced by agricultural historians where the land was as much a part of the classification as its use, rendering it impossible to determine the extent to which the latter was a function of the former (Kerridge 1967; Thirsk 1967). Cluster analysis offers one means of fulfilling this aim, provided that it is recognized that the result obtained will in part at least be a function of the clustering technique chosen. One way of minimizing this limitation is to apply several different clustering techniques and compare the results, since those case differences commonly identified by all solutions are more likely to be products of the data than the methods. Such an approach has already been used to derive robust national classifications of farming enterprise for the consecutive periods 1250–1349 and 1350–1449 using the clustering techniques Ward's, K-Means, and Normix (Power and Campbell 1992; Campbell et al 1996). Since there is historical advantage in producing classifications of farming enterprise and land use which are methodologically directly comparable, Ward's and K-Means have also been employed here. Normix has been replaced by Within Group Linkage due to the operational difficulties of getting the former to run on such a large data set.

Ward's method is an agglomerative hierarchical clustering technique which operates by initially defining each case in the data set as a unique cluster. At successive stages these are combined into larger groups until finally all the cases are placed in a single group. At each stage the increase in the error-sum of squares which would result from the joining of every possible pair of clusters is calculated and those two clusters are combined which result in the smallest error-sum increase. While Ward's method has been found to be reliable in discovering well-separated spherical clusters, it can perform poorly when other configurations exist. K-Means is an optimization technique which begins with the data allocated to a predetermined number of clusters. The number and composition of these clusters may be randomly selected or chosen on some predefined criterion such as the output of another clustering technique. The similarity between each case and the centre of every cluster is calculated and the case assigned to that cluster to which it is most similar. The method involves an iterative process whereby the cluster centres are recalculated after each case relocation and all cases re-examined to check for the necessity of any further relocations. The method terminates when no further relocations are required. While the relocation process should improve upon the initial partitioning of the data the final result may be a local optimum. Only if the same solution is obtained from different starting configurations can there be confidence that a global optimum has been reached. Within Group Linkage is an agglomerative hierarchical technique which combines clusters so that the average distance between all cases in the resulting cluster is as small as possible. The distance between groups is defined as the average distance between all possible pairs of cases in the group that would be created by combining the two initial groups.

Each of these three techniques was used to generate a range of cluster solutions. To create a set of orthogonalized variables, a principal components analysis was conducted on the 12 designated variables. The first seven principal component scores, accounting for 80 per cent of the variance, were used in the cluster analysis, thereby decreasing the computational time and memory resources required to run the cluster analysis. The result which yielded the greatest consensus between the three techniques was a six-cluster solution. This was accepted as the optimum result. Those cases commonly classified by all three cluster techniques were then used to define the core characteristics of each land-use type (Table 1). How to classify the residual of non-core cases was determined by the application of Fisher's linear discriminant functions (Figure 5) (Hand 1981; Power and Campbell 1992, 232; Bartley 1996). This has the further merit that it can be used to assign a 'second choice' classification to the non-core cases, thereby bringing out further inherent differences within the data and lending greater subtlety to the result (Figure 6). For example, all non-core cases assigned to land-use type 1 can be further subdivided into those with land-use type 2 as their second choice, those with land-use type 3 as their second choice, and so on. For a description of land-use types see Section 5.

This method relies on matching the results from three clustering techniques to establish a consistently differentiated core of well-defined clusters. The possibility of the final result being biased by a single clustering technique is thereby minimized. Other mechanisms for clustering the data could be used in conjunction with, or instead of, the methods outlined above, for example Normix (from the Clustan suite of pro grams, University of St Andrews, Computing Laboratory) or Modeclus (from SAS). Nevertheless, the key to this method of classification lies less in the precise choices of clustering technique than in the principle that core areas are only defined when there is a clear consensus between cluster solutions. The strength of the method therefore lies in the creation of a set of stereotypes, each of which emerges strongly from the data.

A prerequisite for the use of cluster analysis as a classification tool is the existence of a data matrix of cases. Creating such a matrix from a set of 12 potential maps – one for each chosen land-use variable – is again only possible using GIS. A net of 47,741 regular hexagons, each with sides 1 km in length and a total area of 2.6 km^2, was imposed to cover the entire country. The mean land-use characteristics of each hexagon were then determined by sampling each of the component potential maps at the centre of each hexagon and storing the results. Due to the idiosyncrasies in recording already noted in Cornwall and the north of England, a submatrix of 34,489 cases was established for England east of the Tamar and south of the Trent. To counteract prohibitive computational times and, more critically, memory requirements, a further 25 per cent subsample of 8,623 locations was selected for the application of cluster analysis. The 3,187 locations identically classified by all three cluster techniques were used to map the distribution of the six 'core' land-use types, treating each location as the centroid of an enlarged hexagon of 10.4 km^2. From this core classification, a set of linear discriminant functions was derived and used to assign all remaining unclassified areas (including those west of the Tamar and north of the Trent) to their most approximate land-use type on the basis of the original net of 2.6 km^2 hexagons. In this way, surface values were translated into location-specific matrices of data, classified, and then converted back into a surface map which depicts land use *across*

Table 1. Core demense land-use types, 1300–49: variable means and standard deviations

Land-use variable	Weighting	Land-use type						All land use
		1	2	3	4	5	6	
Value of pasturage as % of (arable + pasturage)	2.0	32.0 *13.2*	36.7 *12.6*	31.3 *8.4*	34.5 *8.9*	38.9 *9.5*	22.3 *7.3*	32.4 *10.8*
% of *IPM*s recording WFCP	2.0	36.8 *12.6*	13.9 *8.2*	45.8 *14.0*	55.8 *10.8*	30.9 *12.2*	25.8 *14.1*	36.9 *18.0*
Unit value of grassland : unit value of arable	2.0	6.4 *2.8*	2.7 *1.0*	5.6 *1.5*	1.8 *0.8*	4.0 *1.2*	2.6 *0.8*	4.3 *2.1*
Area of arable as % of (arable + grassland)	1.5	85.1 *16.2*	81.6 *15.7*	93.2 *6.8*	69.8 *8.5*	86.3 *11.9*	91.5 *7.8*	87.1 *12.9*
Value of meadow as % of all pasturage	1.5	61.2 *24.2*	76.5 *11.9*	65.1 *14.6*	28.6 *13.6*	67.2 *15.5*	44.3 *15.2*	60.3 *20.8*
Unit value of meadow : unit value of arable	1.5	7.0 *2.6*	3.1 *1.1*	6.7 *1.5*	3.7 *1.3*	4.5 *1.4*	3.3 *1.2*	5.2 *2.2*
Unit value of meadow : unit value of pasture	1.5	12.3 *3.0*	3.3 *3.0*	2.9 *1.0*	5.9 *2.9*	4.2 *2.4*	2.5 *0.8*	4.3 *3.4*
Value of (pasture + herbage) as % of (all pasturage - meadow)	1.2	64.4 *32.9*	74.3 *28.3*	92.9 *11.7*	67.8 *29.5*	90.6 *13.7*	58.0 *25.0*	80.3 *25.5*
% of WFCP *IPM*s recording wood	1.2	93.7 *7.8*	98.6 *4.4*	95.3 *4.9*	92.4 *8.4*	73.3 *14.6*	98.6 *3.0*	92.5 *11.2*
% of WFCP *IPM*s recording FCP	1.2	15.9 *12.5*	4.0 *6.6*	24.0 *13.9*	21.2 *13.3*	60.0 *14.9*	22.2 *17.2*	25.2 *20.9*
% of *IPM*s recording 'other' land-uses	1.2	2.4 *2.7*	2.3 *2.3*	2.0 *1.9*	2.2 *1.8*	2.0 *2.1*	19.3 *6.6*	4.2 *6.4*
Value per acre of meadow (d.)	1.2	13.6 *3.4*	15.9 *3.2*	22.1 *3.4*	19.4 *5.7*	15.3 *3.3*	15.5 *2.8*	18.3 *4.9*
Number of cases		302	447	1,243	332	468	395	3,187

Notes
Roman = mean
Italics = standard deviation
grassland = meadow, pasture
pasturage = grassland, 'land', forelands, verges, herbage, waste, heath, broom, moor, marsh
other land-uses = bog, turbary, bracken, reeds, rushes, warrens, heronries
FCP = forest, chase, park
WFCP = wood, forest, chase, park

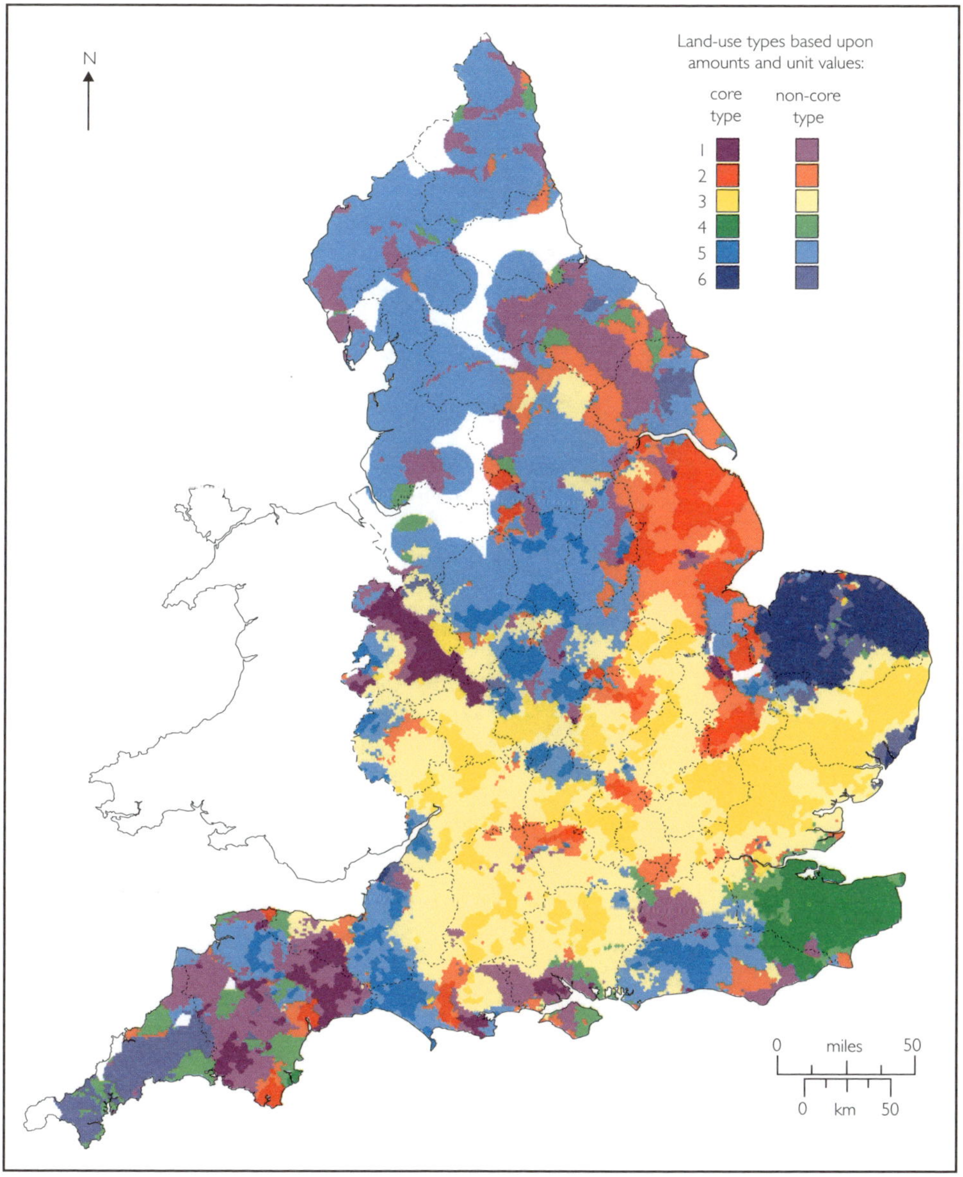

Figure 5. National classification of demesne land use, 1300–49, first choice only.

England, differentiated according to whether core land-use characteristics were strongly or weakly defined.

As can be seen from Figure 5 and as was expected, the results for Cornwall and England north of the Trent appear poorly focused and are the least robust. Elsewhere such limitations are mostly less apparent, especially in those parts of central, eastern,

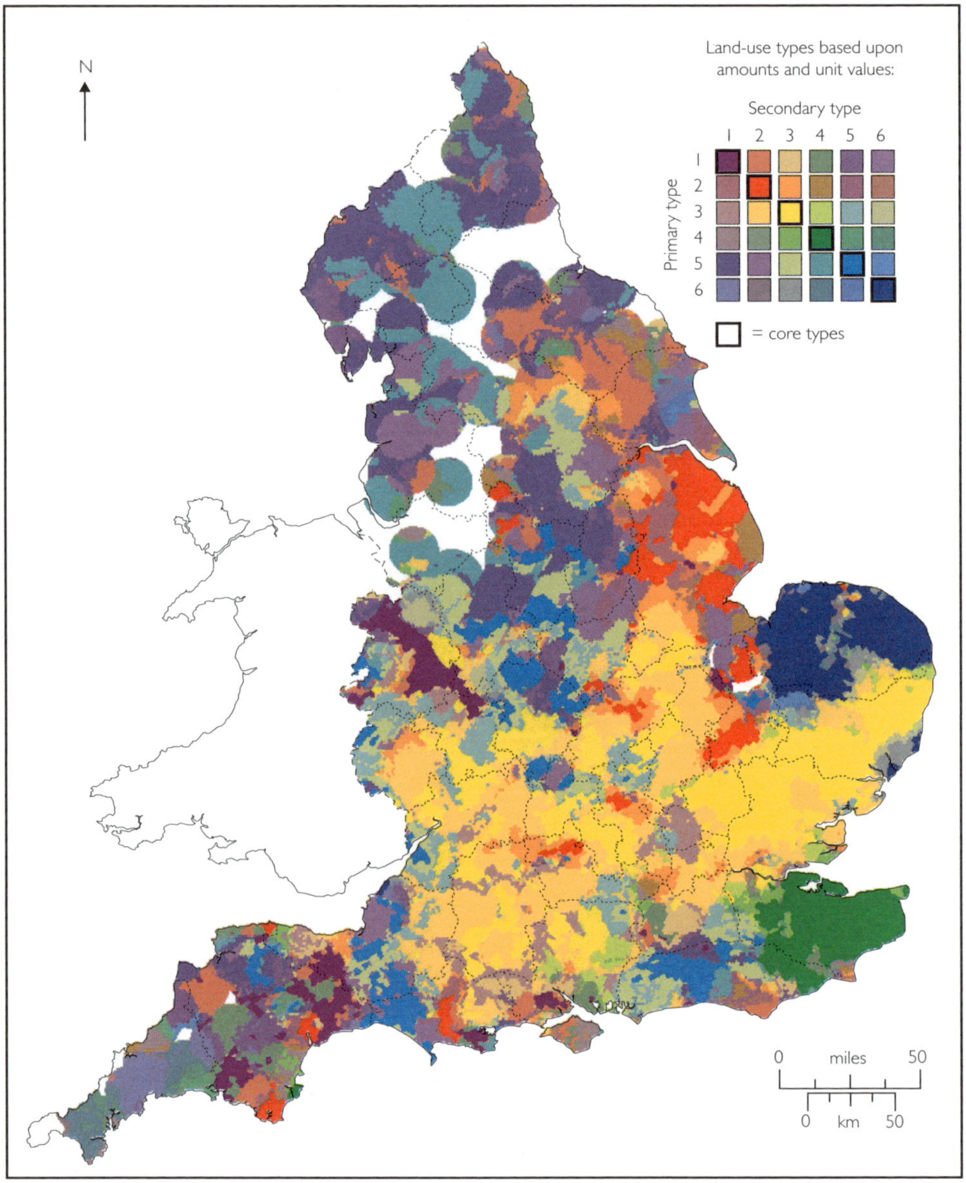

Figure 6. National classification of demesne land use, 1300–49, first and second choice.

and southern England where the density of *IPM* coverage is at its greatest (Figure 2). Land-use variations within these regions come into even sharper focus when discriminant functions are used to reclassify non-core hexagons according to their first- and second-choice land-use categories. To the six core land-use categories are now added 30 subcategories (Figure 6). Clearly, at demesne level there were many subtle gradations of land use which differentiated one locality from another. Nevertheless, several broad regional contrasts in land-use do emerge. These are all the more significant because the method adopted did not preordain a result characterised by pronounced spatial differentiation. The implication is that the patterns revealed by Figures 5 and 6 encapsulate real geographical differences in land-use.

5. The land-use classification

The mean attributes of the six most sharply differentiated demesne land-use types present on the eve of the Black Death are set out in Table 1. Collectively, they encompass 37 per cent of the classified area east of the Tamar and south of the Trent. The remaining 63 per cent of the area is made up of land uses intermediate to these diagnostic types. This is consistent with the naturally kaleidoscopic character of land use and emphasizes the tendency for minor rather than major gradations in land use to be the norm in this period, at least at the level of demesnes. Each of the core land-use types may be broadly defined as follows:

5.1 Land-use type 1: Poor land with low unit land values

This occurred on a smaller scale than any other land-use type and was confined to a few highly specific but quite widely separate locations, notably south-western Surrey, the New Forest in southern Hampshire, the Isle of Purbeck in Dorset, the Brendon Hills in south-western Somerset, south-eastern and south-western Devon, the upper Severn valley in Shropshire, and the Huntingdonshire fen edge. A version of the same land-use type also prevailed in the East and North Ridings of Yorkshire and several other localities in the north of England. Unit values of meadow, pasture, and arable well below the national average were its most distinguishing feature, with, nevertheless, an exceptionally wide differential between the unit values of meadow and arable. These characteristics are consistent with the known physical limitations of most of these areas, especially with regard to arable husbandry.

5.2 Land-use type 2: Open country with little differentiation of unit land values

This land-use type was most widely represented in Lincolnshire, the Isle of Ely, and the closely-settled countryside of south Cambridgeshire. Other occurrences include the limestone country of central Northamptonshire and mid Buckinghamshire, the Vale of the Whitehorse in Berkshire, the Dorset Downs, the immediate environs of Exeter, and southernmost tip of Devon. The most striking features of this land-use type were the scarcity of woods and private hunting grounds, the relatively narrow differential between the unit values of meadow and arable (indicative of the relative scarcity or abundance of one and profitability of the other), and the dominance of pasturage by grassland and particularly by meadow.

5.3 Land-use type 3: Arable country with limited but valuable grassland

This was the characteristic land-use type of much of the heavier land of lowland England, especially inland and away from the coast. No other land-use type was as well or as widely represented. It was the predominant land use of the closely settled East Anglian boulder-clay plateau stretching in a broad swathe from south-east Norfolk across Suffolk and Essex into eastern Hertfordshire. Other notable occurrences included the lower Welland valley of Northamptonshire and adjacent Rutland, northern Bedfordshire, north-eastern Buckinghamshire and south-western Northamptonshire, the Avon valley of south Warwickshire and Worcestershire, the mid Severn valley, eastern Shropshire, the Vale of Oxford, and substantial portions of the Hampshire and Wiltshire Downs. It was the typical land-use type of much of commonfield England as well as of several extensive areas outside that zone. Arable was everywhere present in above average quantities while grassland, partly because it was scarce and carefully managed, was valuable, especially the meadow. A higher than average number of demesnes also possessed some woodland, although as with the grassland the quantities were mostly small.

5.4 Land-use type 4: Superior arable with several pasture and wood

This minority land-use type was associated almost exclusively with central and eastern Kent and neighbouring portions of Sussex. Elsewhere there are hints of it in south-eastern Hampshire and parts of Devon and Cornwall. These were all areas where, 'waste' apart, the bulk of all land was held in severalty. In the absence of much or any common pasture, demesnes in these areas were endowed with well above average quantities of several pasture. No doubt for similar reasons, woods, forests, chases, and parks were also exceptionally well represented. Equally distinctive was the high unit value of arable, resulting in a narrower differential between the unit values of arable and grassland than in any other land-use type.

5.5 Land-use type 5: Inferior arable and pasturage with private hunting grounds

This land-use type was well represented in several widely separate areas: the West-Sussex Weald, a belt of country extending from the Mendip Hills southwards across mid Somerset into west and south Dorset, the hilly country of north Devon, north Oxfordshire and the Stour valley of south Warwickshire, a scattering of locations along the Welsh border, and much of the north midlands. Unit values of arable, meadow and pasture all tended to be below average. Grassland, present in above average quantities, was complemented by several other forms of pasturage. Private hunting grounds, enclosed from the waste and wood, were also quite a typical feature of these areas. A version of this demesne land-use type was also widespread throughout much of the north of England.

5.6 Land-use type 6: Open arable country with assorted lesser land-uses

This was the predominant land-use type of Norfolk and was scarcely ever encountered outside that county, the only notable exceptions being the immediately neighbouring portion of the Suffolk Breckland and the Sandlings of south-eastern Suffolk, which shared many of the same physical characteristics. Within Norfolk this land-use type was pre-eminent everywhere apart from on the stiff boulder-clay soils of the south-east, the Norfolk Fenland in the extreme west, and along the boulder-clay watershed which separated east from west Norfolk. It was therefore most characteristic of soils which were either light or medium and free draining. In both parts of the county demesne land-use was dominated by arable. Several grassland was present in below average quantities and mean land values suggest that what there was was often of inferior quality. The differential between the unit values of meadow and pasture was smaller than in any other land-use type. Woodland was also under-represented by national standards and it is clear that much of the countryside wore a very open aspect. Hence the relative importance of such minor land-use resources as heath, turbaries, and rushes, which feature more prominently in this land-use type than in any other. The latter no doubt explains why an analogous land-use type shows up in Cornwall at the opposite end of the country.

These six core land-use types encapsulate a range of agrarian 'landscapes' which will be familiar to many historians of the period. Their clearly articulated regional characters are genuine products of the data and were not pre-ordained by the method (Figure 5). Many were deeply etched and long enduring. The distinctive character of land use in Kent, where common rights were weakly developed, emerges strongly. Norfolk, too, whose farming enterprise early set it apart from the rest of the country, possessed a land-use character all of its own (Campbell et al 1996: 143–54). Dorset, in contrast, displayed a range of land uses which matched the marked environmental contrasts within the county. Significantly, it was around England's margins and at its extremities that the greatest variety of land uses was to be found. Here, after all, was where environmental and economic differences were at their most pronounced. In contrast, throughout the arable heartland of central and eastern England homogeneity rather than heterogeneity was the land-use norm. From Suffolk to Worcestershire and Rutland to Wiltshire demesne land-use varied within remarkably narrow bounds, reflecting the more muted environmental and economic contrasts of this relatively land-locked area. Because the core classification is defined at a more-or-less national scale it brings out very clearly which land-use distinctions were nationally significant and which were not. Sometimes, as in the case of Dorset, this highlights the importance of local land-use variations, at others, as in the case of Kent, it suggests that local variations were rarely of more than local significance. In this way the critical similarities and dissimilarities between localities and regions are brought into sharper focus. The fact that the results of the classification make such sound historical and geographical sense is one of the most effective tests of the credibility of the method. It also endorses the faith placed in the *IPM*s as a source.

Second-choice classification of the non-core areas brings out much secondary variation within the broader picture (Figure 6). Several small but nonetheless distinctive localities thereby emerge, differentiated from those around them by their land-use: Romney Marsh on the south coast of Kent, the 'high' Chilterns of west Hertfordshire and south Buckinghamshire, the scarp edge of the Berkshire and Marlborough Downs, the Lincoln Edge and Isle of Axholme are some of the more obvious. Nevertheless, it is of considerable interest that only certain areas of reclaimed coastal marshland, some wooded hills, and a few scarp edges stand out in this way. Plainly, not all topographical variations impacted equally strongly upon demesne land-use. It is in determining which were most and which least significant that the kind of systematic and objective classification offered here comes into its own.

6. Historical and methodological implications of the results

Land use was itself only one component of a more complex agrarian scene. Of itself, it reveals much about the combination and quality of the landed resources available to demesne managers but little, directly, about how those resources were managed. The classification offered here therefore provides a framework within which existing case studies of the pre-Black Death countryside may be interpreted. It also helps identify localities and regions which would repay closer investigation. Above all, it helps prepare the way for a more effective regionalization of medieval England.

Methodologically, this exercise in classification illustrates the dividends which can accrue when GIS is harnessed to the analysis of historical data. England is exceptionally rich in sources amenable to this kind of approach. Potential mapping represents a major advance on the point and choropleth mapping upon which historians and historical geographers have hitherto largely relied. It also provides a methodologically effective solution to the common historical problem of how to interpolate missing values from spatially discontinuous or inconsistently-recorded data. Beyond this it constitutes a powerful tool for combining, integrating, and analysing complex data sets and thereby reveals the past in a fresh light.

Acknowledgements

We wish to acknowledge the financial support of the Leverhulme Trust and The Queen's University of Belfast. Research assistance was provided by Roger Dickinson and Marilyn Livingstone. Additional computing advice was supplied by John P. Power. An earlier version of this paper was presented to the medieval seminar in 1996 at All Souls College, Oxford.

References

Bartley, K 1996 Classifying the past: Discriminant analysis and its application to medieval farming systems. *History and Computing*, 8: 1–10

Bonham-Carter, G F 1994 *Geographic Information Systems for Geoscientists: Modelling with GIS*. Kidlington, Pergamon Press

Brunsdon, C 1995 Estimating probability surfaces for geographical point data: An adaptive kernel algorithm. *Computers and Geosciences*, 21: 877–94

Campbell, B M S and Bartley, K [forthcoming] 2006 *England on the Eve of the Black Death: An Atlas of Lay Lordship, Land, and Wealth, 1300–49*, Manchester, Manchester University Press

Campbell, B M S, Bartley, K and Power, J 1996 The demesne-farming systems of post-Black Death England: A classification. *Agricultural History Review*, 44: 131–79

Campbell, B M S, Galloway, J A, Keene, D J and Murphy, M 1993 *A Medieval Capital and its Grain Supply: Agrarian Production and its Distribution in the London Region c.*1300. Historical Geography Research Series 30

Campbell, B M S, Galloway, J A and Murphy, M 1992 Rural land use in the metropolitan hinterland, 1270–1339: The evidence of *Inquisitiones Post Mortem*. *Agricultural History Review*, 40: 1–22

Clanchy, M T 1979 *From Memory to Written Record: England 1066–1307*. London, Edward Arnold

Darby, H C 1977 *Domesday England*. Cambridge, Cambridge University Press

Hand, D J 1981 *Discrimination and Classification*. Chichester, John Wiley & Sons Ltd

Holmes, G A 1957 *The Estates of the Higher Nobility in Fourteenth-Century England*. Cambridge, Cambridge University Press

Kerridge, E 1967 *The Agricultural Revolution*. London, Allen and Unwin

Kosminsky, E A 1956 *Studies in the Agrarian History of England in the Thirteenth Century*. Oxford, Blackwell

Power, J P and Campbell, B M S 1992 Cluster analysis and the classification of medieval demesne-farming systems. *Transactions of the Institute of British Geographers*, 17: 232–42

Prestwich, M 1980 *The Three Edwards: War and State in England 1272–1377*. London, Weidenfeld and Nicolson

Stevenson, E R 1947 The escheator. In Morris, W A and Strayer, J R (eds) *The English Government at Work, 1327–36*. volume 2. Cambridge, MA, Mediaeval Academy of America:109–167

Thirsk, J 1967 The farming regions of England. In Thirsk J (ed) *The Agrarian History of England and Wales, volume 4, 1500–1640*. Cambridge, Cambridge University Press: 1–112

Wattel, E and Van Reenen, P 1995 Visualisation of extrapolated social-geographical data. In Boonstra, O, Collenteur, G and Van Elderen, B (eds) *Structures and Contingencies in Computerized Historical Research*. Hilversum, Uitgeverij Verloren: 253–62

INDEX